Statistical Thermodynamics

Statistical Thermodynamics

Revised Printing

CHANG L. TIEN University of California, Berkeley
JOHN H. LIENHARD University of Kentucky

HEMISPHERE PUBLISHING CORPORATION
New York Washington Philadelphia London

STATISTICAL THERMODYNAMICS

1 2 3 4 5 6 7 8 9 0 BRBR 8 9 8

Library of Congress Cataloging in Publication Data

Tien, Chang L., date–
Statistical Thermodynamics.

Rev. print.
Includes bibliographical references and index.
1. Statistical thermodynamics. I. Lienhard, John H., date. II. Title.
[QC311.5.T54 1985] 536'.7'015195 84-27910
ISBN 0-89116-828-1 Hemisphere Publishing Corporation

Preface

This book has been developed to meet the needs of advanced engineering undergraduate and beginning graduate students for an introduction to statistical thermodynamics. The student that we have envisioned has had at least an introductory course in thermodynamics; he may or may not have had a course in modern physics; and he has had no physics beyond that level. Nor do we presuppose a course in statistics or a mathematics background beyond that which he normally would have developed for his upper-division engineering work. Such a student, although he is capable of learning fairly rapidly, must start at the beginning level as far as the concepts of quantum mechanics and modern physics are concerned.

During the past seven years we first developed class notes, and subsequently organized them into the present text. This text has been developed in close conjunction with classes at the University of California at Berkeley, Washington State University at Pullman, and the University of Kentucky at Lexington.

What we have evolved is an approach that escalates the pedagogical level as it proceeds. The twelve chapters are arranged into a sequence of basic subject groups presented on three pedagogical levels as shown.

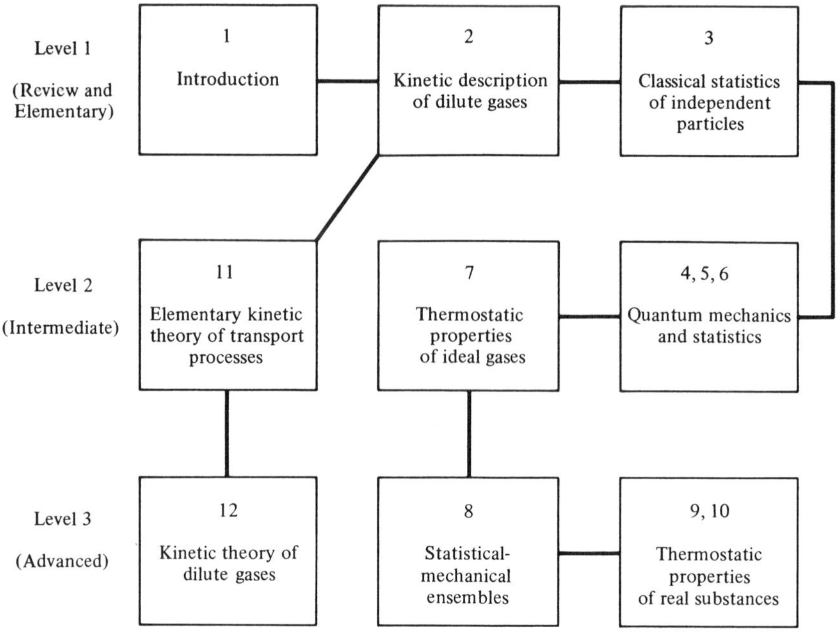

The first three chapters are intended to establish an elementary background, and a language, on which the material in the subsequent chapters is based. Our experience has shown that these should not be covered too rapidly. The fourth and fifth chapters largely contain the needed material that would have been covered in an elementary modern physics course. The detail in which they should be treated will vary according to the background of the class. Chapters 6 and most of 7 and 11 would complete an introduction to the methods of statistical thermodynamics. Depending upon the level of the students, a one-semester course can encompass some or all of the additional material and applications developed in Chapters 8, 9, 10, and 12.

The exposition of the subject is developed with the following objective in mind: We want to bring the student to an appreciation of the role of statistical-thermodynamics methods so that he will understand why they are important, what they can do, and how they can be applied to various engineering systems. Therefore, we have attempted to stay relatively close to the historical evolution of the subject and we have sought to carry this evolution to its fruition in applications. In this connection we have also tried to keep the unity of the subject clear in the student's mind by showing respect for the axiomatic structure of the subject.

In establishing methods for computing the physical properties of substances in equilibrium, for example, we have made strong use of the notion of a *fundamental equation* or generating relation for physical properties. This idea is first developed in Chapter 1 from a strictly macroscopic viewpoint. In subsequent chapters we take care to show how the microscopic generating functions (partition function, q potential, grand canonical partition function, etc.) relate to the macroscopic ones (entropy, energy, the free energies, etc.).

We have attempted, wherever it is appropriate to do so, to demonstrate the usefulness and strength of statistical thermodynamics through simple applications that are important in modern engineering problems. Falling into this category are, for instance, the treatment of the Lighthill dissociating gas and the singly ionized gas, the emphasis on the statistical-mechanical basis of the law of corresponding states in evaluating both thermostatic and transport properties of gases, the statistical-thermodynamic description of the solid and liquid states, and the calculation of thermal and electrical transport in solids.

We owe a great debt of gratitude to students — too many to name — who have generously helped us to improve the successive drafts of the text in the form of class notes. Professors Creighton A. Depew of the University of Washington and Ernest G. Cravalho of the Massachusetts Institute of Technology each provided very helpful and extensive commentaries on the semifinal manuscript. The University of Kentucky contributed heavily to the mechanical burden of preparing the book; Mrs. Linda Boots carried the major task of typing the manuscript; and Mrs. Mardell Haydon and Mrs. Bonnie Turner completed the final revision. We are also grateful to the University of California and Washington State University for many material contributions to completion of the work.

During the course of this effort we have discovered why authors inevitably thank their wives. Their contributions are real, as it turns out. We are very grateful to Di-hwa and to Carol for helping us to find enough peace and quiet within the normal demands of our households to get the task done.

Chang L. Tien
John H. Lienhard

Contents

Statistical Thermodynamics

1 introduction

1.1 SUBJECT OF THERMODYNAMICS

Thermodynamics is commonly thought of as that subject which treats the transformation of energy and the accompanying changes in the states of matter. Surely there are few, if any, disciplines that bear more pretentious definitions than this. In its primal concern with energy and matter, thermodynamics lays claim to the attention of all technical people. The scientist and engineer alike must either come to grips with the subject or suffer the severest limitations upon their professional lives.

The name *thermo-dynamics* calls to mind the idea of thermal energy in transition. The subject is so named because it describes the effects of the dynamic phenomena of heat and work upon systems. The name is misleading insofar as the methods of classical thermodynamics can only describe systems in equilibrium. When a system undergoes a process, classical thermodynamics can do no more than describe the action in terms of static end conditions. In particular, it discloses nothing about the rate of real processes.

The fact that our subject has grown up under the name of *thermodynamics* is related to an accident of history that profoundly influenced its subsequent development — a matter we shall say more about in Sec. 1.3. Lately, the names *thermostatics, equilibrium thermodynamics,* and *reversible thermodynamics* have been proposed as more descriptive and accurate titles. Their use has been encouraged by the development of the subject of *irreversible thermodynamics,* or *nonequilibrium thermodynamics,* or simply *thermodynamics* (in contrast to thermostatics). On the macroscopic level this latter subject is concerned with coupled transport processes such as the simultaneous flows of heat and electrical current in a thermocouple.

But our concern is with learning the gross thermostatic and thermodynamic behavior of systems in terms of the microscopic phenomena in which such behavior has its origins. The present study of statistical thermodynamics will accordingly give greatest emphasis to those manifestations of the *general microscopic behavior* of matter that are thermodynamic in nature.

1.2 NATURE OF STATISTICAL THERMODYNAMICS

The laws of macroscopic thermodynamics clearly must arise out of the microscopic action of myriads of atoms or other particles. In seeking to determine how these laws arise, we might be tempted to embark upon the straightforward prediction of the action of the individual molecules in a group of molecules in the following way.

Let a large number, N, of particles occupy a box. For each particle we must write a second-order differential equation of motion,

$$\sum_j \mathbf{F}_{ij}(\mathbf{r}_i, \mathbf{r}_j, t) = m_i \frac{d^2\mathbf{r}_i}{dt^2}$$

where $\mathbf{r}_i$ and $\mathbf{r}_j$ are the position vectors of the ith and jth particles, m_i is the mass of the ith particle, and t is time. $\mathbf{F}_{ij}(\mathbf{r}_i, \mathbf{r}_j, t)$ is the complicated force of interaction exerted on the ith particle under consideration by the jth particle and the container walls. Each of these N coupled equations requires two initial conditions,

$$\mathbf{r}_i = a \quad \text{and} \quad \dot{\mathbf{r}}_i = b \qquad \text{when } t = 0$$

If N were small, this problem could be done on a computer. However, N is on the order of 10^{15} to 10^{30} molecules in real physical systems of interest. Even the most modern digital computers could not begin to solve a problem of such complexity.[1]

[1] Another impediment to this attack is the fact that the *Heisenberg Uncertainty Principle* precludes the precise simultaneous knowledge of $\mathbf{r}_i$ and $\dot{\mathbf{r}}_i$ at the instant $t = 0$. This is discussed in Sec. 4.5.

But there is a far more important reason why such a straightforward attack upon the problem cannot be fruitful. Such a complete *description of individual particle action would not disclose the gross thermodynamical behavior in which we are interested.* It would reveal only the trees, not the forest.

Clearly, the detailed information about particle behavior must be blurred in some way if a meaningful description of gross behavior is to be obtained. This is what our physical senses do for us automatically when we experience any physical behavior. The physical size of organized animal life is, in fact, determined on this basis. The smallest neural component of an animal must be large with respect to molecular dimensions if it is to respond to average molecular behavior. Otherwise it would respond to the action of individual molecules and behave in a chaotic way.[2] Conversely, we can be sure that any physical system that we *experience directly* is composed of an immense number of particles.

The problem is then for us to develop analytical methods for describing physical systems in the way that our senses describe the physical world to us. We must develop techniques for averaging the behavior of small particles so that, by ignoring detail, we can discover the gross effects of detail. Just as an insurance company need be concerned only with the most probable vital statistics of its policyholders and can ignore their individual characteristics, we need only determine the most probable behavior of large groups of molecules and can ignore their individual dynamics.

Statistical thermodynamics is the study of the techniques for doing this. It is usually regarded as being composed of two subdivisions: *statistical mechanics* and *kinetic theory*. This division is not sharp because both subdivisions are founded upon similar axiomatic structures.

Statistical mechanics is based on the idea that the equilibrium state of a thermodynamic medium is the macroscopic state that corresponds with the most probable microscopic state. The problem of statistical mechanics is that of determining what microscopic state is most probable; and the results of statistical mechanics, like those of classical thermodynamics, are applicable only to equilibrium configurations.

The kinetic theory employs a somewhat more direct attempt to average the behavior of individual particles. It takes into account definite molecular models and the mechanical details of the motion of individual particles. It is characteristically more complicated, but less abstract, than statistical mechanics. The great advantage of

[2]This matter is discussed and amplified by E. Schrödinger in *What Is Life? and Other Scientific Essays*, Doubleday & Co., Inc., Garden City, N.Y., 1956, "What Is Life?."

the kinetic theory over statistical mechanics lies in its applicability to nonequilibrium behavior and the prediction of transport properties.

1.3 HISTORICAL DEVELOPMENT

MACROSCOPIC THERMODYNAMICS

Thermodynamics has evolved fitfully and anachronistically over the past two and a half centuries. Most of the phenomena embraced by the subject can now be explained by applying the laws of classical dynamics and probability to atoms and molecules, using a minimum of additional physical principles. On the gross level, however, a whole structure of new physical laws must be erected to describe thermostatic behavior adequately. These laws cannot be understood as easily in terms of human experience as can, for example, the laws of mechanics. Consequently, the task of putting them in order has been difficult and perplexing.[3,4]

Prior to the eighteenth century, nothing whatever had been done in the field of thermodynamics. In 1695 G. W. Leibniz anticipated the first law of thermodynamics by showing that the sum of kinetic and potential energies remains constant in an isolated mechanical system. But it remained for the development of the chemical and heat engine technologies of the eighteenth century to stir a real interest in heat phenomena.

Several developments in the late eighteenth century laid the foundation for statements of the first and second laws of thermodynamics that were finally made in the early nineteenth century. These included Joseph Black's presentations of the ideas of specific heats and latent heats of phase change (about 1770) and the enunciation of a plausible caloric theory in 1779 by his student William Cleghorn. The caloric theory held that heat was a "subtle fluid" with the following properties: It was elastic; its *particles* repelled one another; it was attracted by ordinary matter in varying degrees; it was indestructible and uncreatable; it was either sensible or latent, and in the latent form could combine with solids to form liquids or with liquids to form vapors; and it was possessed of *weight*.

The indestructibility of caloric caused some difficulty. John Locke in 1772, for example, had observed that axle trees were heated by friction with the wheels, and Albrecht Haller in 1747 had

[3]J. H. Keenan, *Mech. Eng.* **80,** No. 5, 79 (1958).

[4]P. S. Epstein, *A Textbook of Thermodynamics*, John Wiley & Sons, Inc., New York, 1937, pp. 27–34.

(erroneously) explained the heating of blood in the lungs as resulting from frictional dissipation. Even Black had acknowledged that frictional heating could occur. The caloric theory was further challenged by the experiments of Count Rumford and Sir Humphrey Davy in about 1799, who made observations of frictional heating which indicated that, if there were a caloric fluid, it surely could be created.

Carnot's formulation[5] of the second law of thermodynamics in 1824 appeared in turn to strengthen the caloric theory. He reasoned that a constant amount of caloric "falls" through a heat engine in much the same way as water falls through a turbine, degrading its potential for doing additional work as it does so. The quantitative experimental work of James Joule from 1843 to 1849, however, placed the law of conservation of mechanical and thermal energy — the first law of thermodynamics — on firm footing.

Despite Joule's work, the caloric theory persisted for a few more years while an argument was waged over which of the two thermodynamic laws — Joule's or Carnot's — was correct. It appeared that Joule's theory of heat contradicted Carnot's law, which had been formulated in terms of the caloric theory. This question was carried forward by the young physician H. Helmholtz in 1847 and by Lord Kelvin in 1848, and it was finally resolved by R. Clausius in 1850. Clausius showed that Joule's work did not require that heat and work be mutually interchangeable under all circumstances and that Carnot's second law was correct despite his use of an erroneous theory of heat, because the second law and the first law are necessarily independent of one another.

The subsequent organizational work of Clausius, Max Planck, and J. H. Poincaré, and extensions of the subject by J. W. Gibbs, essentially completed the structure of classical thermodynamics by the beginning of the twentieth century.

There has been an important historical flaw in the growth of this subject. Thermodynamics — with its powerful generality — was born into a time when physics was not ready for it. The philosophy of Immanuel Kant (1724–1804) had included a respect for empiricism. A number of his disciples — J. G. Fichte (1762–1814), F. W. Schelling (1775–1859), and G. W. Hegel (1770–1831) — carried the great influence of Kant into the nineteenth century. However, they turned toward a more a priori way of thinking, characterized by Schelling's "Naturphilosophie," a kind of quasi-theological naturalism. This direction in philosophy collapsed with the death of Hegel, and the second quarter of the nineteenth century found physicists

[5]S. Carnot, *Reflections on the Motive Power of Heat* (translated by R. H. Thurston), American Society of Mechanical Engineers, New York, 1943.

rejecting the preceding period, which had fostered so much fruitless generalization and not enough experimentation. Now they wanted to make nonabstract statements about physical behavior on a sound empirical basis.

Thermodynamics was thus erected by pragmatic people — engineers and physiologists — and it was hesitantly accepted by physicists. This acceptance was only given to laws that were *explicit* statements of the human experience upon which the subject was based. The laws spoke of dynamic phenomena, such as working and heating as they occurred in engines.

The name thermo*dynamics* was accordingly given to a subject that is stated in terms of the dynamic processes that bridge the static states it describes. This kind of formulation has proved to be a shortcoming because it renders the axiomatic structure more complicated than it needs to be (something students have told their thermodynamics teachers for nearly 100 years). Abstract reformulations of classical mechanics, and of electricity and magnetism, have long since been made, but only as late as the 1960s did thermodynamics textbooks appear in which the axiomatic structure has been improved. The particular formulation advanced by Callen[6] is discussed in Secs. 1.4 and 1.5.

Attempts to develop clean formulations of thermodynamics, based upon abstract axioms, actually predate Callen by more than half a century. C. Carathéodory partially succeeded in this objective in 1909, and others have worked on such formulations subsequently. During the past few years a number of other thermodynamics textbooks have advanced axiomatic (or partially axiomatic) formulations. Although many of these differ from one another, they all seek to describe the same basic physical behavior. The axioms of any one system are always derivable from those of any other formulation.

STATISTICAL THERMODYNAMICS

Statistical thermodynamics has necessarily developed only since the laws of macroscopic thermodynamics were stated. This is the case because the subject is tied in at each step to the macroscopic limits of microscopic behavior. Nevertheless, well over 2000 years of argument underlay the birth of the subject in the latter nineteenth century.[7]

[6]H. B. Callen, *Thermodynamics*, John Wiley & Sons, Inc., New York, 1960.
[7]Sir James Jeans, *The Dynamical Theory of Gases*, 4th ed., Cambridge University Press, New York, 1925, pp. 11–13.

The argument began in about 400 B.C., when Leucippus (fifth century B.C.), Democritus (460–370 B.C.), and others began to espouse the view that all matter was ultimately composed of indivisible atoms. This was in conflict with the view of Empedocles (484–424 B.C.) that the four *continuous* elements — earth, air, fire, and water — were the basic stuff for all matter. Greek atomism held sway until Aristotle (384–332 B.C.) placed his great authority against it. Since much of Aristotle's philosophy proved an excellent vehicle for Christian thought, the early Christian Church embraced and promulgated his ideas. The Roman poet and philosopher, Lucretius, upheld atomism during the first century B.C., but the atomic view was given little other attention until the Renaissance.

Atomism reemerged after the first great mechanist descriptions of the world were made by Copernicus (1473–1543), Kepler (1571–1630), Galileo (1564–1642), and others. In 1658 P. Gassendi claimed that all material phenomena can be referred to the indestructible motion of atoms and can therefore be described as *kinetic*. Twenty years later R. Hooke independently advanced a similar view, but credit for the first quantitative contribution belongs to Daniel Bernoulli, who in 1738 used a kinetic description of gas molecules to derive Boyle's law.[8] This derivation, despite all its relative crudities and simplifications, provides a clear illustration of the way in which the behavior of individual particles is averaged to give a gross result. It may be paraphrased as follows.

A cylinder contains a gas that is ideal insofar as its molecules are very small and exert no forces-at-a-distance upon one another. It is sealed with a free-floating piston of variable weight w, where w is directly proportional to the pressure p that it exerts (see Fig. 1.1).

Fig. 1.1 Bernoulli's thought model.

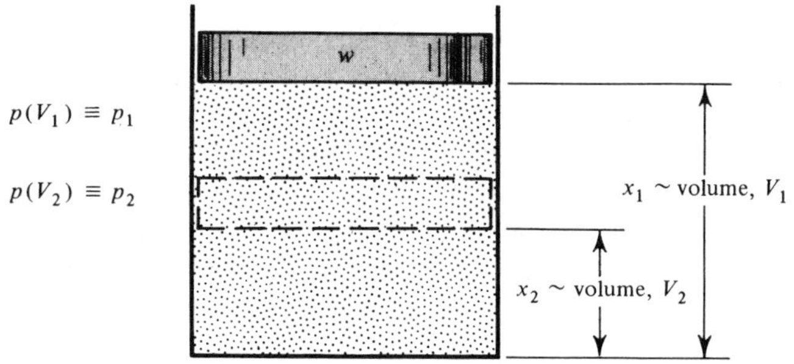

[8]D. Bernoulli, *Hydrodynamica*, Argentoria, 1738, sectio decima. A partial translation is given in J. R. Newman, *The World of Mathematics*, Simon and Schuster, Inc., New York, 1956, vol. 2.

Bernoulli noted that, as the equilibrium position of the piston is changed from x_1 to x_2 by an increase of weight, the distance between molecules, d, will decrease as the cube root of the volume ratio, V_1/V_2. But the number n of particles adjacent to the piston will only increase as the area ratio, which can be expressed as $(V_1/V_2)^{2/3}$.

The force on the piston then increases as the number of impacts of particles against it, or as n/d. Thus

$$\frac{p_1}{p_2} = \frac{n_1}{n_2}\frac{d_2}{d_1} = \left(\frac{V_2}{V_1}\right)^{2/3}\left(\frac{V_2}{V_1}\right)^{1/3} = \frac{V_2}{V_1}$$

It has been implicitly assumed in obtaining this result that the average molecular velocity is constant. If we accept for the moment that this implies an unchanging gas temperature, the result is exactly Boyle's law.

Bernoulli's brilliant foresight was, however, the last contribution to the kinetic theory before the development of macroscopic thermodynamics a century later. The foundations of the kinetic theory were laid by J. C. Maxwell, Clausius, L. Boltzmann, and others between 1857 and 1881. In 1859 Maxwell developed the distribution law of molecular velocities in a uniform gas in equilibrium; and later, in 1866, he first formulated the discussion of a nonuniform gas in a proper mathematical way. In search of an improvement of Maxwell's works, Boltzmann in 1872 established the H theorem and the famous integrodifferential equation (the Boltzmann equation) that the velocity distribution function must satisfy.

Maxwell's work on a uniform gas in equilibrium also spurred discussions of his results by many others, which eventually led to the recognition that these results did not depend on special molecular models. The science of statistical mechanics was then formed. Virtually all that is fundamental to statistical mechanics was completed by Gibbs at the beginning of this century.[9]

The years from 1900 to 1927 have given us still another chapter in the history of atomism. This period saw Planck's invention of the quantum concept and its later clarification by E. Schrödinger, W. K. Heisenberg, and others. In fairness to the Aristotelian view of matter in terms of *qualities* it should be remarked that the ideas of quantization and indeterminacy have led us a long way back from the Victorian belief that matter can be described in terms of ultimately indivisible components. These ideas have also required important changes in the Victorian formulation of kinetic theory and statistical mechanics, which are of concern to us.

[9]J. W. Gibbs, *Elementary Principles in Statistical Mechanics*, Yale University Press, New Haven, 1902; Dover Publications, Inc., New York, 1960.

1.4 POSTULATES OF MACROSCOPIC THERMODYNAMICS

In Sec. 1.3 we suggested that it might be convenient to abandon the relatively cumbersome phenomenological laws of thermodynamics in favor of a simpler set of abstract postulates. Callen[10] has combined the physical information that is included in the laws of thermodynamics into a set of four postulates. These postulates, and the system of thermodynamics based on them, are not only simple but they take a form very close to the form of the results of statistical thermodynamics. It will therefore be a great help to us, subsequently, if we review macroscopic thermodynamics from this vantage point. The postulates are given as axioms of a logical system and are considered justified a posteriori when the system proves to conform with physical experience. By choosing axioms that lead readily to the important thermodynamic relations, instead of axioms that correspond directly with human experience, Callen is able to set forth a very simple structure of thermodynamics.

The postulates are, for the sake of convenience, stated in terms of the behavior of *simple systems*. A simple system is one that is macroscopically homogeneous, isotropic, and not subject to the effects of electrical charge, chemical reactions, electrical force fields, or surface effects. The first postulate relates to the existence of equilibrium states.[11]

POSTULATE 1 There exist certain states (called equilibrium states) of simple systems that, macroscopically, are characterized completely by the internal energy, U, the volume, V, and the mole numbers, $N_1, N_2, \ldots, N_i, \ldots, N_r$ of the r chemical components.

The first postulate is borne out in the example of a simple fluid system that is initially nonuniform by virtue of, say, a temperature or concentration gradient. If the system does not interact with its surroundings, spontaneous processes will occur within the system until the nonuniformities disappear. The state the system reaches

[10] H. B. Callen, *op. cit.*

[11] The first postulate implicitly includes the *state principle* of phenomenological thermodynamics, which is often assumed tacitly in that discipline. The state principle is the basis upon which we know that two (or fewer) independent extensive variables fix the state of a simple compressible substance.

after sufficient time is unique and specifiable in terms of U, V, and the N_i's alone.

The use of the extensive properties[12] U, V, and the N_i's as independent variables, which according to the first postulate fix the equilibrium state of a system, is based upon the tacit assumption that these variables are *measurable*.[13] The postulate is thus meaningful only if operational means can be proposed to measure these variables. The volume is clearly measurable, and knowledge of the number of moles of the chemical constituents is part of the specification of the given system. That the internal energy is also measurable has to be demonstrated in the following two steps:

1. It is possible to manufacture walls that come as close to being impervious to energy loss as one might wish to have them: An ordinary Thermos bottle, for example, can be filled with a fluid; the stopper can be removed, and a stirrer through which work might enter can be introduced. A known amount of work can then be done upon the fluid to change its state. The limit of the perfectly impervious Thermos bottle (or wall) will be that which yields a maximum change of state.

2. The change of energy of the system between any reference state and the state of interest can then be measured as long as the latter state can be reached from the former by transferring work to or from the system. To do this one need only measure the work done on or by the system while it is encased in energy-impervious walls.

Notice that the terms *work* and *energy* are considered to be understood intuitively (or from the independent study of mechanics).

Still another point of great interest in the first and in the remaining postulates is that the term "heat transfer" never appears. It is instead a defined quantity. If work is delivered to or received by a system in a lesser amount than the change of energy of the system, the walls are then *not* impervious to energy transfer by means other than work. We call such means *heat;* thus: *The difference between the work transferred to a system and the resulting increase in its energy shall be called heat.* This definition is the same as the first law of phenomenological thermodynamics.

The second and third postulates give the answer to a basic problem that cannot be answered with the aid of the first postulate

[12]An *extensive* property is one that increases in direct proportion to the number of moles of the system for which the property is written. Properties that are independent of the number of moles are called *intensive*.

[13]We see, subsequently, that from the microscopic viewpoint a more compelling reason for using U, V, and the N_i's is that these quantities have fundamental microscopic meaning.

alone. Suppose that an isolated system[14] is made up of several subsystems which may or may not be isolated from one another. The problem is then that of determining the final state of this composite system when any such internal constraints as might exist (e.g., walls of any kind) are removed. The first postulate implies that such a final state will be unique but gives no way of determining it.

The second and third postulates provide means for solving this problem, which is the archetype of all problems in thermodynamics.

POSTULATE 2 There exists a function called the entropy, S, of the extensive parameters of any composite system, defined for all equilibrium states and having the following property: The values assumed by the extensive parameters in the absence of an internal constraint are those that maximize the entropy over the manifold of constrained equilibrium states.

POSTULATE 3 The entropy of a composite system is additive over the constituent subsystems. The entropy is continuous and differentiable and is a monotonically increasing function of the energy.

The second postulate is couched entirely in terms of equilibrium states. It says that when a composite system in an equilibrium state departs from this state because a constraint is removed, it seeks *one* of a number of new equilibrium states. It seeks the one for which the entropy is greatest. It would be impossible, however, to write a quantitative expression for the entropy function without the third postulate.

The additivity property in the third postulate means that

$$S(U, V, N_1, \ldots, N_r) = \sum_{\alpha} S^{(\alpha)}(U^{(\alpha)}, V^{(\alpha)}, N_1^{(\alpha)}, \ldots, N_r^{(\alpha)}) \quad (1.1)$$

where the superscript α denotes the αth system. This property when applied to λ identical subsystems gives

$$S(\lambda U, \lambda V, \lambda N_1, \ldots, \lambda N_r) = \lambda S(U, V, N_1, \ldots, N_r) \quad (1.2)$$

In other words, the entropy of a simple system is a function of the extensive parameters that is *homogeneous to the first order.*

The monotonically increasing property of the entropy function implies that

$$\left(\frac{\partial S}{\partial U}\right)_{V, N_i\text{'s}} > 0 \quad (1.3)$$

[14]An isolated system is one that is impervious to heat, mass, or work transfer. As such it has rigid, adiabatic, and impermeable walls.

This is later shown to be equivalent to the requirement that temperature is never negative.

The continuity and differentiability properties allow us to make a unique transformation of the equation

$$S = S(U, V, N_1, \ldots, N_r) \tag{1.4}$$

into the form

$$U = U(S, V, N_1, \ldots, N_r) \tag{1.5}$$

That is to say that the energy is a single-valued, continuous, and differentiable function of S, V, and the N_i's.

The combined statement of Postulates 2 and 3, which rigorously defines the entropy function, is equivalent to the second law of thermodynamics.

POSTULATE 4 The entropy of any system vanishes in the state for which

$$\left(\frac{\partial U}{\partial S}\right)_{V,N_i\text{'s}} = 0$$

The fourth postulate is shown in Sec. 1.5 to be equivalent to Planck's statement of the third law of thermodynamics, which requires that entropy vanish at the zero of temperature.

The way in which the postulates answer the basic problem of thermodynamics — that of defining the state to which an isolated system will tend when an internal constraint is removed — is very interesting. The basic or archetypal problem becomes that of evaluating the entropy function, or of writing Eq. (1.4). Therefore, *if we can write this function for a system, we can obtain all the thermodynamic information about the system from it.* This fact is of focal interest in Sec. 1.5.

EXAMPLE 1.1 The following expression was offered as an approximate representation for some thermodynamic data:

$$S = \text{constant}\left(\frac{NU}{V}\right)^{1/3}$$

Can this be a legitimate expression?

To answer this question we must assure ourselves that S is continuous, differentiable, monotonically increasing in energy, that it vanishes when $\partial U/\partial S|_{V,N}$ vanishes, and that it is additive or extensive.

The cube-root function is continuous and differentiable and

$$\left.\frac{\partial S}{\partial U}\right|_{V,N} = \frac{\text{constant}}{3}\left(\frac{N}{U^2 V}\right)^{1/3} = \text{positive}$$

so the first three conditions are met. We also see that

$$\left.\frac{\partial U}{\partial S}\right|_{V,N} \sim U^{2/3} \sim S^2$$

so that this derivative vanishes with S.

However, if the mass of the system is doubled, the extensive properties N, U, and V also double, but the resulting entropy of the system does not double. Therefore, the purported equation does not yield an extensive entropy function and cannot be correct.

1.5 FORMULATION OF MACROSCOPIC THERMODYNAMICS

FUNDAMENTAL EQUATION OF THERMODYNAMICS

Equation (1.4) or its inversion, Eq. (1.5), is called the fundamental equation for a thermodynamic system because, when it is known, all thermodynamic information can be found for the system. However, the explicit formulation of the fundamental equation for any actual system must await the statistical description of that system. Indeed, our primary aim in studying statistical mechanics will be to establish the fundamental equation. Once we have obtained it explicitly, macroscopic thermodynamics will tell us whatever else we might wish to know about the system as a whole.

We begin the formulation of a general theory by writing the fundamental equation in differential form,

$$dS = \left(\frac{\partial S}{\partial U}\right)_{V,N_i\text{'s}} dU + \left(\frac{\partial S}{\partial V}\right)_{U,N_i\text{'s}} dV + \sum_{i=1}^{r} \left(\frac{\partial S}{\partial N_i}\right)_{U,V,N_j\text{'s}} dN_i \quad (1.6)$$

where N_j signifies any mole number other than N_i. The partial derivatives in Eq. (1.6) are important enough to receive special symbols,

$$\left(\frac{\partial S}{\partial U}\right)_{V,N_i\text{'s}} \equiv \frac{1}{T}, \quad \left(\frac{\partial S}{\partial V}\right)_{U,N_i\text{'s}} \equiv \frac{p}{T}, \quad \left(\frac{\partial S}{\partial N_i}\right)_{U,V,N_j\text{'s}} \equiv -\frac{\mu_i}{T} \quad (i \neq j) \quad (1.7)$$

where $1/T$, p/T, and $-\mu_i/T$ are each functions of U, V, and the N_i's, and where T, p, and μ_i are called the *temperature*, the *pressure*, and the *chemical potential* of the ith component, respectively.

Equations (1.7) are only given as definitions in the present axiomatic formulation of thermodynamics. It remains for us to prove that temperature, pressure, and chemical potential (thus defined) have the same meanings that we are accustomed to giving these words.

It is easy to show that T, p, and μ_i are intensive properties. For example, the temperature of a system consisting of λ subsystems is identical to that of the individual subsystems, because

$$\begin{aligned}\frac{1}{T} &\equiv \frac{\partial}{\partial(\lambda U)} S(\lambda U, \lambda V, \lambda N_1, \ldots, \lambda N_r) \\ &= \frac{\partial}{\partial(\lambda U)} \lambda S(U, V, N_1, \ldots, N_r) \\ &= \frac{\partial}{\partial U} S(U, V, N_1, \ldots, N_r) \equiv \frac{1}{T}\end{aligned}$$

Substitution of Eq. (1.7) into Eq. (1.6) then gives

$$dS = \frac{1}{T} dU + \frac{p}{T} dV - \sum_{i=1}^{r} \frac{\mu_i}{T} dN_i \tag{1.8}$$

which can be rearranged into the form

$$dU = T\,dS - p\,dV + \sum_{i=1}^{r} \mu_i\,dN_i \tag{1.9}$$

where T, $-p$, and the μ_i's are each functions of S, V, and the N_i's. On the other hand, Eq. (1.9) could have been derived directly from Eq. (1.5). The extensive variable and its associated intensive variable in each term of Eqs. (1.8) and (1.9) are known as *conjugate variables*. For example, $1/T$ and U are conjugate variables. So are $-p$ and V or μ_i and N_i, and so forth.

In the case of a simple system for which the mole numbers are constrained to be constant,

$$dU = T\,dS - p\,dV \tag{1.10}$$

The integration of Eq. (1.10) is restricted to quasi-static or reversible processes by the equilibrium requirement of the second postulate. The equation can nevertheless be used for any process, because it only relates state points to one another. If two points are actually connected by an irreversible path, all we have to do is to make our calculations along some other reversible path between the same two points. From the definition of heat in Sec. 1.4 we can now write[15]

$$dU = \delta Q + \delta Wk \tag{1.11}$$

where, in this case, we specify that the heat and work must be done reversibly. If the defined quantity, p, is indeed the pressure, as we understand the word, then the reversible mechanical work is given by the laws of mechanics as

$$\delta Wk = -p\,dV \tag{1.12}$$

[15]The symbol δ denotes an imperfect or inexact differential — an infinitesimal change in a quantity that cannot be expressed as a function of state variables.

Equation (1.12) is consistent with a sign convention that calls work done on a system, positive. Then from Eq. (1.11) we obtain for reversible heat transfer

$$\delta Q = T\,dS \tag{1.13}$$

Equation (1.13) is consistent with our experience (as expressed by phenomenological thermodynamics), so we can be satisfied that p is the conventional pressure.

THERMAL EQUILIBRIUM AND THE MEANING OF TEMPERATURE

We must next relate the defined term *temperature* to the intuitive understanding of temperature. Consider an isolated system comprised of two subsystems separated by rigid, impermeable, but diathermal walls. Denoting the subsystems by the subscripts 1 and 2 we can write

$$U_1 + U_2 = \text{constant} \tag{1.14}$$

or

$$dU_1 = -dU_2 \tag{1.15}$$

The system at equilibrium must satisfy the maximum entropy requirement of the second postulate and the additivity requirement of the third postulate. Thus, since the N_i's and V are constant,

$$dS = dS_1 + dS_2 = \left(\frac{\partial S_1}{\partial U_1}\,dU_1 + \frac{\partial S_2}{\partial U_2}\,dU_2\right) = 0 \tag{1.16}$$

Substitution of Eqs. (1.7) and (1.15) into Eq. (1.16) gives, for dS,

$$dS = \left(\frac{1}{T_1} - \frac{1}{T_2}\right) dU_1 = 0 \tag{1.17}$$

It follows that at equilibrium

$$T_1 = T_2$$

Suppose, now, that the subsystems had initially been separated by an adiabatic wall and in equilibrium at temperatures T_1 and T_2 such that $T_1 > T_2$. Then let the adiabatic wall be replaced with a diathermal wall so that the system is no longer in equilibrium. The entropy of the system, in accordance with the second postulate, assumes a larger value at a new equilibrium state such that

$$\Delta S > 0 \tag{1.18}$$

If the difference between the initial values of T_1 and T_2 is small, Eq. (1.17) can be approximated by

$$\Delta S \simeq \left(\frac{1}{T_1} - \frac{1}{T_2}\right) \Delta U_1 \tag{1.19}$$

Since the expression in parentheses is negative, ΔU_1 must also be negative. It follows that spontaneous heat flow must take place from bodies of higher temperature to bodies of lower temperature. This is in agreement with our intuitive understanding of the temperature. Similar considerations of mechanical and chemical equilibrium reveal that the defined pressure and chemical potential are also consistent with our understanding of these words.

It is now clear why the fourth postulate is equivalent to Planck's statement of the third law of thermodynamics — that the entropy of any pure substance vanishes at $T = 0$. But what justification actually exists for the third law or the fourth postulate? Since we actually have no direct experience of any kind at $T = 0$, any claims we make must either be extrapolative or they must be based upon a microscopic formulation.

The original statement of the third law was set down by W. H. Nernst, who generalized certain experiments at low temperatures. What Nernst actually found was that the change $\Delta\mu$ during any spontaneous isothermal process appeared to be such that

$$\lim_{T\to 0} \left.\frac{\partial \Delta\mu}{\partial T}\right|_p = 0 \tag{1.20}$$

But it is easy to show [see Eq. (1.34)] that $(\partial\mu/\partial T)_p = S/N$, whereupon Eq. (1.20) becomes

$$\lim_{T\to 0} (\Delta S) = 0 \tag{1.20a}$$

where ΔS is the increase of entropy in any spontaneous isothermal process.

Nernst accordingly could conclude, on the basis of macroscopic experiments, that the entropy approached a constant value as T approached absolute zero. Planck's subsequent claim that the value of this constant should be zero was based upon Boltzmann's entropy formula, $S \sim \ln W$. We discuss this expression in chapter 3. For the present it suffices to say that W is the number of distinguishable ways that we can distribute the indistinguishable molecules of the substance among the available energy levels. Planck thought that only one arrangement was possible for a pure substance in the zero-temperature state, and that S would have to vanish. The subsequent development of modern quantum theory suggests that some rearrangements might still be possible at absolute zero and that the entropy could possibly remain finite. While the validity of Planck's statement is still being argued, Nernst's hypothesis continues to be an acceptable physical principle that appears to be subject to no exceptions.

EQUATIONS OF STATE

The intensive variables given in Eq. (1.7) are all of the form $(\partial S/\partial X_i)_{X_j\text{'s}}$. The corresponding variables in the energy formulation are

$$P_i = \left(\frac{\partial U}{\partial X_i}\right)_{X_j\text{'s}} = P_i(S, V, N_1, \ldots, N_r) \tag{1.7a}$$

where P_i is any intensive variable and X_i is the conjugate extensive variable. As a consequence of the homogeneity requirement we can, if we wish, introduce $\lambda = 1/N_r$ in the following way:

$$\begin{aligned} P_i &= P_i(\lambda S, \lambda V, \lambda N_1, \ldots, \lambda N_r) \\ &= P_i\left(\frac{S}{N_r}, \frac{V}{N_r}, \frac{N_1}{N_r}, \ldots, \frac{N_{r-1}}{N_r}, 1\right) \end{aligned} \tag{1.21}$$

Each of the $r + 2$ equations for the P_i's now depends upon $r + 1$ new intensive variables. The fact that a simple system of r components has $r + 1$ independent variables or "degrees of freedom" is a special case of the Gibbs phase rule. Gibbs's phase rule, which we do not prove here, states that

$$f = r - \Phi + 2 \tag{1.22}$$

where f is the number of thermodynamic degrees of freedom in a system with r components and Φ phases.

A relation among intensive and extensive properties such as Eq. (1.7a) is called an *equation of state*. The fundamental equation for a simple system can be obtained from $r + 2$ independent equations of state or vice versa. This can be nicely illustrated in the following example. The fundamental equation of a perfect monatomic gas (which we obtain by statistical-mechanical means in chapter 7) is

$$S = \left(\frac{N}{N_0}\right) S_0 + NR^0 \ln\left[\left(\frac{U}{U_0}\right)^{3/2}\left(\frac{V}{V_0}\right)\left(\frac{N_0}{N}\right)^{5/2}\right] \tag{1.23}$$

where the subscript $_0$ refers to a reference state and R^0 is the universal gas constant. Then the $r + 2$, or 3, equations of state are

$$\frac{1}{T} = \frac{\partial S}{\partial U} = \frac{3}{2}\frac{NR^0}{U} \tag{1.24}$$

$$\frac{p}{T} = \frac{\partial S}{\partial V} = \frac{NR^0}{V} \tag{1.25}$$

and

$$-\frac{\mu}{T} = \frac{\partial S}{\partial N} = \frac{S_0}{N_0} + R^0 \ln\left[\left(\frac{U}{U_0}\right)^{3/2}\frac{V}{V_0}\left(\frac{N_0}{N}\right)^{5/2}\right] - \tfrac{5}{2}R^0 \tag{1.26}$$

These equations can be rearranged into the familiar relations for a perfect monatomic gas,

$$U = \tfrac{3}{2} NR^0T \qquad \left(\text{or } c_v \equiv \frac{1}{N}\frac{\partial U}{\partial T} = \tfrac{3}{2} R^0\right) \tag{1.27}$$

$$pV = NR^0T \tag{1.28}$$

and

$$\mu = R^0T[\phi(T) + \ln p] \tag{1.29}$$

where

$$\phi(T) \equiv \frac{5}{2} - \frac{S_0}{N_0R^0} - \ln \frac{N_0R^0T^{5/2}}{V_0{T_0}^{3/2}} \tag{1.30}$$

EULER AND GIBBS–DUHEM EQUATIONS

The extensive nature of internal energy enables us to write, as we did for entropy,

$$U(\lambda S, \lambda V, \lambda N_1, \ldots, \lambda N_r) = \lambda U(S, V, N_1, \ldots, N_r) \tag{1.31}$$

Differentiation of this expression with respect to λ gives

$$\frac{\partial U(\lambda S, \ldots)}{\partial(\lambda S)} S + \frac{\partial U(\lambda S, \ldots)}{\partial(\lambda V)} V + \cdots = U(S, V, \ldots)$$

or

$$U = TS - pV + \sum_{i=1}^{r} \mu_i N_i \tag{1.32}$$

This can be rearranged as

$$S = \frac{1}{T}U + \frac{p}{T}V - \sum_{i=1}^{r} \frac{\mu_i}{T} N_i \tag{1.33}$$

We will make use of this result later to evaluate the absolute entropy of systems. Equations (1.32) and (1.33) are called the Euler equations in the energy representation and in the entropy representation, respectively.

Differentiating either form of the Euler equation and combining it with the differential fundamental equation gives the Gibbs–Duhem equation. Thus from Eqs. (1.32) and (1.9) we obtain

$$S\,dT - V\,dp + \sum_{i=1}^{r} N_i\,d\mu_i = 0 \tag{1.34}$$

and from Eqs. (1.33) and (1.8) we obtain

$$U\,d\!\left(\frac{1}{T}\right) + V\,d\!\left(\frac{p}{T}\right) - \sum_{i=1}^{r} N_i\,d\!\left(\frac{\mu_i}{T}\right) = 0 \tag{1.35}$$

Either of these forms of the Gibbs–Duhem equation gives the relationship among the intensive parameters in differential form.

If all the equations of state are known for a system, their substitution into the Euler equation yields the fundamental equation for the system. In other words, knowing all the equations of state of a system is equivalent to having all conceivable thermodynamic information about the system. If any one equation of state is lacking, it can be obtained for use with the Euler equation (within a constant of integration) by integrating the Gibbs–Duhem equation.

LEGENDRE TRANSFORMS

Up to this point the dependent variable, S or U, has been expressed in terms of extensive parameters. But such intensive parameters as T, p, and μ_i are often more easily measured than the extensive parameters. In most systems the use of a combination of intensive and extensive parameters as independent variables is particularly convenient. Thus we should like to be able to transform dependent variables that depend upon extensive parameters into new dependent variables that depend upon intensive parameters or a combination of intensive and extensive parameters. The transformation that accomplishes this without losing any of the essential content of the original system is called the *Legendre transformation.*

Consider, for example,

$$dU(S, V, N_1, \ldots, N_r) = T\,dS - p\,dV + \sum_{i=1}^{r} \mu_i\,dN_i \qquad (1.9)$$

The following simple transformations allow us to exchange the conjugate intensive and extensive parameters. Since

$$d(TS) = T\,dS + S\,dT \qquad (1.36)$$

and

$$d(pV) = p\,dV + V\,dp \qquad (1.37)$$

substituting Eqs. (1.36) or (1.37), or both in Eq. (1.9) gives

$$d(U - TS) = -S\,dT - p\,dV + \sum_{i=1}^{r} \mu_i\,dN_i \qquad (1.38)$$

or

$$d(U + pV) = T\,dS + V\,dp + \sum_{i=1}^{r} \mu_i\,dN_i \qquad (1.39)$$

or

$$d(U + pV - TS) = -S\,dT + V\,dp + \sum_{i=1}^{r} \mu_i\,dN_i \qquad (1.40)$$

The newly generated functions are, in this particular example, the *Helmholtz function*,

$$F(T, V, N_1, \ldots, N_r) \equiv U - TS \tag{1.41}$$

the *enthalpy function*,

$$H(S, p, N_1, \ldots, N_r) \equiv U + pV \tag{1.42}$$

and the *Gibbs function*,

$$G(T, p, N_1, \ldots, N_r) \equiv U + pV - TS = H - TS \tag{1.43}$$

Thus we have created three new forms of the fundamental equation. While they contain the same thermodynamic information as before, they are now expressed in terms of new independent variables. In different situations it is often more convenient to deal with different sets of independent (or *natural*) variables. In general, we can define a new fundamental parameter, $\psi(P_0, X_1)$, in terms of the old fundamental parameter $Y(X_0, X_1)$ as follows

$$\psi \equiv Y - \left(\frac{\partial Y}{\partial X_0}\right) X_0 = Y - P_0 X_0 \tag{1.44}$$

because if ψ is differentiated we get

$$d\psi = -X_0\, dP_0 + P_1\, dX_1 \tag{1.45}$$

where P is $\partial Y/\partial X$. The Legendre transformation can also be extended to change more than one independent variables as we did in Eq. (1.43).

Another set of functions, and their corresponding independent variables, can be obtained by transforming the entropy instead of the energy. These transforms of the entropy are known as *Massieu functions*.

$$-\frac{F}{T}\left(\frac{1}{T}, V, N_1, \ldots, N_r\right) \equiv S - \frac{1}{T}U \tag{1.46}$$

One of the Legendre transforms, the Gibbs function, has a special meaning. Its definition, Eq. (1.43), combined with the Euler equation, Eq. (1.32), gives

$$G = \sum_{i=1}^{r} \mu_i N_i \tag{1.47}$$

Hence, for a single-component system,

$$\mu = \frac{G}{N} \tag{1.48}$$

Therefore, the chemical potential of a component in a multicomponent system is the molar Gibbs function when the component is left alone.

EXAMPLE 1.2 The energy density (U/V) of blackbody radiation is known to be $(4/c_l)\sigma T^4$, where c_l is the speed of light and σ the Stefan–Boltzmann constant. Determine the form of the fundamental equation whose independent variables are V and T, and obtain expressions for pressure and the specific heats from it.

The energy form of the differential fundamental equation in this case is

$$dU = T\,dS - p\,dV$$

The Legendre transform ψ, which will interchange the variables T and S, is obtained from the same arguments that led to Eq. (1.41) and introduced the Helmholtz function, $F(T, V) \equiv U - TS$.

Now we need $U(T, V)$ and $S(T, V)$ for substitution in $F \equiv U - TS$ to complete the development of the fundamental equation. We already have $U(T, V)$, and to get $S(T, V)$ we can integrate the first of Eq. (1.7),

$$S = \left(\int \frac{1}{T} dU\right)_{T=0}^{T=T} = \int_0^T \frac{V}{T}\frac{16\sigma T^3}{c_l}\,dT = \frac{16\sigma}{3c_l} T^3 V$$

where we have used the fact that $S = 0$ when $T = 0$. Thus

$$F = \left(\frac{4\sigma}{c_l} T^4 - \frac{16\sigma}{3c_l} T^4\right) V = -\frac{4\sigma}{3c_l} VT^4$$

The equations of state consist of the expression for $S(T,V)$, above, and the well-known radiation pressure law,

$$p = -\left.\frac{\partial F}{\partial V}\right|_T = \frac{4\sigma}{3c_l} T^4$$

or

$$p = \frac{U}{3V}$$

The specific heats of blackbody radiation are

$$c_v = T \left.\frac{\partial S}{\partial T}\right|_V = \frac{16\sigma}{c_l} T^3 V$$

and, since $p = p(T)$,

$$c_p = T \left.\frac{\partial S}{\partial T}\right|_p \to \infty$$

EXAMPLE 1.3 We discover in chapter 8, in our consideration of the "grand canonical ensemble," that it is sometimes desirable to characterize a single-component system with the independent variables $1/T$, V, and μ/T. What will be the appropriate dependent variable ψ for a system such that $\psi(1/T, V, \mu/T)$ is the fundamental equation for the system?

To obtain this variable we go to Eq. (1.8) and note that we wish to reverse the conjugate variables $1/T$ and U, as well as $-\mu/T$ and N. Since V is already an independent variable we leave it alone. The Legendre transform that will do this is obviously

$$\psi \equiv S - \frac{U}{T} + \frac{N\mu}{T}$$

But from Eq. (1.33) we see that this is exactly $\psi = pV/T$. The result can be divided by R^0 without any loss of generality. Thus we obtain

$$\frac{pV}{R^0 T} = f\left(\frac{1}{T}, V, \frac{\mu}{T}\right)$$

The new dependent variable is a Massieu function and it is equal to the *compressibility factor*, pV/R^0T. The function f must reduce to N in the ideal-gas limit. The evaluation of the equations of state in this case is left as an exercise (Problem 1.10).

1.6 LINEAR TRANSPORT RELATIONS

We have noted that the classical thermodynamics we have been treating does not describe the nonequilibrium behavior of physical systems and that other theoretical frameworks must be called upon to treat such systems. On the microscopic level, kinetic theory can be used. However, much of nonequilibrium behavior can, in fact, be treated effectively with a macroscopic approach that falls outside the framework of macroscopic thermodynamics. This approach should be familiar to us as the one that leads to such expressions as the Navier–Stokes equations, the energy or heat-conduction equation, and the species or mass-diffusion equation. These equations are derived on the basis of three linear transport relations, which are the physical laws (or axioms) upon which the macroscopic linear transport theory is based.

The first is Newton's law of unidimensional viscous shear,

$$\tau = -\mu \frac{du}{dy} \tag{1.49}$$

where τ and u are the shear stress and fluid velocity in the x direction, y is the coordinate perpendicular to the plane of the shear stress, and μ is an empirical coefficient called the viscosity.

The second is Fourier's law of heat conduction,

$$q = -\lambda \frac{dT}{dy} \tag{1.50}$$

where q is the heat flux and λ is an empirical coefficient called the thermal conductivity.

The third is Fick's law of mass diffusion,[16]

$$J_m = -D \frac{dn}{dy} \tag{1.51}$$

where J_m is the mass flux, n the concentration, and D an empirical coefficient called the coefficient of diffusion.

Like many other linear physical laws, these expressions are approximations to a far more complex behavior. Any discussion of

[16] This and Eq. (1.50) can be expressed for three dimensions in vector form as $\mathbf{q} = -\lambda \nabla T$ and $\mathbf{J}_m = -D\nabla n$. We find in chapter 2 that Eq. (1.49) is one of nine components of the tensor which fully represents the state of stress at a point.

the underlying behavior and its complexities must be based on the microscopic study of particle (electronic, atomic, or molecular) motions in the system under consideration. The microscopic kinetic description is discussed in chapters 2, 11, and 12. In particular, we develop the kinetic tools needed to deal accurately with transport processes in dilute gases. One of the major results of the kinetic description is the phenomenon of "coupling." If gradients of n and T exist simultaneously, for example, then J_m depends upon both gradients; q also depends upon both gradients. The phenomenological theory of coupling phenomena constitutes the subject of *irreversible* or *nonequilibrium thermodynamics*.[17]

Equations (1.49), (1.50), and (1.51) are actually quite accurate for processes in which only one gradient is present as long as u, T, and n change very little in a distance equal to the mean travel of molecules between collisions. In chapter 11, we show how these linear laws follow from elementary kinetic considerations when this is the case.

Problems 1.1 The following equations are purported to be fundamental equations of various thermodynamic systems. Find those which are not physically permissible, and indicate the postulate violated by each. The quantities K_1, K_2, and K_3 are positive constants and, in all cases in which fractional exponents appear, only the real positive root is to be taken.

(a) $S = K_1(NVU)^{1/3}$.

(b) $S = K_2(V^3/NU)$.

(c) $U = K_3(S^2/V) \exp (S/NR^0)$.

1.2 Find two equations of state from the fundamental equation for the diatomic ideal gas,

$$S = \frac{N}{N_0} S_0 + NR^0 \ln \left[\left(\frac{U}{U_0}\right)^{7/2} \frac{V}{V_0} \left(\frac{N_0}{N}\right)^{9/2} \right]$$

where S_0 is $[9N_0R^0/2 - N_0\,(\mu/T)_0]$.

1.3 Find the fundamental equation and the equations of state of a monatomic ideal gas in the Helmholtz function representation, $F = F(T, V, N)$.

1.4 Consider an isolated composite system consisting of two simple systems separated by a movable diathermal wall that is impervious to the flow of matter. Determine the condition of mechanical equilibrium.

[17]See, for instance, S. R. DeGroot and P. Mazur, *Non-Equilibrium Thermodynamics*, Interscience Publishers, New York, 1962.

1.5 Consider an isolated composite system consisting of two simple systems separated by a rigid, diathermal wall, permeable to one type of material and impermeable to all others. Determine the condition of chemical equilibrium (no matter flow).

1.6 Find the Legendre transform of the entropy function with $1/T$, V, and N as independent variables. Identify U, p/T, and μ/T as derivatives of the transformed function.

1.7 Two systems of monatomic ideal gas are separated by a diathermal wall. In system 1 there are 2 moles initially at 250°K; in system 2 there are 3 moles initially at 350°K. Find U_1, U_2, and the common temperature after equilibrium has been reached.

1.8 One thousand liters of ideal He gas occupy a tank at $\frac{1}{2}$ atm and 20°C; 1000 liters of ideal He gas occupy a second tank at 1 atm and 80°C. If the tanks are rigid and adiabatic, find the final temperature and pressure in them after a connecting valve has been opened. (The total energy is, of course, constant throughout the process.)

1.9 One mole of a monatomic ideal gas is in a cylinder with a movable piston. The surrounding pressure is 1 atm. How much heat must be transferred to raise the volume from 20 to 50 liters?

1.10 Identify U, p/T, and N as derivatives of the fundamental equation in terms of ψ in Example 1.3. Discuss the equations in the ideal-gas limit.

1.11 Determine the Legendre transform $\psi(T, V, \mu)$ of the fundamental equation $F(T, V, N) = -NkT \ln [Z(T, V)/N]$, where k is a constant and Z is a function of T and V.

1.12 Prove the c_p and c_v must vanish for any substance as $T \to 0$.

1.13 Given that $c_v(T) = A + BT$ for a particular van der Waals gas (see Eq. (9.27)), obtain the fundamental equation for the gas in terms of v and T.

2 kinetic description of dilute gases

2.1 INTRODUCTION

The explanation of ideal-gas behavior in terms of molecular action is an exercise that demonstrates many of the basic strategies of statistical thermodynamics. We use this exercise to develop techniques and to clarify definitions that are useful throughout this book. Before going to the more restrictive consideration of ideal gases it is well to set down three very general hypotheses.

1. The molecular hypothesis states that matter is composed of small discrete particles (atoms or molecules) and that any macroscopic volume contains a vast number of particles. This hypothesis is, for example, satisfied in a *very* low density gas only if we consider a "macroscopic volume" to be one that is rather large. However, a cubic centimeter of ideal gas at 0°C and 1 atm contains the staggering number of 2.7×10^{19} molecules, so we seldom encounter circumstances in which the hypothesis fails.

2. The statistical hypothesis states that any macroscopic observation takes place over a length of time that is much longer than any characteristic time scale of molecular motion. Such a time scale

might be the average time elapsed between two subsequent collisions of a given molecule. This would be on the order of 10^{-10} sec for most gases at 0°C and 1 atm — much less time, indeed, than is needed for any kind of macroscopic observation.

3. The kinetic hypothesis embodies whatever we choose to say about the motions of molecules. Although by its nature it is more restrictive than the other two, it usually includes the assumption that particles (atoms and molecules) obey the laws of classical mechanics.

The first two of these hypotheses are basic to both kinetic theory and statistical mechanics. They are true physical hypotheses in the sense that they explain certain qualities of microscopic behavior using an intuitive notion of "macroscopicness." Because our technical backgrounds are such that we know at the outset all they have to tell us about microscopic behavior, we might do better to regard them as clarifications of the adjective "macroscopic."

Such emphasis may appear to be unnecessarily pedantic, but considerable confusion can follow failure to distinguish among *definitions*, *empirical facts*, and the *analytical results* of combining definitions and facts. This caution must be emphasized again in the very important problem of fixing the notion of an ideal gas.

An "ideal gas" can be defined either in terms of the properties of the molecules that comprise it, or in terms of the macroscopic behavior of the gas. Either definition must be consistent with the other, and neither definition should include requirements that do not somehow appear in the other. For example, to include in the microscopic definition of an ideal gas the requirement the atoms are blue would be wrong, because there would be no macroscopic manifestation of this blueness.

On the macroscopic level we define an ideal gas only by requiring that it obey Boyle's and Charles's laws:

$$\text{Boyle's law:} \quad pv = f_1(T) \tag{2.1}$$

$$\text{Charles's law:} \quad \frac{v}{T} = f_2(p) \tag{2.2}$$

The unknown functions f_1 and f_2 can be obtained by eliminating v from Eqs. (2.1) and (2.2):

$$\frac{f_1(T)}{p} = Tf_2(p) \tag{2.3}$$

It is obvious from the form of Eq. (2.3) that

$$f_2(p) = \frac{R^0}{p} \qquad \text{or} \qquad f_1(T) = R^0T \tag{2.4}$$

Then, from either Eq. (2.1) or (2.2), we obtain

$$\frac{p}{T} = \frac{R^0}{v} \tag{2.5}$$

and the constant R^0 is determined experimentally to be 1.986 Btu/lb_m mole-°R or 1.986 cal/g mole-°K.

Equation (2.5) is a particular case of Eq. (1.21) and is, in fact, the same as Eq. (1.25); it is one of the three equations of state that are needed to describe any substance. The other two,

$$\frac{1}{T} = \left(\frac{\partial S}{\partial U}\right)_{V,N} \quad \text{and} \quad -\frac{\mu}{T} = \left(\frac{\partial S}{\partial N}\right)_{U,V}$$

depend upon specific-heat information and must await specification of the particular kind of ideal gas before they can be worked out. [Recall Eqs. (1.23) through (1.26), which were claimed to apply to ideal *monatomic* gases.]

We do not use the macroscopic definition directly in the present study. However, the fact that Boyle's and Charles's laws, later, are derived from the microscopic definition will give us confidence in the statistical methods.

The microscopic definition of an ideal gas imposes two restrictions upon its molecules. These constitute the kinetic hypothesis for the gas:

1. The gas is dilute. Its molecules are so widely spaced that
   (a) The combined volume of the particles is much less than the total volume that they occupy.
   (b) Intermolecular forces are only important during the brief instant of collision and can be neglected at all other times.
   (c) The average duration of collision as defined above is negligible with respect to the average time between collisions.

Entries (b) and (c) tell us that the molecules are *independent* of one another. They exert no influence upon one another except in the brief instant of collision.

2. No energy is dissipated in the collisions.

2.2 TERMINOLOGY AND BASIC CONCEPTS

Before we can put the microscopic definition of an ideal gas to use in the computation of macroscopic properties, it will be necessary to establish a vocabulary with which to talk about molecular behavior.

POSITION VECTOR

The position vector **r** is (see Fig. 2.1)

$$\mathbf{r} \equiv \mathbf{i}x + \mathbf{j}y + \mathbf{k}z$$

and

$$r \equiv |\mathbf{r}| = \sqrt{x^2 + y^2 + z^2}$$

MOLECULAR VELOCITY AND SPEED

The absolute *velocity* of a molecule **c** is

$$\mathbf{c} \equiv \mathbf{i}u + \mathbf{j}v + \mathbf{k}w$$

where u, v, and w are the component speeds in the x, y, and z directions, respectively (see Fig. 2.1). The *speed* c is the magnitude of the vector **c**:

$$c \equiv |\mathbf{c}| = \sqrt{u^2 + v^2 + w^2}$$

MOMENTUM OF A PARTICLE

The momentum of a molecule or other moving object of mass m is expressed either as a vector **p** or a scalar p:

$$\mathbf{p} \equiv m\mathbf{c} \qquad \text{or} \qquad p \equiv mc$$

LOCAL NUMBER DENSITY

The local number density n of particles of a gas is defined as the number of particles in ΔV, divided by ΔV, where ΔV is a very small increment of volume surrounding the point of interest. In any real system ΔV would have to be extremely small before molecular motions could cause n to be perceptibly nonuniform. In the limit as $\Delta V \rightarrow 0$, however, n has no meaning. The definition is therefore couched in terms of a ΔV that shrinks to a very small, but still finite, value.

Fig. 2.1 Velocity and position vectors.

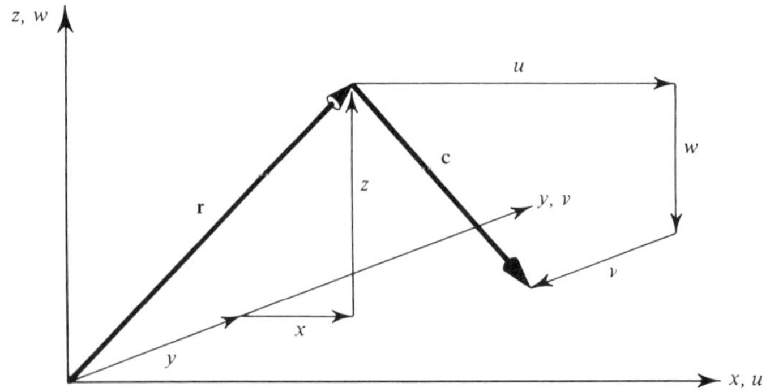

LOCAL MASS DENSITY

The local mass density ρ of a gas must also be defined on the basis of a volume sufficiently large to contain a statistically homogeneous sample of particles. Thus ρ is the mass in a very small increment of volume ΔV divided by ΔV.

MASS OF A MOLECULE

The mass of a molecule m can be calculated from

$$m = \frac{M}{N_A} \tag{2.6}$$

where M is the molecular weight of the gas and N_A is Avogadro's constant, $N_A = 6.02 \times 10^{23}$ molecules/g mole. Thus, for example, for oxygen (O_2)

$$m_{O_2} = \frac{32 \text{ g/g mole}}{6.02 \times 10^{23} \text{ molecules/g mole}} = 5.3 \times 10^{-23} \text{ g/molecule}$$

Throughout this book we view N_A as a conversion factor that converts N moles to N molecules, or m g/mole to m g/molecule. The distinction need not be made explicitly because the requirement of dimensional homogeneity dictates the correct units. We also note that

$$\rho \text{ g/cm}^3 = m \text{ g/molecule} \times n \text{ molecules/cm}^3 \tag{2.7}$$

PHASE SPACE

We now propose a method for simultaneously characterizing both the position and the velocity of particles. The method simply consists of identifying the six coordinates of position and momentum (or velocity) of a particle at a given instant. The hyperspace defined by these six coordinates was given the name phase space[1] by Gibbs, who found that it was expedient to deal with this space mathematically.

Before we develop the notion of phase space further, it is well to recollect some notions relating to position space. Figure 2.2 depicts an infinitesimal volume element in position space. The element located at position $\mathbf{r}$ has a volume[2] δV, such that $\delta V = \delta x\, \delta y\, \delta z$.

Any coordinate scheme serves to display $\mathbf{r}$ and δV. Figure 2.3, for example, shows a volume element in spherical coordinates which will prove useful in our later work. In this case δV can be expressed (see Fig. 2.3) as $\delta V = r^2 \sin\theta\, \delta\phi\, \delta\theta\, \delta r$.

[1]Technically the name "phase space" is given only to momentum-position space, not to velocity-position space.
[2]The free *variation* δ of x, y, z, or V is employed here in place of the differential d because there is no functional restriction upon the variables and the changes are wholly arbitrary.

Fig. 2.2 Elemental volume in Cartesian position space.

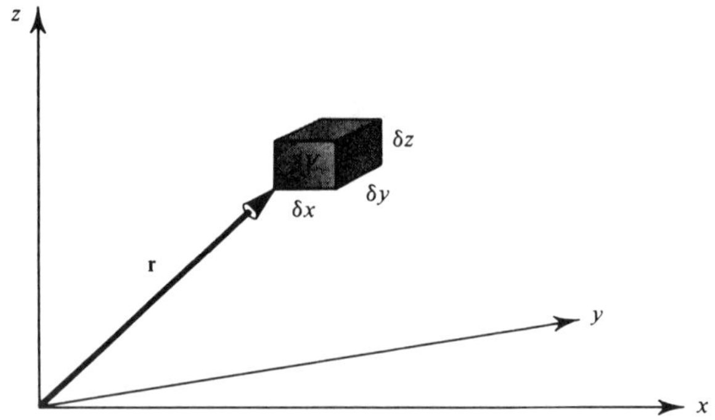

Fig. 2.3 Elemental volume in spherical coordinates.

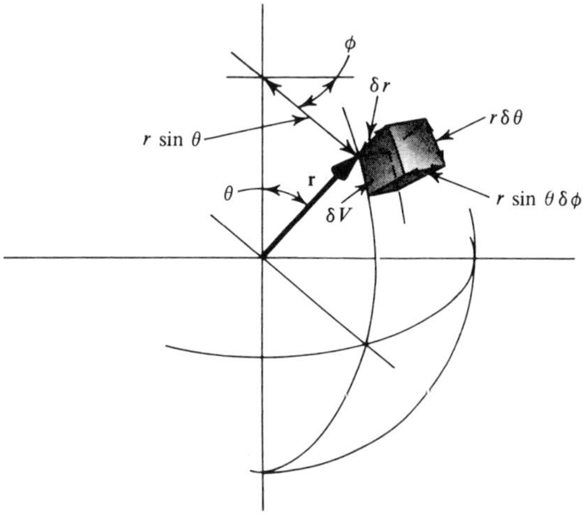

In a completely analogous fashion we can describe a velocity or momentum "space" in the coordinates u, v, and w or p_u, p_v, and p_w. Just as δV has the physical significance of being an infinitesimal collection of adjacent positions that a particle might assume, an element in velocity or momentum space is a collection of adjacent velocities or momenta that it might assume. Accordingly, for Cartesian velocity coordinates we define

$$\delta\Omega \equiv \delta u\, \delta v\, \delta w$$

and

$$\delta\Phi \equiv \delta p_u\, \delta p_v\, \delta p_w = m^3\, \delta\Omega$$

It follows that for spherical velocity coordinates

$$\delta\Omega = c^2 \sin\theta\, \delta\theta\, \delta\phi\, \delta c$$

and

$$\delta\Phi = p^2 \sin\theta\, \delta\theta\, \delta\phi\, \delta p$$

Phase space is a six-coordinate hyperspace in which a "volume" element $\delta\tau$ has the physical significance of embracing an infinitesimal collection of adjacent states of position and momentum which a particle might assume: $\delta\tau \equiv \delta x\, \delta y\, \delta z\, \delta p_u\, \delta p_v\, \delta p_w$. This element is usually represented schematically by a sketch such as is shown in Fig. 2.4.

An insight into the conceptual utility of the phase-space representation might be gained by reconsidering the relatively clumsy representation that had to be used in its absence in Fig. 2.1.

The reader acquainted with the Heisenberg uncertainty principle (see Sec. 4.5) might well wonder about the frequent implication that we intend to specify the exact simultaneous position and momentum of particles. There are, as it happens, circumstances in which we can deal with particles as though they had exact locations in phase space. In the many cases for which we cannot, however, we will speak of particles as "occupying" small finite elements in phase space at given instants.

DISTRIBUTION FUNCTION

The concept of a distribution function f is among the most important notions in the field of statistical thermodynamics. It is the basis upon which, time and again, we shall make the transition from the behavior of single particles to the gross behavior of molecular aggregates. To gain an insight into its meaning let us forget molecular aggregates and consider a large basket filled with equal numbers of red and black marbles.

Imagine that a blindfolded man reaches into the basket and withdraws samples of 10 marbles. After each such observation, he returns the sample to the basket and mixes it in with the rest. Figure 2.5(a) shows the results of 1024 of his drawings. The number of occurrences of different percentages of red marbles in the

Fig. 2.4 Elemental volume in Cartesian phase space.

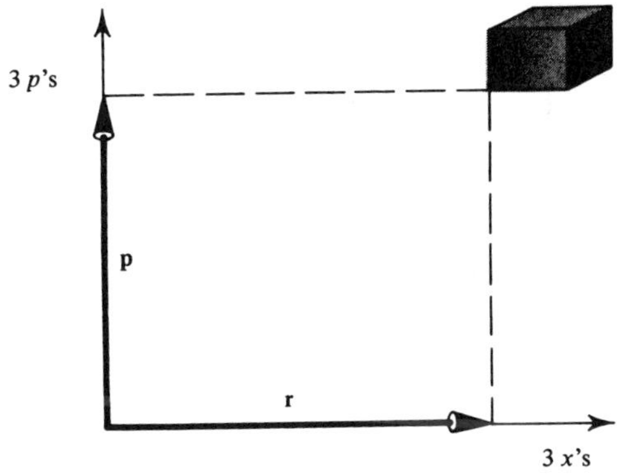

Fig. 2.5 Distribution histograms: (a) a posteriori distribution and (b) a priori distribution

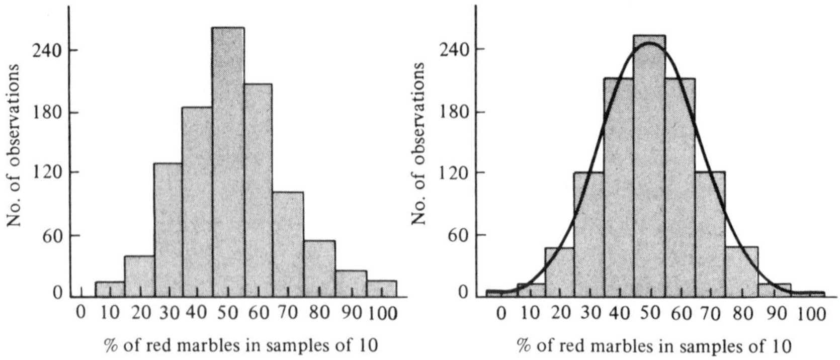

(a) (b)

samples is plotted against these percentages. Such a plot is called a *histogram*.

We can now obtain a posteriori or "empirical" probabilities from Fig. 2.5(a). For example, in 130 drawings three marbles were red; hence we would conclude that the likelihood of drawing three red marbles in a sample of 10 would be 130/1024, or 12.7 percent.

This problem is well known in applied statistics, and the distribution of percentages can be predicted by the normal distribution. The results would be a priori or "theoretical" probabilities, and they would correspond with the a posteriori values under two conditions: (1) if the number of experimental trials approached infinity, and (2) if the experiment were absolutely unbiased. Figure 2.5(b) shows the resulting a priori histogram, and in this case we note that the probability of drawing three red marbles is actually only 120/1024, or 11.7 percent.

Our interest will lie with continuous random variables such as molecular velocities instead of discrete ones such as the percentage of red marbles in a small sample. Suppose that we wish to describe the statistical behavior of a random variable x which can take on any value. In this case a histogram would have to be formed by lumping together all values of the variable within given ranges (e.g., 1 to 2^-, 2 to 3^-, 3 to 4^-, etc.), because given values are not apt to recur. Figure 2.6 shows such a histogram.

The probability that an event will take place in a given range of the independent variable[3] $[x_{i-1}, x_i)$ is then

$$\mathcal{P}[x_{i-1}, x_i) = \frac{N_i}{N} \tag{2.8}$$

[3]The notation $[a, b)$ designates a range from a to b, including the point a but excluding the point b.

Fig. 2.6 Typical histogram.

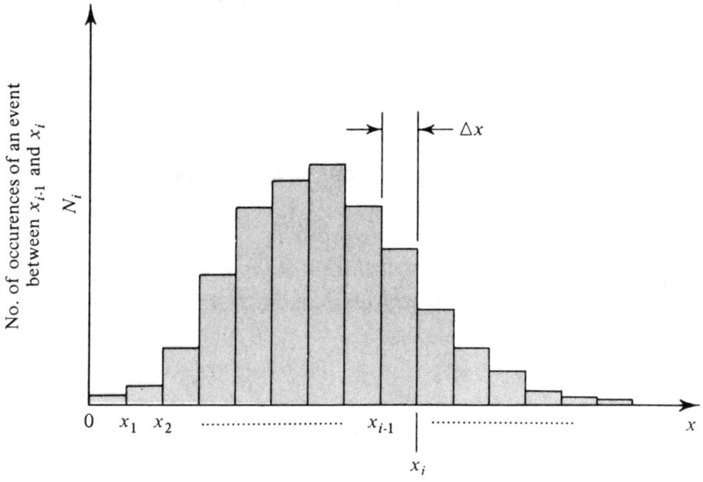

where N_i is the number of observations in $[x_{i-1}, x_i)$ and N is the total number of observations, or $\sum_{i=1}^{\infty} N_i$. We should now like to define a function $f(x)$ which has the property that

$$\int_{x_{i-1}}^{x_i} f(x)\, dx = N_i \tag{2.9}$$

or[4]

$$\int_{-\infty}^{\infty} f(x)\, dx = N \tag{2.10}$$

With the aid of the mean-value theorem and Eq. (2.9) it is easy to show that

$$f(x_i) = \lim_{\Delta x \to 0} \left(\frac{N_i}{\Delta x} \right) \tag{2.11}$$

or, if we denote as N_x all observations in the range $[-\infty, x)$, and consider N_x to be a continuous quantity,

$$f(x) = \frac{dN_x}{dx} \tag{2.12}$$

The function $f(x)$ is called the *distribution function*. It provides a continuous representation of the distribution of an event in a continu-

[4]The lower limit of minus infinity applies for variables that assume all real values. If x can assume only positive values, the lower limit can be taken as zero.

ous random variable. For the marble-sampling example it would be [see the solid line in Fig. 2.5(b)] the *normal distribution*,

$$f(x) = \frac{N}{\sqrt{2\pi}\,\sigma} \exp\left(\frac{-x^2}{2\sigma^2}\right) \tag{2.13}$$

where x is the difference between the number of red marbles and the average number, 5, and is treated as a continuous variable. The symbol σ denotes the standard deviation (or root-mean-square deviation) of x from 5.

We also define a *normalized distribution function*, or *probability density function* $F(x)$ such that

$$F(x) = \frac{d}{dx}\mathcal{P}[-\infty, x) = \frac{1}{N}\frac{dN_x}{dx} \tag{2.14}$$

Using Eq. (2.10) we find that $F(x)$ has the property that

$$\int_{-\infty}^{\infty} F(x)\,dx = \frac{N}{N} = 1$$

Thus the probability that x will always be observed in the range $-\infty < x < \infty$ is certain.

MOLECULAR DISTRIBUTION FUNCTION

A very general molecular distribution function $f(\mathbf{r}, \mathbf{c}, t)$ describes the time-dependent distribution of molecules in phase space (actually in velocity-position space). It contains the complete statistical description of the gas. For example, once it is known, expressions can be formed for the number of particles N in a finite volume V,

$$N(t) = \underbrace{\iiint}_{\text{over volume, } V} \int_{w=-\infty}^{+\infty}\int_{v=-\infty}^{+\infty}\int_{u=-\infty}^{+\infty} f(\mathbf{r}, \mathbf{c}, t)\,du\,dv\,dw\,dx\,dy\,dz$$

$$= \int_V \int_\Omega f(\mathbf{r}, \mathbf{c}, t)\,d\Omega\,dV \tag{2.15}$$

for the number density n of the gas,

$$n(\mathbf{r}, t) = \int_\Omega f(\mathbf{r}, \mathbf{c}, t)\,d\Omega \tag{2.16}$$

and for the density ρ,

$$\rho(\mathbf{r}, t) = m\int_\Omega f(\mathbf{r}, \mathbf{c}, t)\,d\Omega \tag{2.17}$$

Much of the time we shall work with gases in equilibrium. In this case there can be no variation in the gross state of the gas from position to position and the gross state will be uniform in time. Thus

$$f_{\text{equilibrium}} = f(\mathbf{c}\ \text{only}) \tag{2.18}$$

The equilibrium density and number density will accordingly be constant.

The physical nature of the equilibrium molecular distribution function is such that it is always *finite, positive, continuous,* and *the limit of f as* $\mathbf{c} \to \infty$ *is zero.*

MOLECULAR AVERAGES

Average values of such molecular properties as kinetic energy, velocity, and momentum are very important to the prediction of gross behavior. The properties vary greatly from particle to particle at a given instant, and from instant to instant (by virtue of collisions) for a given particle. If $\phi(\mathbf{r}, \mathbf{c}, t)$ is some molecular property, the average value of ϕ will be the sum of the ϕ's for all the particles under consideration divided by the number of these particles. Different averages can be formed by averaging over different sets of particles. In all cases the ponderous summations that would otherwise have to be made are avoided with the help of the distribution function.

The most commonly used average, the local average $\bar{\phi}$, is in general position- and time-dependent. The local average of ϕ is defined as the total amount of ϕ in an element of volume δV divided by the number of particles in δV. Thus

$$\bar{\phi} = \frac{\left[\int_\Omega \phi(\mathbf{r}, \mathbf{c}, t)\, f(\mathbf{r}, \mathbf{c}, t)\, d\Omega\right] \delta V}{\left[\int_\Omega f(\mathbf{r}, \mathbf{c}, t)\, d\Omega\right] \delta V} \tag{2.19}$$

or

$$\bar{\phi} = \frac{\int_\Omega f\phi\, d\Omega}{\int_\Omega f\, d\Omega} \quad \text{or} \quad \frac{1}{n}\int_\Omega f\phi\, d\Omega \tag{2.20}$$

The *ensemble average* $\langle\phi\rangle$ is an average over phase space—over location as well as momentum (or velocity)—and in general it is only time-dependent:

$$\langle\phi\rangle = \frac{\int_V\int_\Omega f\phi\, dV\, d\Omega}{\int_V\int_\Omega f\, dV\, d\Omega} = \frac{1}{N}\int_V\int_\Omega f\phi\, dV\, d\Omega \tag{2.21}$$

Consider, for example, the application of the local average to the property **c**. We define the local average velocity $\mathbf{c}_0 \equiv \bar{\mathbf{c}}$, whence

$$\mathbf{c}_0 = \frac{1}{n}\int_\Omega \mathbf{c}\, f\, d\Omega \tag{2.22}$$

This clearly has the physical significance of being the gross velocity with which the gas is moving locally. In accordance with Fig. 2.7, we also define the velocity of a particle relative to the local element of moving gas. This is called the *peculiar* or *thermal velocity* **C** of a particle and it must satisfy the relation

$$\mathbf{c} = \mathbf{c}_0 + \mathbf{C} \tag{2.23}$$

It follows directly from Eqs. (2.22) and (2.23) that $\bar{\mathbf{C}} = 0$. Consequently,

$$\bar{\mathbf{C}} = \overline{(\mathbf{i}U + \mathbf{j}V + \mathbf{k}W)} = 0 \tag{2.24}$$

where U, V, and W, are the x, y, and z components, respectively, of **C**. It follows that $\bar{U} = \bar{V} = \bar{W} = 0$. Finally, for a stationary gas (one in which $\mathbf{c}_0 = 0$), we have $\mathbf{c} = \mathbf{C}$ and $\bar{\mathbf{c}} = \bar{\mathbf{C}} = 0$. Thus the distinction between **c** and **C** need only be made when there is relative motion in a gas.

EXAMPLE 2.1 Suppose that the molecular speed of N particles is distributed between 0 and c_{max} according to the function

$$f(c) = \frac{Ac(c_{max} - c)}{c_{max}^3}$$

What are A and $\bar{c}$?

The constant A gives us the relative scale of the distribution as expressed by the number density n because

$$n = \int_0^\infty f(c)\, dc = \int_0^{c_{max}} \frac{Ac(c_{max} - c)}{c_{max}^3}\, dc + \int_{c_{max}}^\infty 0\, dc$$

Fig. 2.7 Thermal and gross components of molecular velocity.

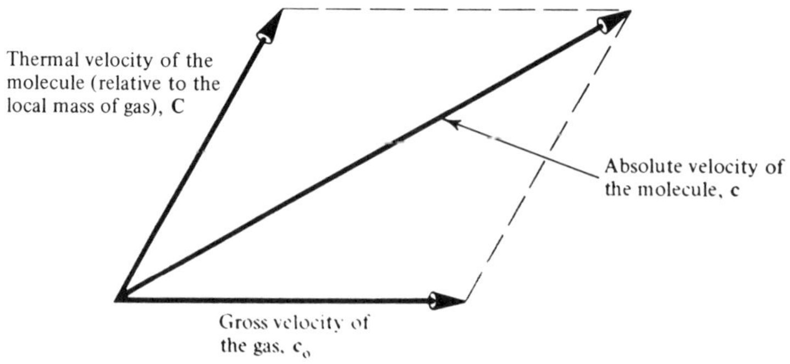

Thus

$$n = A\left(\frac{c^2}{2c_{\max}^2} - \frac{c^3}{3c_{\max}^3}\right)_0^{c_{\max}}$$

or

$$A = 6n$$

The local average can then be written with the help of Eq. (2.20) as

$$\bar{c} = \frac{1}{n}\int_0^{c_{\max}} \frac{6nc(c_{\max} - c)}{c_{\max}^3}\, c\, dc$$

from which we obtain

$$\bar{c} = \frac{c_{\max}}{2}$$

TRANSPORT OF MOLECULAR PROPERTIES

A problem of very great importance in kinetic theory is that of determining the rate at which molecular properties are transported across any plane within the gas. In particular, the pressure is given by the flux of molecular momentum at any point in the gas.

Let there be a fixed elemental area δs in a stationary gas, or an elemental area moving with velocity $\mathbf{c}_0$ in a moving gas. Molecules will then impinge upon the surface with thermal velocity $\mathbf{C}$ in either case. Consider one of the molecules approaching δs with velocity $\mathbf{C}$ and let it be encased in a skewed cylinder of length $C\,\delta t$ with δs forming one end. Furthermore, let δt be selected small enough to preclude collisions within the cylinder (see Fig. 2.8).

Fig. 2.8 Model for counting particles crossing a surface.

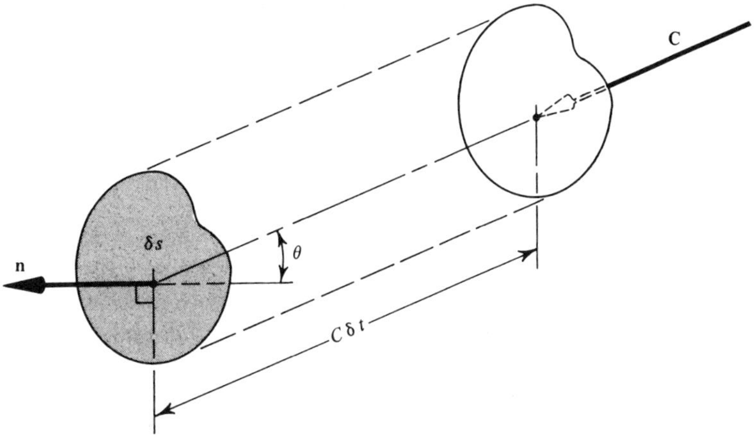

The volume of the cylinder is $\mathbf{C}\cdot(\mathbf{n}\,\delta s)\,\delta t$, where $\mathbf{n}$ is a unit vector perpendicular to δs. Alternatively, the volume can be written as $C\cos\theta\,\delta s\,\delta t$, where θ is the angle between $\mathbf{C}$ and $\mathbf{n}$. The number of particles per unit volume in the velocity range $d\Omega$ is $f\,d\Omega$; hence the number of particles in such cylinders over the velocity range $d\Omega$ is $\mathbf{C}\cdot\mathbf{n}\,f\,d\Omega\,\delta s\,\delta t$. Thus

$$\begin{pmatrix}\text{net transport of } \phi \text{ across an element}\\ \text{of area during an interval of time}\end{pmatrix} = \left(\int_\Omega \phi\mathbf{C}\cdot\mathbf{n}\,f\,d\Omega\right)\delta s\,\delta t \tag{2.25}$$

The net transport of ϕ per unit area and unit time (or more simply, the *flux* of ϕ) J_ϕ is then

$$J_\phi = \int_\Omega \phi\mathbf{C}\cdot\mathbf{n}\,f\,d\Omega \tag{2.26}$$

The flux J_ϕ will be a scalar or a vector depending upon the character of ϕ.

A very simple application of Eq. (2.26) can be made in the case of the *mass flux*. Let $\phi = m$, the molecular mass, so that $J_\phi = J_m$. Then, since m is a constant,

$$J_m = m\int_\Omega \mathbf{C}\cdot\mathbf{n}\,f\,d\Omega \tag{2.27}$$

The integral in Eq. (2.27) can be eliminated if we use Eq. (2.20) to define the average of $\mathbf{C}\cdot\mathbf{n}$,

$$J_m = m(n\,\overline{\mathbf{C}\cdot\mathbf{n}}) = 0$$

This result applies just as well to any other velocity-independent property $\phi(\mathbf{r}, t)$ although it is restricted to single-component systems. In this case it merely expresses the conservation of molecular mass.

The *momentum flux* J_{mom}, or rate of transfer of momentum across an area, is equal to the force that area must exert upon the gas to sustain equilibrium,

$$J_{\text{mom}}\ (\text{g-cm/sec})/\text{cm}^2\text{-sec} = J_{\text{mom}}\ \text{dynes/cm}^2$$

Thus J_{mom} is actually a generalized pressure or fluid stress. This can be illustrated by looking at the way in which molecular transport leads to the *hydrostatic* or *normal pressure*, p. When δs is taken to be an arbitrary imaginary surface element in the midst of a gas, the gaseous regions on either side of it impart momentum to one another by two means. The first of these is by "pushing" molecules into the other side with a component of velocity normal to δs. The second is by absorbing molecules with a normal velocity component from the other side. While momentum is *conserved* in each of the two sides, they exert a mutual normal pressure on one another due to the continuing *interchange* of momentum. This pressure exists at each point through the gas and is finally sustained at the walls of the container, which receive incoming molecules and reflect them back as outgoing molecules.

The use of Eq. (2.26) to evaluate J_{mom} is not entirely straightforward. There are, at any point, three independent orientations that δs may assume, and in each orientation there are three independent components of momentum that can cross δs. Any one of these nine possible momentum fluxes (or stresses) can be represented by the scalar quantity p_{ij}, where

$$p_{ij} = \int_{\Omega} (mu_j)u_i f \, d\Omega$$

or

$$p_{ij} = n(m\overline{u_j u_i}) = \rho \overline{u_j u_i} \tag{2.28}$$

The velocity components u_j and u_i are the components in the directions of the momentum flux and normal to δs, respectively. If we take the double-subscripted quantity p_{ij} as designating the second-rank tensor $\rho\overline{\mathbf{cc}}$, then

$$p_{ij} = \begin{vmatrix} p_{xx} & p_{xy} & p_{xz} \\ p_{yx} & p_{yy} & p_{yz} \\ p_{zx} & p_{zy} & p_{zz} \end{vmatrix} = \begin{vmatrix} \rho\overline{u^2} & \rho\overline{uv} & \rho\overline{uw} \\ \rho\overline{vu} & \rho\overline{v^2} & \rho\overline{vw} \\ \rho\overline{wu} & \rho\overline{wv} & \rho\overline{w^2} \end{vmatrix} \tag{2.29}$$

Figure 2.9 displays the stresses at a point on a cubical element, at that point.

Fig. 2.9 Stresses on a Cartesian fluid element.

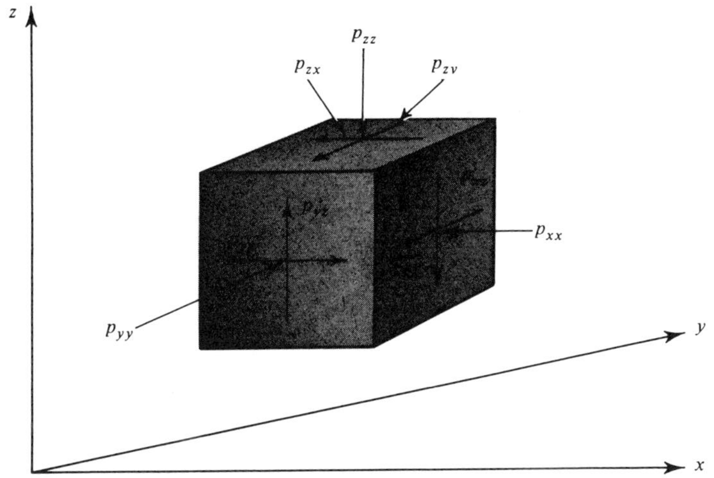

We note that p_{ij} designates the stress in the j direction on a plane normal to the i direction. Common shorthand expresses the stress tensor as

$$p_{ij} = p\,\delta_{ij} + \tau_{ij} \tag{2.30}$$

where p is the normal or hydrostatic pressure, $\rho\overline{u_i^2}$; τ_{ij} is the shear-stress component (equal to zero for $i = j$); and δ_{ij} is the Kronecker delta. δ_{ij} is equal to unity when $i = j$ and zero when $i \neq j$.

2.3 DERIVATION OF THE EQUATION OF STATE FOR AN IDEAL GAS

The first step in the formulation of an equation of state among the variables p, v, and T is the explanation of the variables in terms of microscopic behavior. This has already been done for v inasmuch as we have shown that the specific volume, $v = 1/\rho = 1/mn$.

To evaluate p we observe that τ_{ij} is always zero in a uniform, stationary gas. Although this is easy to see intuitively, it can be proved once we have derived the distribution function for an ideal gas. Then from Eqs. (2.30) and (2.29),

$$p_{ij} = p\delta_{ij} = \begin{vmatrix} \rho\overline{U^2} & 0 & 0 \\ 0 & \rho\overline{V^2} & 0 \\ 0 & 0 & \rho\overline{W^2} \end{vmatrix}$$

or

$$3p = \rho(\overline{U^2} + \overline{V^2} + \overline{W^2}) = \rho\overline{C^2}$$

so

$$p = \frac{\rho}{3}\overline{C^2} \tag{2.31}$$

or

$$pv = \tfrac{1}{3}\,\overline{C^2} \tag{2.31a}$$

Equation (2.31a) is Boyle's law if we can guess (as Daniel Bernoulli did) that the value of $\overline{C^2}$ is fixed uniquely by the temperature of the gas.

KINETIC MEANING OF TEMPERATURE

Temperature is a word that we understand in two ways. Intuitively we perceive it as the intensity of a physical sensation. Quantitatively we understand it as being defined by the magnitude of

$(\partial U/\partial S)_{V,N_i\text{'s}}$ or, for an ideal gas, by pv/R^0.[5] How then do we define T in terms of microscopic properties? Clearly, if the microscopic and macroscopic definitions are to be consistent, we must make this definition in such a way that Eq. (2.31a) becomes the ideal-gas law. Thus

$$\tfrac{1}{3}\,\overline{C^2} = \frac{R^0T}{M} = \frac{N_A kT}{M}$$

Here we have introduced Boltzmann's constant, $k = 1.3805 \times 10^{-16}$ erg/°K, which can be construed as the gas constant based upon a single molecule (i.e., $k \equiv R^0/N_A$). Then

$$T = \frac{M\overline{C^2}}{3N_A k} = \frac{m\overline{C^2}}{2}\,\frac{2}{3k} \tag{2.32}$$

The kinetic energy of a molecule is thus

$$\frac{m\overline{C^2}}{2} = \frac{3}{2}kT \tag{2.32a}$$

and the x, y, and z components of energy are

$$\frac{m\overline{U^2}}{2} = \frac{m\overline{V^2}}{2} = \frac{m\overline{W^2}}{2} = \frac{kT}{2} \tag{2.33}$$

because, on the average, no preference should be given to the kinetic energy in any direction.

Temperature thus has the direct physical significance of specifying the average translational kinetic energy of ideal-gas molecules. We shall find in our subsequent study of statistical mechanics that a somewhat more convenient measure is the "temper," β, defined as

$$\beta \equiv \frac{1}{kT} \tag{2.34}$$

Two other results of some interest can be written down immediately with the aid of Eq. (2.32). One is the root-mean-square molecular speed,

$$\sqrt{\overline{C^2}} = \sqrt{3RT} \tag{2.35}$$

where $R = R^0/M$ is the gas constant on a unit mass basis. The well-known sound speed, a, for an ideal gas is similar to Eq. (2.35),

$$a = \sqrt{\gamma RT}$$

where $\gamma \equiv c_p/c_v$, and c_p and c_v are the specific heats at constant pressure and volume, respectively. It follows that

$$\frac{\text{root-mean-square speed}}{\text{sonic speed}} = \sqrt{\frac{3}{\gamma}}$$

[5] In the phenomenological organization of thermodynamics, T is defined in terms of the Carnot cycle. It is easy to prove that $T = pv/(R^0/M)$ is equivalent to the Carnot-cycle definition.

We find in Sec. 3.7 that γ is bounded between unity and $\frac{5}{3}$, or $1.34 < \sqrt{3/\gamma} < 1.73$; therefore, the rate of propagation of a weak disturbance in an ideal gas is limited (as we might have guessed) to a speed that is a little below the root-mean-square molecular speed.

Finally, we can recast the ideal-gas law, Eq. (2.5), in microscopic terms. Combining Eqs. (2.31a) and (2.32) we obtain

$$pv = \frac{kT}{m}$$

But $v = 1/mn$, so

$$p = nkT \tag{2.36}$$

EXAMPLE 2.2 Two thousand years ago Julius Caesar gasped "Et tu, Brute!" and died. In that dying breath he released a number of inert N_2 molecules which, over the millenia, have diffused uniformly around the world. Do you suppose we have ever breathed any of those same molecules?

To answer the question we need to know the "tidal capacity" of the human lungs and the total volume of air in the world. The former is about 500 cm^3 for normal breathing. The latter can be approximated (very crudely) as the volume at standard conditions of a spherical shell 13 000 km in diameter and 10 km thick, or 5×10^{24} cm^3.

We also need the number density at standard conditions. For air this is

$$n = \frac{p}{kT} \simeq \frac{10^6 \text{ dynes/cm}^2}{(1.38 \times 10^{-16} \text{ erg/°K})(298\text{°K})} \simeq 2.5 \times 10^{19} \text{ molecules/cm}^3$$

and for N_2 it would be 79 percent of this, or 2×10^{19} molecules/cm^3.

Thus each breath corresponds to $(2 \times 10^{19})(500)$, or 10^{22}, molecules of N_2, but it also corresponds to $500/5 \times 10^{24}$, or $1/10^{22}$, of the N_2 molecules in the world. On the average, then, we breathe one N_2 molecule of Julius Caesar's dying breath *each time we inhale!*

2.4 MAXWELL'S DERIVATION OF THE MOLECULAR-VELOCITY DISTRIBUTION FUNCTION

As we have noted earlier, the distribution function provides the link through which we obtain gross thermodynamic properties from microscopic information. The most basic molecular distribution function is the equilibrium distribution developed by Maxwell,[6] and his derivation is of interest to us both historically and for its intuitive value. It cannot be regarded as rigorously correct, however, for reasons that will be apparent in a moment.

DERIVATION

Maxwell considered a spatially and temporally uniform stationary ideal gas. An expression for $f(\mathbf{C})$ can be developed quite easily once Maxwell's two physical assumptions have been accepted. These assumptions are that (1) the molecular velocities are isotropic in the absence of external fields (i.e., there are no preferred directions), and (2) the distribution of molecular speeds in any one component of velocity is independent of that in any other component.

Although the first assumption is easy to accept, the second is much harder to see. It says that a molecule moving very rapidly in the x direction is no more likely to have a high y-component speed than is one with a low x-direction speed. We can be fairly confident that it is correct, because subsequent theoretical work and actual measurements have verified the distribution based upon it. The implication of the assumption, although not immediately obvious, is that the molecules are in their most chaotic, or least organized, state. This idea is explored further in chapter 3.

The probability of finding a molecule in a given element of velocity space, $\delta\Omega$, is then

$$\mathcal{P}[(U, V, W), (U + \delta U, V + \delta V, W + \delta W)) = \frac{f(\mathbf{C})\,\delta\Omega}{n} \tag{2.37}$$

and the probabilities of finding a molecule in each of the component elements, δU, δV, and δW, are

$$\mathcal{P}[U, U + \delta U) = \frac{1}{n}\left[\int_{-\infty}^{\infty}\int_{-\infty}^{\infty} f(\mathbf{C})\,dV\,dW\right]\delta U \equiv F(U)\,\delta U$$

$$\mathcal{P}[V, V + \delta V) = G(V)\,\delta V \tag{2.38}$$

and

$$\mathcal{P}[W, W + \delta W) = H(W)\,\delta W$$

The second assumption says that the component probabilities given by Eq. (2.38) are *independent* of one another. Therefore, the probability given by Eq. (2.37) can be expressed as the product of the probabilities of the three independent events, $[U, U + \delta U)$ $[V, V + \delta V)$, and $[W, W + \delta W)$:

$$\frac{f(\mathbf{C})\,\delta\Omega}{n} = F(U)G(V)H(W)\,\delta U\,\delta V\,\delta W \tag{2.39}$$

[6] *The Scientific Papers of James Clark Maxwell*, Dover Publications, Inc., New York, 1952, vol. 2, pp. 43–47.

From the first assumption, that of isotropic behavior, we know that F, G, and H should all be the same function. Thus Eq. (2.39) becomes

$$f(\mathbf{C}) = nF(U)F(V)F(W)$$

The next problem is that of determining what kind of function of **C** can be decomposed into a product of functions of components of **C**. It will help us to note that

$$\mathbf{C}^2 = \mathbf{C}\cdot\mathbf{C} = U^2 + V^2 + W^2 = C^2$$

The choice $F(U) = F(V) = F(W) = a \exp (bU^2)$ gives the appropriate form,

$$f(\mathbf{C}) = a^3 n e^{b(U^2+V^2+W^2)} \tag{2.40}$$

If this function satisfies the two obvious constraints,

$$n = \int_\Omega f(\mathbf{C})\, d\Omega \tag{2.41}$$

and

$$T = \frac{m}{3k}\overline{C^2} = \frac{m}{3kn}\int_\Omega \mathbf{C}^2 f(\mathbf{C})\, d\Omega \tag{2.42}$$

it should be correct.

Substitution of Eq. (2.40) into Eqs. (2.41) and (2.42) and integration over all three component velocities gives two equations in a and b. Solving these we obtain

$$a = \sqrt{\frac{m}{2\pi kT}} \qquad \text{and} \qquad b = -\frac{m}{2kT}$$

so that

$$f(\mathbf{C}) = n\left(\frac{m}{2\pi kT}\right)^{3/2} \exp\left(-\frac{1}{kT}\frac{mC^2}{2}\right) \tag{2.43}$$

This is the well-known Maxwell molecular-velocity distribution function. We note that it is a special case of the *normal* or *Gaussian distribution function*, Eq. (2.13).

The component distributions $F(U)$, $F(V)$, and $F(W)$ can also be written down immediately as

$$F(U) = \left(\frac{m}{2\pi kT}\right)^{1/2} \exp\left(-\frac{mU^2}{2kT}\right) \tag{2.44}$$

$$F(V) = \left(\frac{m}{2\pi kT}\right)^{1/2} \exp\left(-\frac{mV^2}{2kT}\right) \tag{2.44a}$$

and

$$F(W) = \left(\frac{m}{2\pi kT}\right)^{1/2} \exp\left(-\frac{mW^2}{2kT}\right) \tag{2.44b}$$

Because the velocity components are normally distributed about zero, the most probable component is zero, and the most probable velocity is zero as well. This does not really imply that we should expect to find any stationary particles in a gas, however. The number of particles in a vanishingly small range of velocity about zero is, in fact, zero:

$$\lim_{\mathbf{C}\to 0} f(\mathbf{C})\,\delta\Omega = \lim_{\mathbf{C}\to 0}\left[n\left(\frac{m}{2\pi kT}\right)^{3/2}\exp\left(-\frac{m\mathbf{C}^2}{2kT}\right)\right] C^2 \sin\theta\,\delta\theta\,\delta\phi\,\delta C = 0$$

The most probable speed, on the other hand, is not zero — a fact that might seem odd at first glance. The reason is that velocity requires far more specification than speed. While $C = \sqrt{U^2 + V^2 + W^2}$ for any number of choices of U, V, and W, $\mathbf{C} = \mathbf{i}U + \mathbf{j}V + \mathbf{k}W$ for one unique set of U, V, and W. We find in chapter 5 that the values which U, V, and W can assume are uniformly distributed in velocity space, in accordance with the dictates of quantum mechanics. Thus the larger C is, the more possible sets of velocity components will correspond to it. Thus a given speed is far more likely than any of its corresponding velocities, with the single exception of $C = 0$, for which U, V, and W have to be zero.

The speed distribution $f(C)$ can be determined by regarding $f(\mathbf{C})$ as $f(C, \theta, \phi)$ and integrating out the angular dependence:

$$\mathcal{P}[C, C+\delta C) = \frac{1}{n} f(C)\,\delta C = \frac{1}{n}\left[\int_\theta \int_\phi f(C, \theta, \phi) C^2 \sin\theta\, d\theta\, d\phi\right]\delta C$$

or

$$\mathcal{P}[C, C+\delta C) = \left(\frac{m}{2\pi kT}\right)^{3/2} 8\left[\int_{\theta=0}^{\pi/2}\int_{\phi=0}^{\pi/2} C^2 \exp\left(-\frac{mC^2}{2kT}\right)\sin\theta\, d\theta\, d\phi\right]\delta C \quad (2.45)$$

The factor of eight arises in Eq. (2.45) because the integration is being made over only one-eighth of a sphere. The result is the *Maxwell speed distribution*,

$$f(C) = \left[4\pi\left(\frac{m}{2\pi kT}\right)^{3/2}\right] nC^2 \exp\left(-\frac{mC^2}{2kT}\right) \quad (2.46)$$

The most probable speed is clearly not zero; in fact, the distribution function vanishes as C goes to zero.

The velocity and speed distributions [Eqs. (2.43) and (2.46)] have both been plotted against the nondimensional speed in Fig. 2.10.

EXPERIMENTAL VERIFICATION OF MAXWELL'S DISTRIBUTION

A variety of experimental verifications of the Maxwell velocity distribution law have been made. The speed distribution was measured in the early 1930s using the method illustrated in Fig. 2.11.

Fig. 2.10 Maxwell velocity and speed distributions.

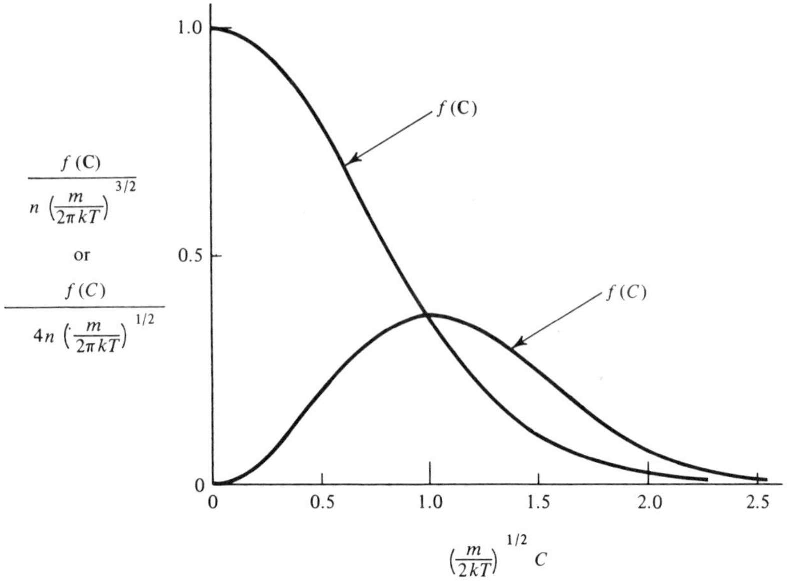

Fig. 2.11 Typical apparatus for measuring a molecular-speed distribution.

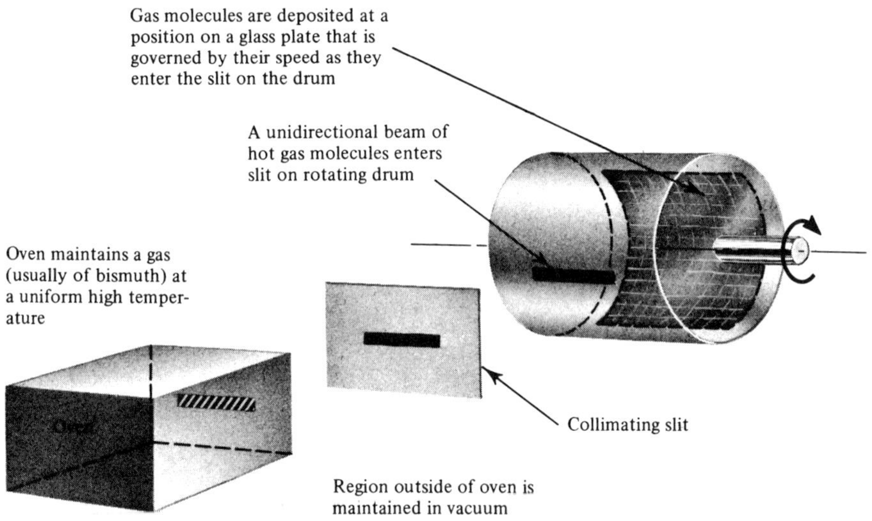

Bismuth vapor was heated to a high-temperature equilibrium condition and a beam of escaping molecules was collimated to facilitate a measurement of the speed C instead of a velocity component. The molecular beam entered a slit in a rotating drum each time it intercepted the beam. The molecules passed across the inside of the moving drum at differing speeds and distributed themselves accordingly upon a glass surface on the inside of the drum. The number of molecules arriving at any position was then determined by removing the glass surface and measuring the darkening that had resulted with a recording microphotometer.

Improved molecular-beam devices, similar to the one described above, have subsequently been developed and more refined measurements have been made. In all cases the Maxwell distribution was found to represent the experimental data extremely well.

An interesting "mathematical experiment" has been carried out[7] to show the existence of the Maxwell velocity distribution in the equilibrium state. With the aid of a computer the motions of 100 particles in a cubic box were treated mathematically by solving the equations of motion with arbitrarily prescribed initial positions and velocities of all particles. After a rather modest number of collisions, the particles approached an equilibrium state and it was found that the velocities of particles were distributed according to the Maxwell velocity distribution.

MAXWELL MOMENTUM DISTRIBUTION

In mechanics it has been established that the momentum is a more fundamental parameter than the velocity, although they are linearly related ($\mathbf{p} = m\mathbf{c}$), and the velocity is a quantity that is easy to understand. Thus the Maxwell distribution of vector momentum $f_m(\mathbf{p})$ can be obtained from the Maxwell velocity distribution $f_v(\mathbf{c})$ in the following way:

$$\mathcal{P}(\mathbf{c}, \mathbf{c} + d\Omega) = \frac{1}{n} f_v(\mathbf{c})\, d\Omega = \frac{1}{n} f_v\left(\frac{\mathbf{p}}{m}\right) \frac{d\Phi}{m^3} = \frac{1}{n} f_m(\mathbf{p})\, d\Phi$$

Therefore,

$$f_m(\mathbf{p}) = \frac{1}{m^3} f_v(\mathbf{c})$$

and substitution of this result in Eq. (2.43) gives the momentum distribution $f_m(\mathbf{p})$ for an equilibrium gas. Dropping the unnecessary subscript m, we have

$$f(\mathbf{p}) = n(2\pi mkT)^{-3/2} \exp\left(-\frac{\mathbf{p}^2}{2mkT}\right) \tag{2.47}$$

The Maxwell distribution for the magnitude of momentum can likewise be obtained from either Eqs. (2.43) or (2.46). It is

$$f(p) = n(4\pi p^2)(2\pi mkT)^{-3/2} \exp\left(-\frac{p^2}{2mkT}\right) \tag{2.47a}$$

[7]See B. J. Alder and T. E. Wainwright, *J. Chem. Phys.* **33**, 1439 (1960), and references cited therein.

The results of averages based on the velocity and the momentum distributions are, of course, identical. For example, the molecular mean speed obtained from the speed distribution is

$$\bar{c} = 4\pi\left(\frac{m}{2\pi kT}\right)^{3/2} \int_0^\infty c^3 \exp\left(-\frac{mc^2}{2kT}\right) dc = \left(\frac{8kT}{\pi m}\right)^{1/2}$$

while the momentum distribution gives

$$\bar{c} = \left(\frac{\bar{p}}{m}\right) = \frac{4\pi m^{1/2}}{(2\pi m^2 kT)^{3/2}} \int_0^\infty p^3 \exp\left(-\frac{p^2}{2mkT}\right) dp = \left(\frac{8kT}{\pi m}\right)^{1/2} \tag{2.48}$$

EXAMPLE 2.3 Why does the earth keep an atmosphere while the moon does not?

The answer to this question can be obtained without working to any great precision. Therefore, let us consider the temperature of an atmosphere on either planet to be 300°K and consider an air atmosphere. The root-mean-square speed in either case would be

$$C_{rms} = \sqrt{\frac{3kT}{m}} = \left[\frac{3(1.3805 \times 10^{-16})(300)}{4.8 \times 10^{-23}}\right]^{1/2} = 50{,}900 \text{ cm/sec}$$

But the velocity required to escape the earth's gravitational field is 1.1×10^6 cm/sec, or $22 \times C_{rms}$. Using the result of Problem 2.13 and taking $C_{rms} \simeq C_m$ we find that the fraction of particles with speeds in excess of the escape velocity is

$$\frac{2}{\sqrt{\pi}}\sqrt{\frac{3}{2}}(22) \exp\left[-\tfrac{3}{2}(22)^2\right] = 30e^{-726}$$

and this is an incomprehensibly small number. On the moon, however, the escape velocity is much less—about 2×10^5 cm/sec, or $4 \times C_{rms}$—and the fraction of molecules with sufficiently high velocity to escape is

$$\frac{2}{\sqrt{\pi}}\sqrt{\frac{3}{2}}(4) \exp\left[-\tfrac{3}{2}(4)^2\right] \simeq 2 \times 10^{-10}$$

There would therefore be a small but steady attrition of the moon's atmosphere to outer space, if it had an atmosphere. It would continue to lose about 1 five-billionth of its atmosphere in the time required for the distribution function to readjust to the loss of particles.

2.5 FLUX OF MOLECULES

The flux of molecules *in one direction*, at a point in a gas, is given by Eq. (2.26) in the form

$$J_{\text{molecules}} = \int \phi\, \mathbf{c} \cdot \mathbf{n}\, f(\mathbf{c})\, d\Omega \tag{2.49}$$

In this case ϕ will be unity and the integration will be made over all angles on one side of the plane only. Accordingly, we can introduce $\mathbf{c}\cdot\mathbf{n} = c\cos\theta$ and write

$$J_{\text{molecules}} = \int_{c=0}^{\infty}\int_{\theta=0}^{\pi/2}\int_{\phi=0}^{2\pi} f(\mathbf{c})c^3 \sin\theta\cos\theta\, d\phi\, d\theta\, dc$$

$$= \pi\int_{c=0}^{\infty} c^3 f(\mathbf{c})\, dc \tag{2.50}$$

Had the integration been over all space ($0 \leq \theta \leq \pi$) the *net* flux, at least for a uniform gas, would have been zero. The result above gives the flux from one side only.

Equation (2.50) might seem to imply that $J_{\text{molecules}}$ can be evaluated only when the distribution function for the gas is known. Actually it is possible to circumvent the need for $f(\mathbf{c})$. We can write

$$n\bar{c} \equiv \frac{n}{n}\int_{\Omega} cf(\mathbf{c})\, d\Omega = \int_0^{\infty}\int_0^{\pi}\int_0^{2\pi} f(\mathbf{c})c^3 \sin\theta\, d\theta\, d\phi\, dc$$

$$= 4\pi\int_0^{\infty} c^3 f(\mathbf{c})\, dc \tag{2.51}$$

Substitution of Eq. (2.51) into (2.50) then gives

$$J_{\text{molecules}} = \frac{n\bar{c}}{4} \tag{2.52}$$

or, if we use Eq. (2.48),

$$J_{\text{molecules}} = n\left(\frac{kT}{2\pi m}\right)^{1/2} \tag{2.53}$$

Equation (2.52) is not restricted to Maxwellian gases, but Eq. (2.53), of course, is. If $\bar{c}$ is replaced by $\bar{C}$ in Eq. (2.52), the result will give the flux at a point that is moving with the gas as a whole. Equation (2.53) applies only to points at rest with respect to $\mathbf{c}_0$.

Thus far we have made no attempt to relate fluxes to gradients of any kind, and, strictly speaking, Eqs. (2.50), (2.52), and (2.53) are only applicable when it is reasonable to talk about an *equilibrium* distribution function, $f(\mathbf{c})$. For example, only a very small hole could exist in a pressurized vessel without causing gross changes in the local distribution function which would invalidate these equations.

The subject of molecule fluxes receives more detailed consideration when we take up the kinetic theory of gases. It will become clear that such a hole would have to be smaller than the mean free path l (or average distance traveled by molecules between collisions) if a gross velocity distribution $\mathbf{c}_0(\mathbf{r})$ were not to be induced in its vicinity.

EXAMPLE 2.4 How do the mass fluxes of an ideal gas from a tank differ between the case of a large hole and that of a very small hole?

If the hole is much larger than the mean free path, a gross motion of the gas will be induced and Eq. (2.53) is not valid. Actually, kinetic-theory methods for dealing with the situation do exist and will be of interest to us in chapter 12. However, the very elementary Bernoulli equation applies here. It gives

$$\dot{m}_{\text{large}} = \rho\sqrt{2\left(\frac{\Delta p}{\rho}\right)}$$

where the orifice coefficients have been assumed approximately equal to unity. If the mean free path is larger than the hole we must use Eq. (2.53),

$$\dot{m}_{\text{small}} = mJ_{\text{molecules}} = n\left(\frac{kTm}{2\pi}\right)^{1/2}$$

The use of Eq. (2.52) presupposes a zero back pressure, in that no returning component of flow is considered. Thus to make proper comparison we must replace Δp with p, or nkT, in the $\dot{m}_{\text{large}}$ expression. Then if we note that $\rho = nm$,

$$\frac{\dot{m}_{\text{large}}}{\dot{m}_{\text{small}}} = 2\sqrt{\pi} \simeq 3.54$$

Actually the advantage of the large hole is not really this great, because if there is no back pressure the flow will be sonic and

$$\dot{m}_{\text{large}} = \rho\sqrt{\gamma RT} = \rho\sqrt{\frac{\gamma k N_{\text{A}} T}{M}}$$

so a more appropriate ratio would be

$$\frac{\dot{m}_{\text{large}}}{\dot{m}_{\text{small}}} = \sqrt{2\pi\gamma} \simeq 2.5 \quad \text{to} \quad 3.23$$

where we have used the fact that $1.0 < \gamma \leq \frac{5}{3}$.

In either case the mass motion of fluid through an aperture is clearly a much more effective means of gas removal than the individual escape of particles represented by Eqs. (2.52) and (2.53).

We now leave considerations that belong in the domain of kinetic theory until chapters 11 and 12. Chapters 3 through 10 deal specifically with the more general subject of statistical mechanics and those aspects of quantum mechanics that are needed to discuss statistical mechanics.

Problems 2.1 Show that for a two-dimensional gas, $p = \rho\overline{C^2}/2 = K$, where ρ is the mass per unit area and K is the mean translational kinetic energy per unit area.

2.2 Suppose the speed distribution function $f(c)$ takes the form given in Example 2.1. Evaluate the mean-square speed and the most probable speed. Can you compute $\langle c \rangle$? What is its value?

2.3 How many molecular collisions are made per second on each square centimeter of a surface exposed to air at a pressure of 1 atm and at 300°K? The mean molecular weight of air is 29.

2.4 A closed vessel contains liquid water in equilibrium with its vapor at 100°C and 1 atm. One gram of water vapor at this temperature and pressure occupies a volume of 1670 cm^3. The heat of vaporization at this temperature is 2250 Joules/g.

(a) How many molecules are there per cubic centimeter of vapor?
(b) How many vapor molecules strike each square centimeter of liquid surface per second?
(c) If each molecule that strikes the surface condenses, how many evaporate from each square centimeter per second?
(d) Compare the mean kinetic energy of a vapor molecule with the energy required to transfer one molecule from the liquid to the vapor phase.

2.5 A pressure vessel contains V_p cm^3 of an ideal gas initially at pressure p_i. The vessel, which is located in a vacuum chamber of volume V_c has a very small hole (a leak) of known area in its side. Derive expressions for the pressure in the vessel and in the chamber as a function of time after the leak occurs. Assume that there is enough heat transfer to keep the temperatures in both containers at the same uniform value.

2.6 Two chambers of the same gas at the same temperature, but at different pressures, are connected through a membrane with very small holes. Determine the net molecular flow rate per unit area of the holes. (This is the phenomenon of *effusion*.)

2.7 Plot, to scale, the motion of the simple harmonic oscillator, shown in Fig. 2.12, in phase space, (a) for $c_f = 0$ lb$_f$/(ft/sec) and (b) for $c_f = \sqrt{8}$ lb$_f$/(ft/sec). What would happen to the phase-space trajectories if time appeared explicitly in the equation of motion—if, for example, there were a time-dependent forcing function?

2.8 Prove that division by N normalizes the normal distribution function.

2.9 A vertical cylinder is fitted with a piston of mass M which is a distance h above the cylinder head (the bottom of the cylinder) and which can follow the influence of gravity without friction. The cylinder contains a sphere of mass $m \ll M$, which moves only up and down with a speed c. The sphere is elastically reflected by the piston and the cylinder head. The influence of gravity upon the *sphere* is negligible.

Fig. 2.12 Damped, simple harmonic oscillator.

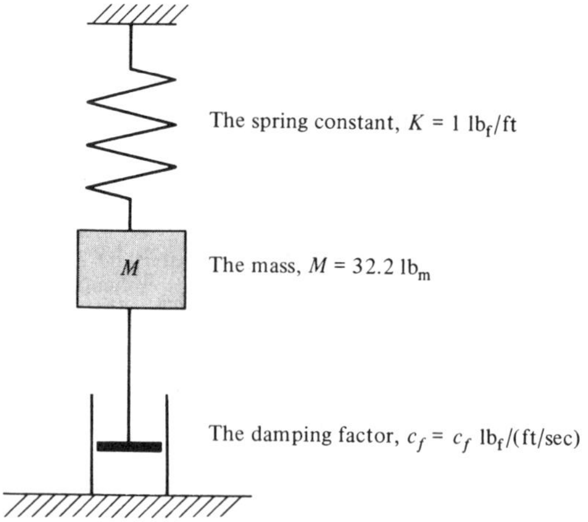

(a) Establish the condition of equilibrium for the piston and compare it with the ideal-gas law. Ignore the size of the sphere. Comment upon your result in terms of the principle of equipartition of energy (see Sec. 3.7).

(b) If the piston is withdrawn at a velocity V_p, where $V_p \ll c$, compare the loss of energy of the sphere with the work of an equivalent gas ($\delta W = p\,dV$).

2.10 A monatomic gas is expanded from a stagnation condition defined by T_0, through a converging–diverging nozzle, to the maximum speed attainable in such a nozzle. Compare this speed with the root-mean-square speed at T_0. Under what condition would these speeds be equal? What would be the physical meaning of this condition?

2.11 On the basis of the Maxwell distribution show that the most probable speed C_m is $\sqrt{2kT/m}$; the root-mean-square speed is $\sqrt{3kT/m}$; and the mean speed is $\sqrt{8kT/\pi m}$. Show that

$$C_m : \bar{C} : \sqrt{\overline{C^2}} :: \sqrt{2} : \sqrt{\frac{8}{\pi}} : \sqrt{3} :: 1 : 1.13 : 1.22$$

2.12 Obtain $\overline{U^2}$ from the Maxwell distribution for the U-component velocity. How could you have obtained this result far more simply?

2.13 Show that if the fraction of molecules with speeds between C and $C + \delta C$ is written $\Psi(C)\,\delta C$, then the probability density is

$$\Psi(C) = \frac{4}{C_m{}^3\sqrt{\pi}} C^2 \exp(-\alpha^2)$$

where $\alpha \equiv C/C_m$ and C_m is the most probable speed. Prove that the fraction of the molecules having speeds in excess of C is

$$1 + \frac{2\alpha}{\sqrt{\pi}} \exp(-\alpha^2) - \operatorname{erf}(\alpha)$$

Finally, show that when $\alpha > 1.5$ this fraction is well approximated by

$$\frac{2}{\sqrt{\pi}} \sqrt{\frac{mC^2}{2kT}} \exp\left(-\frac{mC^2}{2kT}\right)$$

2.14 Derive the expression for $f(E)$, where E is the translational energy of a particle $mC^2/2$. Use your result to evaluate $\overline{E}$ in terms of temperature. Fission neutrons have an average energy of $2(10)^6$ eV. What is their "temperature"?

2.15 One thousand H_2 molecules are in thermal equilibrium at 0°C. With the aid of the result from Problem 2.13, choose 100-m/sec increments of velocity between 0 and 3000 m/sec, and plot the number of molecules in each increment. Show C_m, $\overline{C}$, and C_{rms} on your curve. (You had best use a digital computer for this problem.)

2.16 Complete the missing steps in the derivation of Eq. (2.43), Maxwell's velocity distribution.

2.17 Obtain Eq. (2.53) directly, using Eq. (2.50).

2.18 Show whether or not there can be shear stress in a Maxwellian gas, using Eq. (2.28). Can there be viscosity in a Maxwellian gas? Discuss.

2.19 Derive the general expression for $J_{\text{molecules}}$ for a two-dimensional gas. Your result should not be restricted to a Maxwellian gas.

2.20 A particle moves away from a point $x = 0$ in a straight line. It is influenced by a force field such that $F = -a/x^2$. Plot the motion of the particle in phase space. Assume that it has insufficient energy to escape from $x = 0$.

2.21 Compute the maximum possible heat flux that could ever occur in a condensation process, by assuming that all molecules of saturated vapor impinging on a surface were to condense on that surface. What would this heat flux be for steam at atmospheric pressure?

3 classical statistics of independent particles

The preceding developments of the ideal-gas law and other relationships began in fairly concrete descriptions of molecular behavior. These developments belong in that area of statistical thermodynamics which is generally called kinetic theory. Now we embark upon a more abstract strategy for determining gross effects of microscopic behavior. The methods are those of elementary statistical mechanics and we find that they lead to results based upon surprisingly little information about molecular behavior.

Consideration is restricted to particles that are independent in the sense that the energy of each is independent of the energy of the others.[1] Actually energy *can* be exchanged in the brief instant of collision. However, we find in chapter 12 that collisions cease to influence the distribution of energy at equilibrium. Our statistical method therefore envisions an equilibrium array of particles without making any reference to collisions.

[1]This independence is not complete. The total energy of an equilibrium group of particles is constrained to be a value that can, for all practical purposes, be considered constant.

3.1 MACROSTATES AND MICROSTATES

The starting point of our statistical-mechanical strategy is the idea of macrostates and microstates. The following example serves to illustrate the meaning of this idea. Imagine for a moment that a full salt shaker is attached at its open end to an open, full pepper shaker. Together they constitute a tube with pure salt in one half and pure pepper in the other. Now imagine that they are shaken; the mixture will assume an intermediate gray. Shake them 1000 times; the original black-and-white configuration will not recur, nor would one (on the basis of his experience) *expect* it to. Nevertheless, we *know* that it would be possible for it to recur, because it corresponds with certain of the possible rearrangements of particles.

This paradoxical situation holds a vital key to the behavior of molecular aggregates. In the salt and pepper shakers there are millions upon millions of rearrangements that will appear as a uniform gray to our undiscerning eyes. There are far fewer, however, that will appear black and white.

We give the name *microstates* to the detailed configurations, or rearrangements, that can exist on the microscopic level. *Macrostates* are those manifestations of the microstates that we can distinguish from one another on the gross level. In the salt-and-pepper-shaker example, each of the vast number of arrangements of salt and pepper particles corresponding with the macrostate "gray" was a different microstate.

The thing that we tend to accept intuitively about the salt-and-pepper example is that all the microstates are equally likely. Thus we cannot hope to regain the *rare* black–white configuration in a reasonable number of shakes. Similarly, the marble-sampling illustration employed in Sec. 2.2 showed that there were 252 times as many ways of drawing the most probable sample of 10 marbles (i.e., 5 red and 5 black) than there were ways of drawing 10 red marbles. If the sample were increased to 100 marbles, 50 red and 50 black marbles would then be 6.65×10^{28} times as likely to occur as 100 red marbles and about 3000 times more likely than even a 30 : 70 mixture.

It is therefore clear that there is at play a phenomenon that has been called "the tyranny of large numbers." Deviations from those microstates that correspond with the most probable macrostate become "impossible" when sample sizes are at all large.

The task that we undertake here is that of deriving a general distribution function that will include the Maxwell distribution as a special case. We accomplish this by determining the number of molecular microstates corresponding with each macrostate and then finding how the molecules are distributed in phase space when this number is maximized. The general result is named the Boltzmann distribution (or the Maxwell–Boltzmann distribution) after Ludwig Boltzmann, who formulated it in 1871.[2]

PRINCIPLE OF EQUAL A PRIORI PROBABILITIES

The fundamental assumption in statistical mechanics is the principle of equal a priori probabilities. In slightly restrictive form it says: *All microstates of motion occur with equal frequency.*

The principle implies that all arrangements of a system of molecules in phase space are equally probable — a fact that we have to use in writing the probability of a given macrostate. There is no definite information or logical reason to favor one particular arrangement, so this principle is indeed the only alternative. The validity of this principle, however, can only be supported by the a posteriori success of statistical mechanics in predicting the macroscopic behavior of thermodynamic systems.

The principle of equal a priori probabilities includes as a special case the principle of molecular chaos. The latter principle says that there is no order in molecular motion, and it generally takes the form of Maxwell's second assumption (Sec. 2.4). Since three components of velocity have to be specified independently to identify the microstate of a particle, Maxwell's assumption clearly follows the principle of equal a priori probabilities.

3.2 WAYS OF ARRANGING OBJECTS

Following Davidson[3] we digress from the problem at hand to enumerate the solutions of five combinational problems that we must subsequently use in writing the probabilities of macrostates. Some of these results are not used until chapter 6, but it is convenient and easy to develop them now along with the results that we need immediately.

PROBLEM 1. *How many ways can we arrange N distinguishable objects?* Let us suppose that, for example, we wish to arrange N books in various ways on a shelf. When the first book is placed in any one of N ways there remain only $N-1$ ways of placing the second. There are then $N(N-1)$ ways of placing the first two, and,

[2]L. Boltzmann, *Lectures on Gas Theory*, English translation by S. G. Brush, University of California Press, Berkeley, Calif., 1964, chap. IV.

[3]N. Davidson, *Statistical Mechanics*, McGraw-Hill, Inc., New York, 1962 Sec. 5.2.

when they are in position, $N-2$ ways of placing the third. Thus there are

$N!$ ways of arranging N distinguishable objects

PROBLEM 2. *How many ways can we put N distinguishable objects into r distinguishable boxes (without regard to order within the boxes) such that there are N_1 objects in the first box, N_2 in the second, . . . , and N_r in the* r*th box?* Once more we have N books but now there are r shelves, and the order within any shelf is unimportant. There are again a total of $N!$ arrangements of books, but the arrangements that result from changing books on a given shelf must be divided out, because they are not relevant. We must divide $N!$ by $N_1!$, $N_2!$, . . . , $N_i!$, . . . , $N_{r-1}!$, and $N_r!$ to account for the meaningless rearrangements. There are thus

$$\frac{N!}{\prod_{i=1}^{r} N_i!}$$ *ways of putting $N_1, N_2, \ldots, N_r$ distinguishable objects into r distinguishable boxes (without regard to order)*

PROBLEM 3. *How many ways can we select N distinguishable objects from a set of g distinguishable objects?* This is nothing more than putting N books on one of two shelves and $g-N$ books on the other. The distinguishable "books" are simply divided in two groups within which there is no concern for order. The answer to this problem can be written immediately, as a special case of Problem 2, in the form

$$\frac{g!}{N!\,(g-N)!}$$ *ways of selecting N distinguishable objects from g distinguishable objects*

PROBLEM 4. *How many ways can we put N indistinguishable objects into g distinguishable boxes, if there is no limit on the number of objects in any box?* This is the first problem in which we must cope with the indistinguishability of objects. We might liken it to the problem of placing N copies of the same book among g shelves. The problem can be solved very simply if we first reduce it to a more abstract form. Let us designate the books with identical dots, and the separators between shelves with slashes:

$$\cdots/\cdots\cdot\cdot/\cdot/\cdots\cdot/\cdot\cdot$$

There are N dots and $g-1$ slashes, denoting a total of g boxes. The "boxes" are distinguishable by virtue of their location, and the apparent indistinguishability of the slashes is irrelevant to the problem. The problem can now be asked in the form: How many ways can we arrange the $N+(g-1)$ *distinguishable* locations which contain dots and slashes in the sketch above?, or, more simply, How many ways can we select N distinguishable dot locations and $g-1$ distinguishable slash locations from $N+(g-1)$ dot and slash locations? The problem is now reduced to Problem 3 and the answer is

$$\frac{(N+g-1)!}{N!\,(g-1)!}$$ *ways of putting N indistinguishable objects into g distinguishable boxes*

PROBLEM 5. *How many ways can we put N distinguishable objects into g distinguishable boxes?* Each of N different books can be put on any of g shelves. The first book can be placed in g ways, the second book can also be placed in g ways, and so on. There are thus

g^N *ways that N distinguishable objects can be placed in g distinguishable boxes*

3.3 ON THE SPECIFICATION OF MOLECULAR MICROSTATES

The meaningful specification of a finite array of microstates can only be made after means have been devised for breaking the energy of particles into finite increments. Quantum-mechanical considerations will subsequently provide the appropriate means for doing this. Until we have taken the trouble to develop these means in chapters 4 and 6, it suffices to subdivide energy[4] in the following arbitrary way.

An array of N particles has a total energy U. The energies of the individual particles are assumed to take on the discrete values $\epsilon_0, \epsilon_1, \ldots, \epsilon_i, \ldots$. The number of particles with energy ϵ_0 is N_0, the number with ϵ_1 is N_1, and so on. Thus

$$\sum_{i=0}^{\infty} N_i = N \tag{3.1}$$

and

$$\sum_{i=0}^{\infty} \epsilon_i N_i = U \tag{3.2}$$

The word "macrostate" can now be applied to the gross (or observable) state that corresponds with a given set of numerical values $N_1, N_2, \ldots, N_i, \ldots$, and thus satisfies the two constraints. The

[4]The salt-and-pepper example (Sec. 3.1) dealt with the arrangement *in space* of particles, as identified by color. Now we talk about the arrangement *in energy* of identifiable particles.

number of microstates for each macrostate will be equal to the number of ways in which we can choose these N_i's from the N particles.

We wish to take the view that particles can be distinguished from one another when we count the microstates of the system that comprises them. But to make such a statement requires that we be exceedingly careful in understanding what the claim means. If two particles are truly indistinguishable, the operation of interchanging them is meaningless. When we have finished interchanging two indistinguishable particles we have, in fact, *done nothing at all.*

To our classical way of thinking the last sentence might seem like nonsense. The reason is that indistinguishability is an idea that really makes sense only within the framework of a quantum description of particles.[5] In the classical view we feel that given a large enough microscope we could always devise some means for distinguishing, or identifying, particles.

The number of microstates, W, corresponding to a given macrostate can be written down immediately. It is given by the answer to Problem 2 in Sec. 3.2, which tells the number of ways in which we can divide N distinguishable objects into groups of $N_1, N_2, \ldots, N_i, \ldots,$

$$W = N! \prod_{i=0}^{\infty} \frac{1}{N_i!} \tag{3.3}$$

This W is an enormous number—a number comparable with $N_A!$. If, on the other hand, the particles were to be viewed as indistinguishable, it is clear that no rearrangements within a macrostate would be meaningful and W would simply be *unity.* The statistical description that we now formulate is self-consistent with a classical view. It will also bear a strong resemblance to the revised quantum-statistical description that is developed in chapter 6.

3.4 THERMODYNAMIC PROBABILITY

The probability of encountering any macrostate, $\mathcal{P}$(macrostate), and the probability of encountering any microstate, $\mathcal{P}$(microstate), can be written in accordance with the principle of equal a priori probabilities as

$$\mathcal{P}(\text{macrostate}) = \frac{\text{number of microstates corresponding to the given macrostate}}{\text{total number of microstates}} \tag{3.4}$$

and

$$\mathcal{P}(\text{microstate}) = \frac{1}{\text{total number of microstates}} \tag{3.4a}$$

[5]This idea is discussed by D. ter Haar in *Elementary Statistical Mechanics,* Holt, Rinehart and Winston, Inc., New York, 1960, p. 72.

respectively. The number of microstates consistent with a given macrostate W is given the name *thermodynamic probability*. It is not a true probability in that its sum over all macrostates is not equal to unity. Instead

$$W = \frac{\mathcal{P}(\text{macrostate})}{\mathcal{P}(\text{microstate})} \tag{3.5}$$

Another name for W, sometimes used in place of the thermodynamic probability, is the *disorder number*.

Since $\mathcal{P}$ (microstate) is fixed in any system, W is proportional to $\mathcal{P}$ (macrostate). The fact that it is vastly bigger than the true probability will not affect the type of computation we now wish to make.

EXAMPLE 3.1 A total of 40 people in a small precinct vote for two candidates — 16 for the Tory and 24 for the Whig. Evaluate W, $\mathcal{P}$ (macrostate), and $\mathcal{P}$ (microstate) for the vote distribution.

We must first recognize that what really interests us is the gross outcome and that the individual action of voters is to be viewed as microscopic detail that we shall ignore. Thus the distribution, 24 Whigs to 16 Torys, is a macrostate whose thermodynamic probability is given in Problem 3, Sec. 3.2, as

$$W = \frac{40!}{16!\,24!} = 6.285 \times (10)^{10}$$

But the total number of ways in which 40 individuals can cast their votes in either of two ways is given by the answer to Problem 5 as

$$(2)^{40} = 1.0995 \times (10)^{12}$$

Thus

$$\mathcal{P}\ (\text{microstate}) = (2)^{-40} = 9.091 \times (10)^{-13}$$

and

$$\mathcal{P}\ (\text{macrostate}) = \mathcal{P}\ (\text{microstate}) \times W = 0.0571$$

Thus there are an immense number of ways of achieving the given macrostate and a negligible probability of guessing how each of the 40 voters will decide. However, the specific defeat suffered by the Tories was not unreasonably surprising.

3.5 MAXWELL-BOLTZMANN STATISTICS

The equilibrium distribution of the N_i's will be assumed to be the one for which the thermodynamic probability is maximum. The fact that *the maximum W is overwhelmingly greater than W's corre-*

sponding to macrostates that differ appreciably from the equilibrium macrostate is subject to strictly mathematical proof. It is shown to be true for large ensembles in more detailed treatments of statistical mechanics. We accept the fact in the present discussion, on the basis of its intuitive appeal.

It is simpler to maximize ln W than to maximize W itself. Thus our problem will be that of maximizing ln W as given by Eq. (3.3),

$$\ln W = \ln N! - \sum_{i=0}^{\infty} \ln N_i! \tag{3.6}$$

subject to the two constraints given in Sec. 3.3,

$$\sum_{i=0}^{\infty} N_i = N \qquad \text{and} \qquad \sum_{i=0}^{\infty} \epsilon_i N_i = U$$

This maximization is accomplished using the method of Lagrangian multipliers and Stirling's approximation to the logarithm of large factorials.[6] The method of Lagrangian multipliers is a technique for picking the values of the independent variables (in this case, the N_i's) that maximize a function of those variables (in this case, W).

The introduction of Stirling's approximation in Eq. (3.6) and differentiation of the result gives

$$d \ln W = -\sum_{i=0}^{\infty} (\ln N_i)\, dN_i = 0 \tag{3.7}$$

The constraints (3.1) and (3.2) yield

$$\alpha \sum_{i=0}^{\infty} dN_i = 0 \qquad \text{and} \qquad \beta \sum_{i=0}^{\infty} \epsilon_i\, dN_i = 0 \tag{3.8}$$

where α and β are undetermined multipliers. The subtraction of Eq. (3.7) from the sum of Eqs. (3.8) gives

$$\sum_{i=0}^{\infty} (\ln N_i + \alpha + \beta\epsilon_i)\, dN_i = 0 \tag{3.9}$$

Since the coefficients of the dN_i's must vanish identically,

$$\ln N_i + \alpha + \beta\epsilon_i = 0$$

or

$$N_i = e^{-\alpha} e^{-\beta\epsilon_i} \tag{3.10}$$

[6]Stirling's approximation, which says that for large n, $\ln (n!) \simeq n \ln n - n$, is developed in Appendix A. The method of Lagrangian multipliers is described in Appendix B.

The first constraint (3.1) and Eq. (3.10) can be combined to show that[7]

$$\frac{N_i}{N} = \frac{e^{-\alpha} e^{-\beta\epsilon_i}}{\sum_{i=0}^{\infty} e^{-\alpha} e^{-\beta\epsilon_i}} \tag{3.11}$$

But α does not contain the summation index, i; thus

$$\frac{N_i}{N} = \frac{e^{-\beta\epsilon_i}}{\sum_{i=0}^{\infty} e^{-\beta\epsilon_i}} \tag{3.12}$$

which is the Boltzmann or Maxwell-Boltzmann distribution. The problem of evaluating α has conveniently been avoided by considering the *proportion* of particles at the ith energy level instead of the absolute number. We can thus anticipate that the multiplier α (which arises from the constraint of constant N) should be related in some fashion to the absolute number of particles present.[8] The constant β, on the other hand, is related in some way to the general level of particle energy. To evaluate β we must first propose a microscopic meaning for entropy.

MICROSCOPIC MEANING OF ENTROPY

Figure 3.1 shows an isolated system composed in turn of two subsystems, (1) and (2). The entire system is in equilibrium; hence the entropy function has a unique value. The maximum thermo-

Fig. 3.1 Two subsystems of an isolated system.

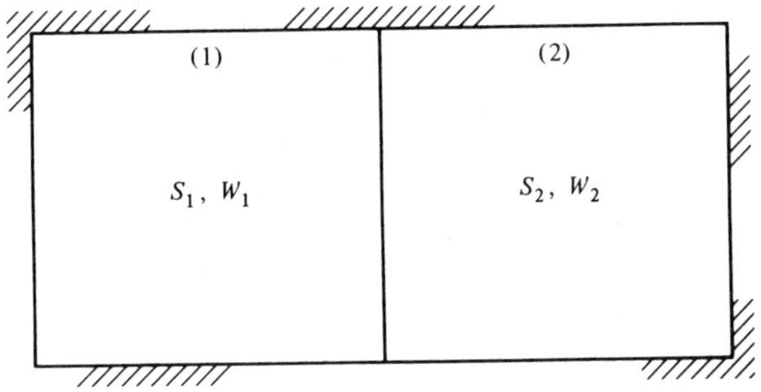

[7]The particular set of N_i's for which W is maximum should be given a special symbol — say N^*_i. We do not wish to complicate our notation in this way. Henceforth N_i is simply used to designate this particular value.

[8]It is shown in Sec. 6.1 that the multiplier, α, is related to the chemical potential, μ, the conjugate of the mole number, which is, in turn, the absolute number of particles divided by Avogadro's number. The exact relation is $\alpha = -\mu/kT$.

dynamic probability is also uniquely determined for the given system by the fact that it is in equilibrium. There should, therefore, be a functional relationship between S and W such as

$$S_1 = f(W_1) \qquad \text{and} \qquad S_2 = f(W_2) \tag{3.13}$$

where f denotes the same function in either case.

The entropy function is additive, however. Therefore, the entropy, S, of the combined system is

$$S = S_1 + S_2 \tag{3.14}$$

The thermodynamic probabilities of these independent systems are, on the other hand, multiplicative:

$$W = W_1 W_2 \tag{3.15}$$

To satisfy Eqs. (3.13), (3.14), and (3.15), the entropy function must take the form

$$S = f(W_1 W_2) = f(W_1) + f(W_2) \tag{3.16}$$

We must define S so that it will fit the narrow confines of Eq. (3.16). The form of the function, f, that will satisfy Eq. (3.16) is

$$S(W) = k \ln W \tag{3.17}$$

The multiplying constant is given the symbol k because it is possible to identify it with Boltzmann's constant later in this section. It is also shown that the entropy so defined is wholly consistent with macroscopic entropy.

Equation (3.17) lends insight to the meaning of entropy. Imagine, for example, an isolated system in equilibrium, with certain internal constraints. These might include adiabatic, rigid, and/or impermeable walls separating regions of differing temperature, pressure, and/or chemical potential. When the constraints are removed, spontaneous processes take the system to a new equilibrium state. The entropy of the new state exceeds that of the old one in accordance with either the second law of thermodynamics or the second and third thermodynamic postulates.

We also know that when the constraints are removed, if there is a more probable equilibrium state among the subsystems than formerly existed, the system will tend toward it. Equation (3.17) expresses the resulting maximization of entropy, not as a physical principle, but as a consequence of the laws of probability.

The salt-and-pepper-shaker example used in Sec. 3.1 can now be viewed in another light. Prior to mixing, the pure substances comprised subsystems of an equilibrium system. The system included one constraint—an impermeable wall—which was subsequently removed. Agitation of the system upset the metastable equilibrium of the unconstrained separate states and mixing occurred. As a consequence there was a finite increase of entropy, resulting from mixing alone. This phenomenon is also familiar in macroscopic thermodynamics.

Consider an isothermal, isobaric system consisting of two subsystems separated by an impermeable wall — the system shown in Fig. 3.1, for example. One subsystem consists of a mole of an ideal gas, No. 1; the other consists of a mole of another ideal gas, No. 2. The membrane is removed and the gases are allowed to mix. The change in molar entropy is well known [see, e.g., Eq. (7.66a)], and for this case it is

$$\Delta S = -R^0[1 \ln x_1 + 1 \ln x_2]$$

where the x_i's are the mole fractions in the mixture. In this case $x_1 = x_2 = \frac{1}{2}$, so

$$\Delta S = R^0 \ln 4 \tag{3.18}$$

This increase of entropy results solely from mixing the molecules and has meaning only as long as we can propose operations by which the molecules can be distinguished from one another. An entropy of mixing would thus arise if we mixed two very similar isotopes of oxygen but not if we mixed two samples of pure oxygen.[9]

An extremely simple microscopic argument also results in Eq. (3.18): When the wall is removed, there are exactly twice as many cells that each particle can occupy in position space. $W_{\max}$ is accordingly doubled for each particle present and Eq. (3.17) gives,

$$\Delta S = k \ln \frac{W_f}{W_i} = k \ln \left(\frac{W_{1_f}}{W_{1_i}} \frac{W_{2_f}}{W_{2_i}} \right)$$

or, in this case,

$$\Delta S = k \ln [(2)^{N_A}(2)^{N_A}] = (N_A + N_A) k \ln 2$$

so

$$\Delta S = R^0 \ln 4 \tag{3.18}$$

This result vindicates the selection of k as the multiplying constant in the entropy function, even though it is a restrictive example. It shows that if the calculated entropy change is the same whether it is based upon macroscopic or microscopic arguments — and it *must be* the same — then k must indeed be Boltzmann's constant.

[9]The apparent discontinuity in the entropy function as gases approach indistinguishability was noted by Gibbs and is called *Gibbs's paradox*. We return to this point in chapter 7.

PARTITION FUNCTION

Equation (3.12) can be rewritten as

$$\frac{N_i}{N} = \frac{e^{-\beta\epsilon_i}}{Z} \tag{3.12a}$$

where

$$Z \equiv \sum_{i=0}^{\infty} e^{-\beta\epsilon_i} \qquad \textit{partition function} \tag{3.19}$$

The partition function, Z, is so named because it expresses the *partition* or *distribution* of energies over the various energy levels. The German word for Z is *Zustandssumme*, the *sum of states* of the system, and is equally descriptive of the meaning of Z. The great value of the partition function will be found to lie in its role as a *generating function* for the macroscopic thermodynamic properties.

The partition function is a thermodynamic property itself in the sense that it is a unique function of the state of the gas. It depends upon T through β, and it depends upon V through ϵ_i, because the translational energy of particles rises or drops when the system is compressed or expanded adiabatically.[10]

The entropy for a system of distinguishable particles can be expressed in terms of the partition function with the help of the thermodynamic probability. Thus

$$S = k \ln W = k\left(N \ln N - \sum_{i=0}^{\infty} N_i \ln N_i\right)$$

The substitution of Eqs. (3.12) and (3.19) yields

$$\begin{aligned} S &= kN \ln N - k \sum_{i=0}^{\infty} N_i \ln\left(N \frac{e^{-\beta\epsilon_i}}{Z}\right) \\ &= kN \ln N - k \sum_{i=0}^{\infty} N_i \ln N + k \sum_{i=0}^{\infty} N_i \ln Z + k \sum_{i=0}^{\infty} N_i \beta\epsilon_i \\ &= kN \ln N - kN \ln N + kN \ln Z + k\beta U \end{aligned}$$

or

$$S = kN \ln Z + k\beta U \tag{3.20}$$

Equation (3.20) will serve as the basis for expressing the remaining macroscopic properties in terms of Z, k, N, and T. Beta is

[10]This point requires more attention than we have given it here. Equation (5.3) shows the dependence of the energy on container size, and Eq. (5.28) indicates the role of V in the partition function.

eliminated from Eq. (3.20) with the help of the definition of temperature,

$$\left(\frac{\partial S}{\partial U}\right)_{V,N} \equiv \frac{1}{T} = \left[\frac{\partial}{\partial U}(kN \ln Z + k\beta U)\right]_{V,N}$$

This differentiation gives

$$\frac{1}{T} = k\beta + k\frac{\partial}{\partial \beta}\left(N \ln \sum_{i=0}^{\infty} e^{-\beta\epsilon_i} + \beta U\right)\frac{\partial \beta}{\partial U} \tag{3.21}$$

but

$$\frac{\partial}{\partial \beta}\left(N \ln \sum_{i=0}^{\infty} e^{-\beta\epsilon_i} + \beta U\right) = -N\frac{\sum_{i=0}^{\infty} e^{-\alpha}\epsilon_i e^{-\beta\epsilon_i}}{\sum_{i=0}^{\infty} e^{-\alpha}e^{-\beta\epsilon_i}} + U$$

where $e^{-\alpha}$ has arbitrarily been multiplied and divided into the first term. Then

$$\frac{\partial}{\partial \beta}\left(N \ln \sum_{i=0}^{\infty} e^{-\beta\epsilon_i} + \beta U\right) = -\frac{N\sum_{i=0}^{\infty} N_i\epsilon_i}{\sum_{i=0}^{\infty} N_i} + U$$

or

$$\frac{\partial}{\partial \beta}\left(N \ln \sum_{i=0}^{\infty} e^{-\beta\epsilon_i} + \beta U\right) = -U + U = 0$$

Therefore, Eq. (3.21) reduces to

$$\beta = \frac{1}{kT} \tag{2.34}$$

The choice of the symbol β for the second Lagrangian multiplier was made in anticipation of the fact that this multiplier is the temper of the distribution. We have now shown that this choice was correct.

The entropy, energy, Helmholtz function, and pressure for a localized distribution are easily shown (Problem 3.5) to be

$$S = kN \ln Z + \frac{U}{T} \tag{3.22}$$

$$U = kT^2N\left(\frac{\partial \ln Z}{\partial T}\right)_{V,N} \tag{3.23}$$

$$F = -kTN \ln Z \tag{3.24}$$

$$p = kTN\left(\frac{\partial \ln Z}{\partial V}\right)_{T,N} \tag{3.25}$$

These results illustrate the way in which the remaining thermodynamic properties can be extracted from Z once it is known. But, if both the parameter, Z, and the fundamental equation of macroscopic thermodynamics give complete thermodynamic information, we ought to be able somehow to obtain a macroscopic parameter equivalent to Z.

The Massieu function given by Eq. (1.48) provides the basis for doing this:

$$-\frac{F}{T} \equiv S - \frac{U}{T} = -\frac{F}{T}\left(\frac{1}{T}, V, N\right) \tag{1.48}$$

From Eq. (3.24), on the other hand, we can write this Massieu function as

$$-\frac{F}{T} = kN \ln Z\left(\frac{1}{T}, V\right) \tag{3.26}$$

Rearrangement of this equation gives the fundamental equation for the system, in the Helmholtz-function form:

$$F(T, V, N) = -kNT \ln Z(T, V) \tag{3.26a}$$

This can also be written as

$$-\frac{F}{NT} = k \ln Z(\frac{1}{T}, V) \tag{3.26b}$$

where $-F/NT$ is a Massieu function on a "per-particle" basis. The equation

$$e^{-F/kNT} = Z \tag{3.27}$$

therefore provides us with a macroscopic generating function equivalent to the partition function.

Further discussion of the evaluation of Z is deferred until Sec. 3.7. Since the second Lagrangian multiplier is now known, the derivation of the Maxwell distribution will now be completed.

3.6 MAXWELL DISTRIBUTION

The molecular-velocity distribution in a stationary dilute gas, Eq. (2.43), can now be obtained from the Boltzmann distribution. The energy at the ith level, ϵ_i is assumed to consist of pure translatory energy and the various internal molecular energies, $\sum_j \epsilon_{ji}$, which do not depend upon the translational velocities (the independence of the translational and internal energy modes is discussed further in Secs. 5.3 and 7.1). These latter energy contributions might arise from rotation or vibration of the atoms within a translating molecule, for example. Then

$$\epsilon_i = \frac{m}{2} C_i^2 + \epsilon_{1_i} + \epsilon_{2_i} + \cdots = \frac{m}{2} C_i^2 + \sum_j \epsilon_{j_i}$$

This energy can now be substituted into Eq. (3.12)—the Boltzmann distribution. After we have divided both N and N_i by the volume to get n and the number density n_i of particles at the ith energy level (or "within the ith energy cell") the Boltzmann distribution becomes

$$n_i = \frac{n}{Z} \exp\left(-\sum_j \frac{\epsilon_{ji}}{kT}\right) \exp\left(-\frac{mC_i^2}{2kT}\right) \tag{3.28}$$

where the i cells represent different locations in velocity space and can all be located in a single large cell of position space, $\Delta V = V$, because the gas is in equilibrium and the distribution does not change in space. Accordingly, N has been replaced by N/V, the number density n of the particles; and n_i denotes the portion of n in $\Delta\Omega$. Finally [recall Eqs. (2.11) and (2.12)], we can drop the subscript i in the continuous limit so that Eq. (3.28) becomes

$$f(\mathbf{C}) \simeq \frac{n_i}{\Delta\Omega} \simeq A \exp\left(-\frac{mC^2}{2kT}\right) \tag{3.28a}$$

where $\Delta\Omega$ is volume-dependent, because $\Delta\tau = \text{constant} = m^3 V\,\Delta\Omega$, and

$$A \equiv \frac{n \exp\left(-\sum_j \epsilon_{j_i}/kT\right)}{Z\,\Delta\Omega} \tag{3.29}$$

The factor A can be eliminated from Eq. (3.28a) with the help of Eq. (2.32a),

$$\tfrac{3}{2} kT = \tfrac{1}{2} m\overline{C^2} = \frac{m}{2n}\int_\Omega C^2 f(\mathbf{C})\, d\Omega$$

so

$$\tfrac{3}{2} kT = A \frac{m}{2n} \iiint_{-\infty}^{\infty} (U^2 + V^2 + W^2) \exp\left(-\frac{mU^2}{2kT}\right) \exp\left(-\frac{mV^2}{2kT}\right) \exp\left(-\frac{mW^2}{2kT}\right) dU\, dV\, dW$$

Appropriate integration by parts leads to

$$\tfrac{3}{2} kT = A \frac{m}{2n}\left[\frac{3}{2}\left(\frac{2kT}{m}\right)\left(\frac{2\pi kT}{m}\right)^{3/2}\right]$$

so that

$$A = n\left(\frac{m}{2\pi kT}\right)^{3/2} \tag{3.29a}$$

and the resulting *Maxwell velocity distribution* is

$$f(\mathbf{C}) = n\left(\frac{m}{2\pi kT}\right)^{3/2} \exp\left(-\frac{mC^2}{2kT}\right) \tag{2.43}$$

3.7 PARTITION FUNCTION AND EQUIPARTITION OF ENERGY

THEOREM OF EQUIPARTITION OF ENERGY

Certain of the arguments used to derive Boltzmann's distribution can be recast to obtain another of the very powerful results of classical statistical thermodynamics. This is the *theorem of equipartition of energy*, which says: *Each contribution to the average energy of a particle that is a quadratic function of one of the coordinates of the particle (e.g., x, y, z, U, V, or W) gives an average energy of kT/2 per particle to the system.*

The theorem can be proved as follows: Let the energy of a particle at the ith level be expressed as a sum of component energies, at least one of which is quadratic in form,

$$\epsilon_i = \epsilon_{1_i} + \epsilon_{2_i} + \cdots + \alpha\xi_i^{\,2} \tag{3.30}$$

where ξ_i is any (position or velocity) coordinate of the ith cell and α is a constant of proportionality. Arguments analogous to those which led to Eq. (3.28a) give, in this case,

$$f(\xi) \simeq B \exp\left(-\frac{\alpha\xi^2}{kT}\right) \tag{3.28b}$$

where

$$B \equiv \frac{N \exp\left(-\sum_j \epsilon_{ji}/kT\right)}{Z\,\Delta\xi \prod \Delta\ (\text{other coordinates})} \tag{3.29b}$$

The average of ϵ_ξ, where $\epsilon_\xi \equiv \alpha\xi^2$, is

$$\overline{\epsilon_\xi} = \alpha\overline{\xi^2} = \frac{\alpha B \int_{-\infty}^{\infty} \xi^2 \exp(-\alpha\xi^2/kT)\, d\xi}{B \int_{-\infty}^{\infty} \exp(-\alpha\xi^2/kT)\, d\xi}$$

$$= \frac{\alpha(\sqrt{\pi}/2)\cdot\frac{1}{2}(kT/\alpha)^{3/2}}{(\sqrt{\pi}/2)\sqrt{kT/\alpha}}$$

or

$$\overline{\epsilon_\xi} = \frac{kT}{2}$$

which completes the proof.

The equipartition theorem is a surprising statement about molecular behavior. It says that whatever might be the various ways in which a particle stores energy, the energy will be (on the average) equally distributed among these modes of storage.[11] If the molecules of a system rotate in two directions as they translate, the average molecule will possess just as much kinetic energy of rotation in each direction as it will due to translation in, for example, the y direction. It should be emphasized, however, that the validity of the equipartition theorem as stated above is subject to the following restrictions:

1. The Boltzmann distribution law is valid. This requires that the gas be at an equilibrium condition.
2. The particles must be independent of one another.
3. The approximation of substituting integrals for summations is allowable. This implies that the spacing between neighboring energy levels must be small. We find out later that this assumption breaks down for rotational and vibrational energies at sufficiently low temperatures.
4. The total energy of a molecule can be split into energies resulting from different modes of motion. This is true if the interaction between different modes of motion is negligible. For most kinds of molecules the rotational and vibrational energies are not really entirely separable.
5. Each component of energy must be a quadratic function of the position or momentum component coordinate. The vibrational energy of a large-amplitude oscillation (anharmonic oscillation) violates this assumption.

If the number of modes of energy storage the particle can employ, or the number of degrees of freedom of the particle, is denoted by D, and if all energy storage is of the form, $\alpha\overline{\xi^2}$, then the specific heat of the particle can be expressed in terms of D. The molar specific heat c_v is, for example,

$$c_v = \left(\frac{\partial U}{\partial T}\right)_V = \left[\frac{\partial}{\partial T}\left(N_A D \frac{kT}{2}\right)\right]_V$$

or

$$c_v = \frac{R^0 D}{2} \tag{3.31}$$

so that for ideal gases

$$c_p = c_v + R^0 = \tfrac{1}{2}R^0(D+2) \tag{3.32}$$

[11] It is helpful to recognize that this theorem is a direct manifestation of the requirement that the system be in its most probable macrostate.

Equation (3.31) is subject only to the restrictions listed above and is not necessarily limited to ideal gases. In chapter 4, for example, we find that it can be adapted to the prediction of the specific heat of a solid.

A stationary monatomic gas that stores energy only in translational motion has an average energy per particle of

$$\bar{\epsilon} = \tfrac{1}{2}m\overline{U^2} + \tfrac{1}{2}m\overline{V^2} + \tfrac{1}{2}m\overline{W^2}$$

Hence it has three degrees of freedom, and

$$\bar{\epsilon} = \tfrac{3}{2}\,kT$$

(something that we already knew), and

$$c_v = \tfrac{3}{2}\,R^0 \qquad c_p = \tfrac{5}{2}\,R^0$$

The ratio of specific heats, is accordingly,

$$\gamma = \tfrac{5}{3} = 1.667$$

which is the same as the experimental value for monatomic argon (for example) at moderate temperatures.

A diatomic gas with two additional modes of rotational motion[12] (see Fig. 3.2) will have energies of rotation about the z and y coordinates equal to $\overline{\epsilon_{\theta_z}}$ and $\overline{\epsilon_{\theta_y}}$, where

$$\overline{\epsilon_{\theta_z}} = \tfrac{1}{2}\,I_z\overline{\dot{\theta}_z{}^2} = \frac{kT}{2}$$

and

$$\overline{\epsilon_{\theta_y}} = \tfrac{1}{2}\,I_y\overline{\dot{\theta}_y{}^2} = \frac{kT}{2}$$

The symbols I_ξ and $\dot{\theta}_\xi$ denote the moment of inertia and the angular velocity about the ξ coordinate, respectively. D is now equal to 5 and

$$c_v = \tfrac{5}{2}\,R^0 \qquad c_p = \tfrac{7}{2}\,R^0$$

so that

$$\gamma = \tfrac{7}{5} = 1.40$$

which is the well-known value for diatomic gases at moderate temperatures.

A diatomic molecule at elevated temperatures will suffer excitation of *two* additional degrees of vibrational freedom. To see

[12]Rotation about the axis is excluded because quantum mechanics shows that this motion can only be excited at unreasonably high temperatures. See the discussion of Θ_r at the end of Sec. 5.3.

Fig. 3.2 Modes of energy storage in a diatomic molecule.

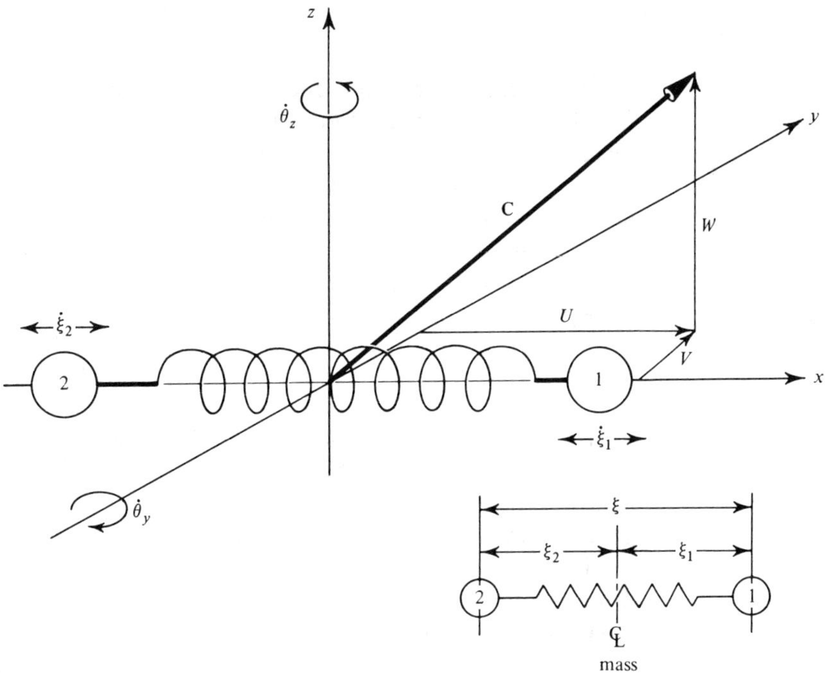

why this should be, consider the inset in Fig. 3.2. The center of mass of the molecule is located so that

$$\xi_1 = \frac{m_2}{m_1 + m_2}\xi \qquad \xi_2 = \frac{m_1}{m_1 + m_2}\xi$$

The total vibrational energy is composed of kinetic energy and potential energy, both of which can be expressed in terms of the square of coordinates. The average kinetic energy of vibration $\overline{\epsilon_{VK}}$ is

$$\overline{\epsilon_{VK}} = \overline{\tfrac{1}{2} m_1\dot{\xi}_1^2 + \tfrac{1}{2} m_2\dot{\xi}_2^2} = \tfrac{1}{2} m_r\overline{\dot{\xi}^2} = \frac{kT}{2}$$

where

$$m_r \equiv \frac{m_1 m_2}{m_1 + m_2}$$

The average potential energy of vibration, $\overline{\epsilon_{VP}}$, can be expressed in terms of the "spring constant" of the molecule K and the equilibrium displacement of the molecule, ξ_e:

$$\overline{\epsilon_{VP}} = \tfrac{1}{2} K\overline{(\xi - \xi_e)^2} = \frac{kT}{2}$$

The total number of degrees of freedom for a vibrating diatomic molecule is thus seven, so

$$\gamma = \tfrac{9}{7} = 1.286$$

The reader might well question the omission of certain additional modes of energy storage in the preceding discussion. These include the third rotational component $\dot{\theta}_x$ and electronic energies. These have been omitted because our intuitive idea that they should contribute little happens to be correct at moderate temperatures. Nevertheless, nothing in the classical theory *tells* us to ignore them. After we have investigated the quantum mechanics of intermolecular energy distributions, we find it possible to account for the internal energy more accurately and to assign a proper role to such contributions as electronic-energy storage. An understanding of quantum-mechanical effects reveals that the excitation of each degree of freedom occurs in a series of very small steps such that $\bar{\epsilon_\xi}$ approaches $kT/2$ asymptotically. The variation of c_v with temperature for molecular hydrogen (for example) thus takes place as sketched in Fig. 3.3 rather than in abrupt steps.

It is of interest to note that the number of coordinates that would be needed to fully describe the motions of all atoms in a molecule is three times the number of atoms. Thus a monatomic particle can have only 3 modes of motion. A diatomic molecule can

Fig. 3.3 Variation of c_v of hydrogen with temperature.

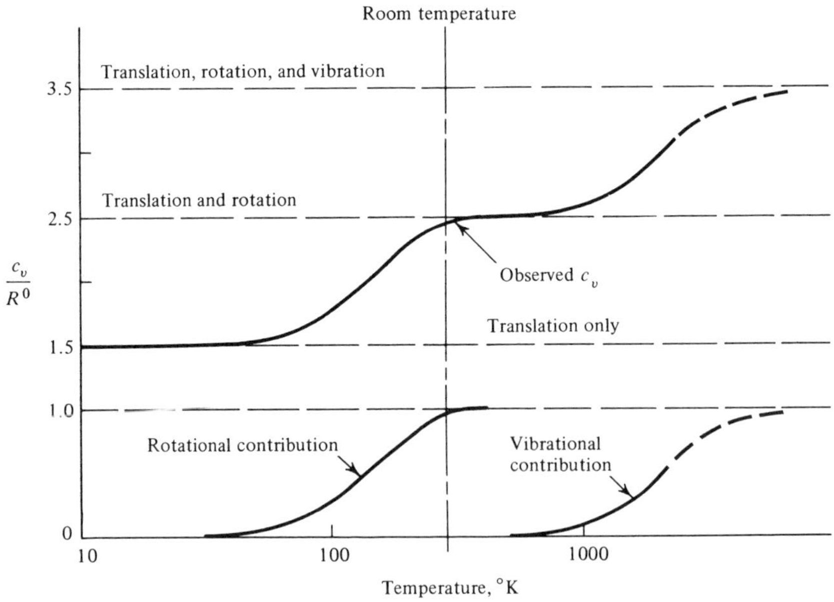

have 3×2, or 6: Three in translation, two in rotation and one in vibration. A planar triatomic molecule has three each (or nine) in translation, vibration and rotation. More complicated molecules will have three possible additional modes of vibration for each atom. Whether or not all the possible motions will all be activated, however, is a quantum mechanical problem.

PARTITION FUNCTION FOR AN IDEAL MONATOMIC GAS

It was shown in Sec. 3.5 that the partition function could be used as a generating function for thermodynamic properties, but the problem of writing it in a useful form was deferred. The partition function for an ideal monatomic gas can now be obtained from Eqs. (3.29) and (3.29a) by noting that the nontranslational energies $\left(\sum_j \epsilon_{ji}/kT\right)$ are zero in this case. Then

$$A = n\left(\frac{m}{2\pi kT}\right)^{3/2} = \frac{Ne^{-0}}{Z\,\Delta\Omega\,V}$$

or, since $N/n = V$,

$$Z = \frac{1}{\Delta\Omega}\left(\frac{2\pi kT}{m}\right)^{3/2} = \frac{V}{\Delta\tau}(2\pi mkT)^{3/2} \tag{3.33}$$

The constant $\Delta\tau$, which can be varied arbitrarily, thus appears to influence the partition function. With reference to S, U, F, and p [Eqs. (3.22) through (3.25)] we see that S and F will be affected by this choice to the extent of an additive constant equal to $-\ln \Delta\tau$, while U and p will not be affected at all. Quantum mechanics will provide means for fixing upon one appropriate value of $\Delta\tau$, the use of which will eliminate this measure of arbitrariness.

EXAMPLE 3.2 Obtain the distribution function, partition function, and specific heat for air molecules in the earth's atmosphere if it is approximately isothermal?

This problem can be solved in either of two ways. We can employ the ideal-gas law, $p = \rho kT/m$, in the hydrostatic equation $dp = -\rho g\,dz$, where z is the vertical coordinate. This gives

$$\frac{dp}{p} = -\frac{mg}{kT}\,dz$$

Integration from $p(z = 0) = p_0$ up to a point of interest gives

$$p = p_0 \exp\left(-\frac{mgz}{kT}\right)$$

But $p \sim n$ in an isothermal atmosphere, so

$$n = n_0 \exp\left(-\frac{mgz}{kT}\right)$$

Although the ideal-gas law used in this derivation was expressed in terms of some microscopic parameters, the derivation has not really used any microscopic concepts. It turns out that we can derive the same relation using Boltzmann statistics without making any reference to the macroscopic notion of a hydrostatic pressure formula. We begin by writing the energy of a particle[13] as

$$\epsilon = \frac{m}{2} C^2 + mgz$$

where rotational and vibrational energy contributions have been neglected.

Thus the Boltzmann distribution of n in z will take the form

$$n = \text{constant} \cdot \exp\left(-\frac{mg}{kT} z\right)$$

where the constant will include such things as Z, the integrated kinetic energy contribution, and cell sizes. We went to the trouble of writing out the constant in an analogous computation in Eq. (3.29), but it is not really necessary to do so. Here we can simply note that when $z = 0$ the constant can be evaluated as n_0. Thus

$$n = n_0 \exp\left(-\frac{mg}{kT} z\right)$$

which is the same result that macroscopic considerations gave.

It is extremely interesting to note that the equipartition theorem does not apply in this case because the gravitational energy is not quadratic in form. Thus the specific heat, for example, cannot be written as $DR^0/2$, with $D = 3 + 1$, or $2R^0$, as we might be tempted to try. Instead

$$c_v = \left.\frac{\partial U}{\partial T}\right|_{V,N} = \frac{\partial}{\partial T}\left(R^0 T^2 \left.\frac{\partial \ln Z}{\partial T}\right|_{V,N}\right)_{V,N}$$

and to write the partition function we ignore rotational and vibrational contributions and begin with

$$Z = \sum_{i=0}^{\infty} \exp\left(-\frac{\epsilon_i}{kT}\right) = \sum_{i=0}^{\infty} \exp\left(-\frac{mC_i^2}{2kT} - \frac{mgz_i}{kT}\right)$$

This can be approximated with integrals[14] as

$$Z = \frac{1}{\Delta\tau} \underbrace{\iiint}_{x,y,z} \underbrace{\iiint}_{\text{mom.}} \exp\left[-\frac{p_x^2 + p_y^2 + p_z^2}{2mkT} - \frac{mgz}{kT}\right] dp_x\, dp_y\, dp_z\, dx\, dy\, dz$$

[13]We do this problem here as though there were no rotational contribution. We see in Sec. 5.3 that this is actually incorrect.

[14]This approximation is discussed with some care in Sec. 5.4.

or

$$Z = \frac{\text{Area}}{\Delta\tau}\left(\frac{kT}{mg}\right)(2\pi mkT)^{3/2}$$

Substituting this in the expression for c_v, we get

$$c_v = \tfrac{5}{2} R^0$$

Thus gravity contributes $2R^0/2$ to the specific heat.

Much of the analytical apparatus that will be used subsequently to obtain thermodynamic information from molecular aggregates has now been established. One important tool — quantum mechanics — is lacking. At several points in the preceding sections arguments had to be terminated short of completion because of this lack. Therefore, we must now digress from the problems of thermodynamics to outline those elements of quantum mechanics that we need, and return to a more complete discussion of the partition function in Secs. 5.3 and 5.4.

Problems 3.1 Suppose you are rolling two dice. What is the probability of rolling any particular microstate? What is the probability of rolling the macrostate 4? What is the thermodynamic probability of the macrostate 4? Discuss the application of the principle of equal a priori probabilities to the dice.

3.2 A playing card lies face down on the table. A picks it up and looks at it but B does not. What entropies do A and B compute for the card? What assumptions must underlie these computations?

3.3 Air at standard conditions circulates freely in a large box subdivided into leaky cells of volume 1 cm^3. The thermodynamic probability, $W_{eq.}$, for the distribution of equal numbers of molecules, $N_{eq.}$, in each of the cells should exceed the value, $W_{ne.}$, for the nonequilibrium situation in which the numbers in the cells differ.

(a) Compute $W_{ne.}$ for the case in which all but two cells contain $N_{eq.}$ particles. One of the remaining cells contains $(N_{eq.} + n)$ and the other, $(N_{eq.} - n)$ particles, where $(n/N_{eq.}) \ll 1$.

(b) Using the power series expansion for $\ln(1 + x)$ in Appendix E, Sec. 2(c), compute $\ln(W_{ne.}/W_{eq.})$ for (a) and use the result to determine $W_{ne.}/W_{eq.}$ for the case in which one cell contains 1 percent more molecules, and one cell contains 1 percent fewer, than $N_{eq.}$. What maximum percent deviation would be expected in a thousand pairs of cells?

(c) What entropy decrease would be associated with the one percent fluctuation described in part (b)?

3.4 Find (a) the partition function for a two-dimensional monatomic gas, and (b) the equations of state of the gas, by regarding the expression for the Helmholtz function as the fundamental equation.

3.5 Derive Eqs. (3.23), (3.24), and (3.25).

3.6 Show the complete derivation of the Maxwell distribution.

3.7 On the basis of the classical theory, determine the values of c_v and γ for a gas composed of molecules with four atoms at the corners of a tetrahedron.

3.8 Use Eqs. (3.33) and (3.25) to derive the ideal-gas law.

4 development of quantum mechanics

By the end of the nineteenth century scientists had generally achieved remarkable advances in discovering what they believed to be the basic mechanisms upon which the universe was ordered. Some men felt that the remaining work of science lay in little more than completing details. So orderly was Victorian science that certain flaws in the structure did not seem to harbor serious trouble. The success of Maxwell–Boltzmann statistics was part of this picture of complacency. When, in 1901, Planck provided a basis for explaining the failure of classical statistics to describe certain phenomena, a revolution was set in motion. By 1927 this revolution had altered not only the structure of science but the very concepts of causality and determinacy upon which it was based.

The revolution originated when Planck succeeded, while classical theory failed, in explaining the energy spectrum of electromagnetic radiation. More questions were raised by his explanation, and the resulting chain of explanations was largely resolved in Schrödinger's mathematical formulation of quantum mechanics in 1926 and Heisenberg's enunciation of the uncertainty principle in 1927. We now introduce the laws of quantum mechanics and treat the statistics based upon quantum concepts in chapter 6.

4.1 PREQUANTUM THEORIES OF BLACKBODY RADIATION

The term "radiation" is applied generally to the process that transmits energy from one body to another in the absence of a material medium between them. This process is often conveniently described in terms of electromagnetic waves. The electromagnetic-wave spectrum ranges from radio waves of extremely long wavelength and low frequency to cosmic rays of extremely short wavelength and high frequency. The thermal radiation emitted by bodies by virtue of their temperature is distributed over wavelengths from about 10^{-5} to 10^{-2} cm in moderate ranges of temperature.

STEFAN-BOLTZMANN LAW

The temperature dependence of the total blackbody energy flux E is given by the Stefan–Boltzmann law,

$$E = \sigma T^4 \tag{4.1}$$

The Stefan–Boltzmann constant of proportionality σ is now known to be 5.6697×10^{-5} erg/cm^2-sec-°K^4. This relation was first suggested in 1879 by Stefan on the basis of experiments. Five years later Boltzmann derived it on the basis of thermodynamic arguments.

It is helpful in following Boltzmann's arguments to take the view that electromagnetic radiation can be interpreted as particle action instead of wave action. From Eq. (2.31), the pressure exerted by particles reflecting from a wall can be expressed as

$$p = \frac{n}{3}(m\overline{C^2}) = \frac{2}{3}\,\frac{\text{kinetic energy}}{\text{unit volume}}$$

Actually, it has been shown that the pressure p_{rad} exerted by radiation is only half this.[1] Thus we have the result noted in Example 1.2:

$$p_{\text{rad}} = \frac{u}{3} \tag{4.2}$$

Equation (2.52) also gives the flux of radiation "particles" as $nc_l/4$, where c_l is the velocity of light. The flux of energy is then the product of this particle flux and the energy per particle, u/n:

$$E = \frac{c_l u}{4} \tag{4.3}$$

[1] For a discussion of this and related matters, see, for example, F. K. Richtmyer, E. H. Kennard, and T. Lauritsen, *Introduction to Modern Physics*, 5th ed., McGraw-Hill, Inc., New York, 1955, chap. 4.

Boltzmann used Eq. (4.2) and the thought model of a Carnot engine driven by radiation to show that

$$u = \text{constant} \cdot T^4 \tag{4.4}$$

Substitution of Eq. (4.3) in Eq. (4.4) then gave the Stefan–Boltzmann law, Eq. (4.1). The Stefan–Boltzmann constant σ remained an empirical parameter in the Boltzmann prediction but Planck's subsequent quantum prediction included a theoretical value of σ.

ENERGY SPECTRUM

The *monochromatic emissive power* or the *spectral emissive power* $E_\lambda(\lambda, T)$ is the distribution function for energy as a function of wavelength. It is defined [recall Eqs. (2.9) and (2.10)] such that

$$E = \int_0^\infty E_\lambda(\lambda, T)\, d\lambda \tag{4.5}$$

We also define a spectral energy density u_λ such that $E_\lambda = c_l u_\lambda/4$. The function E_λ or u_λ has great practical importance, and the determination of its form was the subject of considerable attention in the late nineteenth century.

Wien tried in 1893 to establish the relationship among u_λ, T, and λ and was only partially successful. He was able to prove that the functional relationship was

$$\frac{u_\lambda}{T^5} = \frac{C_1}{(\lambda T)^5} f(\lambda T) \tag{4.6}$$

where C_1 is a constant and the function f was unknown. In 1896 he argued that $f(\lambda T)$ should be exponential in form and obtained what is known as *Wien's distribution*,

$$u_\lambda = C_1 \lambda^{-5} \exp\left(-\frac{1}{C_2 \lambda T}\right) \tag{4.7}$$

where C_2 is a second experimental constant.

Accurate measurements of λ, T, and E_λ were made by Lummer and Pringsheim in 1899. Their data defined curves such as are shown in Fig. 4.1(a). Figure 4.1(b) shows the correlation curve through all these data on $E_\lambda T^{-5}$ versus λT coordinates. It also shows that Wien's distribution law, which works very well for low λT, fails at large λT.

Figure 4.1(b) shows that the maximum E_λ occurs where

$$(\lambda T)_{E_\lambda = \max} = 5216\ \mu\text{-}^\circ\text{R} = 0.2898\ \text{cm-}^\circ\text{K} \tag{4.8}$$

Equation (4.8) is shown as a locus of the maxima of the curves in Fig. 4.1(a). Equation (4.8) is the one empirical fact that Planck had to use in developing the quantum distribution, and we return to it later.

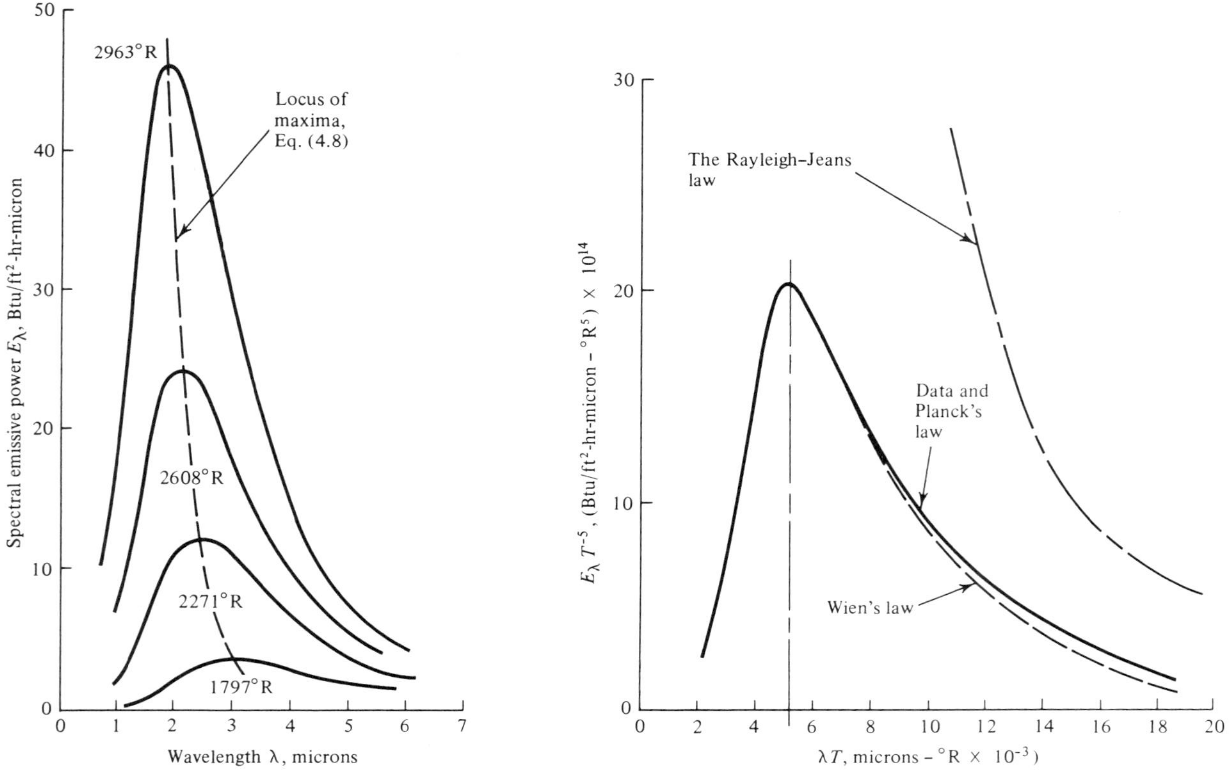

(a) Typical curves defined by the data of Lummer and Pringsheim.

(b) Correlation of data on the basis of Wien's law. Comparison of data with the distributions of Wien, Rayleigh-Jeans, and Planck.

Fig. 4.1 Black-body spectral emissive power — predicted and observed.

RAYLEIGH–JEANS LAW

An important attempt to formulate the distribution law for E_λ was developed by Rayleigh in 1900 and redeveloped in a simpler way by Jeans in 1905. The equation was not accurate in correlating data, but it would be the correct result in a strictly classical world and it has important aspects in common with Planck's derivation. It therefore merits our attention.

Let there be a quantity of radiant energy contained in a box (see Fig. 4.2) of volume $V = abc$. The walls of the box are perfectly conducting and perfectly reflecting. These conditions require that the electromagnetic waves *stand* in the box as shown and that their null points coincide with the wall, which (by virtue of perfect conduction) can sustain no electric potential.

The waves within the box are subject to Maxwell's equations for an electromagnetic field in free space. These can be combined into the following forms:

$$\nabla^2 \mathbf{E} = \frac{1}{c_l^2}\frac{\partial^2 \mathbf{E}}{\partial t^2} \tag{4.9}$$

and

$$\nabla^2 \mathbf{H} = \frac{1}{c_l^2}\frac{\partial^2 \mathbf{H}}{\partial t^2} \tag{4.10}$$

where **H** is the magnetic intensity of the field and **E** the electric intensity.

Equation (4.9) actually consists of three scalar equations of the form

$$\nabla^2 E_x = \frac{1}{c_l^2}\frac{\partial^2 E_x}{\partial t^2} \tag{4.11}$$

Fig. 4.2 Jeans's radiation thought model.

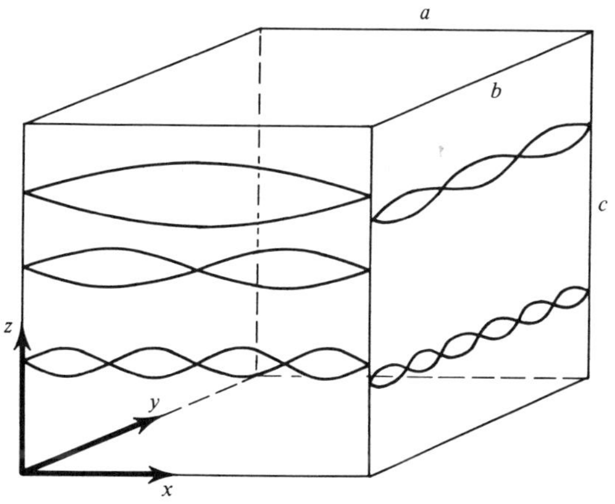

The boundary conditions on $\mathbf{E}(x, y, z, t)$ are

$$\begin{aligned} \mathbf{E}(0, y, z, t) &= \mathbf{E}(a, y, z, t) = 0 \\ \mathbf{E}(x, 0, z, t) &= \mathbf{E}(x, b, z, t) = 0 \end{aligned} \tag{4.12}$$

and

$$\mathbf{E}(x, y, 0, t) = \mathbf{E}(x, y, c, t) = 0$$

The solution for this system is

$$E_x = A \sin 2\pi\nu t \sin\left(\frac{\pi n_x x}{a}\right) \sin\left(\frac{\pi n_y y}{b}\right) \sin\left(\frac{\pi n_z z}{c}\right) \tag{4.13}$$

$$E_y = B \sin 2\pi\nu t \sin\left(\frac{\pi n_x x}{a}\right) \sin\left(\frac{\pi n_y y}{b}\right) \sin\left(\frac{\pi n_z z}{c}\right) \tag{4.14}$$

and

$$E_z = C \sin 2\pi\nu t \sin\left(\frac{\pi n_x x}{a}\right) \sin\left(\frac{\pi n_y y}{b}\right) \sin\left(\frac{\pi n_z z}{c}\right) \tag{4.15}$$

where A, B, and C are constants and ν is the wave frequency,

$$\nu = \frac{c_l}{\lambda} \tag{4.16}$$

The three eigenvalues n_x, n_y, and n_z are integers that have the physical significance of being the number of standing-wave loops in the box. In one-dimensional wave motion, the wave number n is

$$n = \frac{2\nu a}{c_l} = \frac{2a}{\lambda} \tag{4.17}$$

In three-dimensional wave motion, λ (or c_l/ν) has to satisfy wave numbers in three directions, so that

$$\frac{1}{\lambda} = \sqrt{\left(\frac{n_x}{2a}\right)^2 + \left(\frac{n_y}{2b}\right)^2 + \left(\frac{n_z}{2c}\right)^2} \tag{4.18}$$

The Rayleigh–Jeans derivation now proceeds as follows. In accordance with the equipartition theorem, each wave component is treated as a simple harmonic oscillator that is free to vibrate in two directions and that is, therefore, possessed of an average energy

$$\bar{\epsilon} = 2(2 \times \tfrac{1}{2}\, kT) = 2kT \tag{4.19}$$

The two directions of vibration are the two *directions of polarization* of the wave. They are illustrated for an x-directed wave in Fig. 4.3. This ingenious application of the equipartition theorem now pro-

Fig. 4.3 Idealization of doubly polarized wave motion used by Rayleigh–Jeans and Planck.

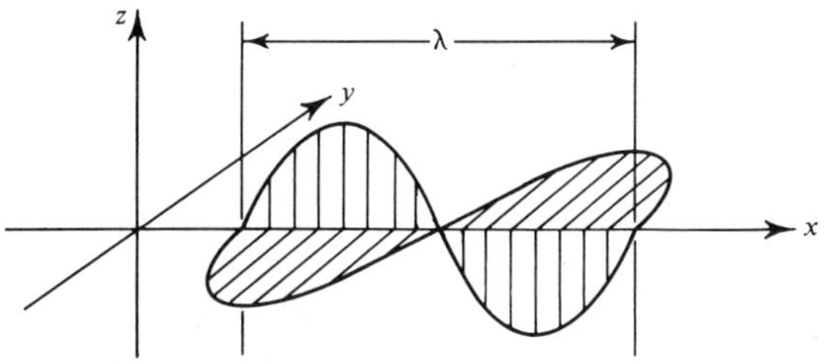

(a) The y and z polarization of an x-directed wave.

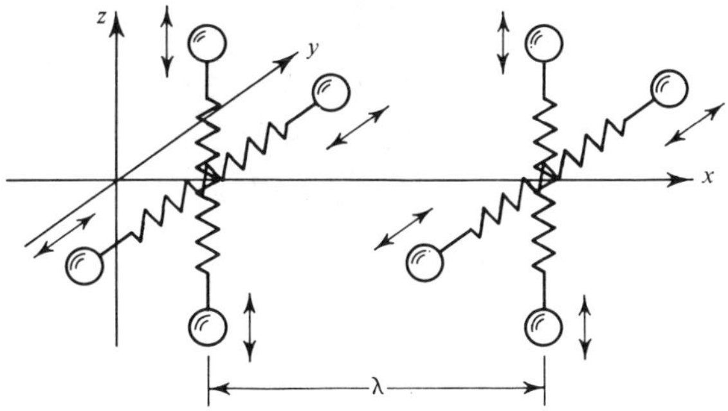

(b) Idealization of the wave as a series of mechanical oscillators.

vides a basis for calculating u_λ (and, with it, E_λ) if we can say how many components of each wavelength are present.

It is expedient to designate as n_0 the number of wave components in the range $0 \leq 1/\lambda \leq (1/\lambda)_{\text{interest}}$. Clearly, if we consider n_0 and λ as continuous,

$$\frac{n_0(2kT)}{V} = \int_\lambda^\infty u_\lambda \, d\lambda = u - \int_0^\lambda u_\lambda \, d\lambda \tag{4.20}$$

or

$$\frac{2kT}{V}\frac{dn_0}{d\lambda} = -u_\lambda \tag{4.21}$$

The number n_0 is obtained with the help of the following simple device. Equation (4.18) defines a different ellipsoid in n_x, n_y, n_z space (see Fig. 4.4) for each value of λ. The total number of points within the λ surface is numerically equal to the volume of the positive eighth of the ellipsoid,

$$n_0 = \frac{1}{8}\frac{4\pi}{3}\frac{2a}{\lambda}\frac{2b}{\lambda}\frac{2c}{\lambda} = \frac{4\pi V}{3\lambda^3} \tag{4.22}$$

Fig. 4.4 Device for counting wave components.

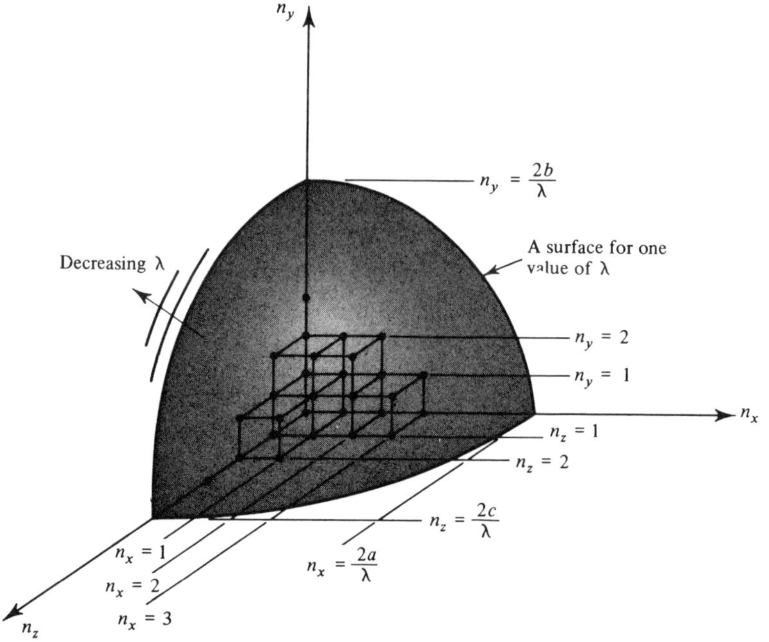

Differentiation of Eq. (4.22) and substitution of the result in Eq. (4.21) gives the Rayleigh–Jeans law directly,

$$u_\lambda = \frac{8\pi kT}{\lambda^4} \quad \text{or} \quad \frac{4\pi}{\lambda^4}\bar{\epsilon} \tag{4.23}$$

or

$$E_\lambda T^{-5} = \frac{2\pi k c_l}{(\lambda T)^4} \tag{4.24}$$

Equation (4.24), which is based upon the classical equipartition theorem, does not provide a good prediction of E_λ except for very large values of λT. A typical comparison of both the Wien and the Rayleigh–Jeans law with experimental data is sketched in Fig. 4.1(b). (The region in which the Rayleigh–Jeans law becomes accurate is off scale in the figure.)

This, then, was the stage for Planck's work in 1901. Wien's semiempirical law offered a convenient representation of the data over most wavelengths, but the classical theory was concurrently providing only a spectacular failure in representing the data.

4.2 PLANCK'S QUANTUM THEORY OF THE ENERGY SPECTRUM

Planck's attack did not differ from the Rayleigh–Jeans development up to the establishment of the relation $u_\lambda = (4\pi/\lambda^4)\bar{\epsilon}$. At this point he replaced the classical equipartition expression for $\bar{\epsilon}$ with the following heuristic proposal:

There is a discrete set of energies that a physical system such as an oscillator can assume. The system cannot assume intermediate energies. The energies $\epsilon_0, \epsilon_1, \ldots, \epsilon_i, \ldots$ are spaced so that[2]

$$\epsilon_i = i\epsilon \qquad i = 0, 1, 2, \ldots \tag{4.25}$$

where, in turn, ϵ is the smallest physically meaningful energy increment. It is directly proportional to the classical frequency ν of the oscillator. Thus

$$\epsilon = h\nu \tag{4.26}$$

where the constant of proportionality h (called *Planck's constant*) will be evaluated subsequently. Planck had originally intended to let $h \to 0$ but found that in doing so he only obtained the Rayleigh–Jeans law, and that violated Eq. (4.1), which is a constraint on the energy.

Planck's proposal is contrary to our intuition as to the "way things are," and indeed they offended the thinking of his contemporaries. The reasonableness of the assumptions will become increasingly clear to us, however, as we trace the developments that followed Planck's work.

The average energy of a wave $\bar{\epsilon}$ is defined as

$$\bar{\epsilon} = 2\frac{\sum_{i=0}^{\infty} N_i\epsilon_i}{\sum_{i=0}^{\infty} N_i} \tag{4.27}$$

where the factor of 2 again accounts for the fact that each wave is being construed as two oscillators by virtue of double polarization.

[2]We find in chapter 5 that, for a harmonic oscillator, Eq. (4.25) should be $\epsilon_i = (i + \frac{1}{2})\epsilon$. This correction, based upon later quantum theory, did not change Planck's results. We also see in chapter 6 that Eq. (4.25) must be abandoned in favor of a different characterization of energy when we take the more sophisticated view that photons form a "degenerate ideal gas."

Using the Boltzmann distribution to describe the energy distribution among the mechanical oscillators that replace the actual waves in this construct, we obtain

$$N_i = \frac{N}{Z} \exp\left(-\frac{\epsilon_i}{kT}\right) \tag{3.12a}$$

in which N_i is the number of oscillators in the ith energy level. The constant N/Z is obviously equal to N_0 for a given distribution, because $\epsilon_0 = 0$; thus

$$N_i = N_0 \exp\left(-\frac{\epsilon_i}{kT}\right)$$

It follows that

$$\sum_{i=0}^{\infty} N_i = N_0 \left\{ 1 + \exp\left(-\frac{\epsilon}{kT}\right) + \left[\exp\left(-\frac{\epsilon}{kT}\right)\right]^2 + \left[\exp\left(-\frac{\epsilon}{kT}\right)\right]^3 + \cdots \right\} \tag{4.28}$$

but the term in braces in Eq. (4.28) is the power-series expansion for $[1 - \exp(-\epsilon/kT)]^{-1}$ when $[\exp(-\epsilon/kT)]^2 < 1$. This is clearly the case, so

$$\sum_{i=0}^{\infty} N_i = \frac{N_0}{1 - \exp(-\epsilon/kT)} \tag{4.29}$$

Likewise,

$$\sum_{i=0}^{\infty} \epsilon_i N_i = \epsilon \sum_{i=0}^{\infty} i N_i = \epsilon N_0 \exp\left(-\frac{\epsilon}{kT}\right) \left\{ 1 + 2\exp\left(-\frac{\epsilon}{kT}\right) + 3\left[\exp\left(-\frac{\epsilon}{kT}\right)\right]^2 + \cdots \right\} \tag{4.30}$$

The term in braces in Eq. (4.30) is the power-series expansion for $[1 - \exp(-\epsilon/kT)]^{-2}$ when $[\exp(-\epsilon/kT)]^2 < 1$, so

$$\sum_{i=0}^{\infty} \epsilon_i N_i = \frac{\epsilon N_0 \exp(-\epsilon/kT)}{[1 - \exp(-\epsilon/kT)]^2} = \frac{\epsilon N_0}{[1 - \exp(-\epsilon/kT)][\exp(\epsilon/kT) - 1]} \tag{4.31}$$

The substitution of Eqs. (4.29) and (4.31) into Eq. (4.27) and the subsequent use of Eq. (4.26) gives

$$\bar{\epsilon} = \frac{2\epsilon}{\exp(\epsilon/kT) - 1} = \frac{2h\nu}{\exp(h\nu/kT) - 1} \tag{4.32}$$

Thus we obtain Planck's distribution,

$$u_\lambda = \frac{4\pi}{\lambda^4}\bar{\epsilon} = \frac{8\pi h\nu}{[\exp(h\nu/kT) - 1]\lambda^4} \tag{4.33}$$

or

$$u_\lambda = \frac{8\pi h c_l}{\lambda^5[\exp(hc_l/\lambda kT) - 1]} \tag{4.34}$$

and

$$E_\lambda = \frac{2\pi h c_l^2}{\lambda^5[\exp(hc_l/\lambda kT) - 1]} \tag{4.34a}$$

Equation (4.34a) predicts values of E_λ that fall within the accuracy of the experimental results. However, its use depends upon the prior evaluation of Planck's constant h. This is achieved by first writing

$$\left.\frac{\partial u_\lambda}{\partial \lambda}\right|_{\lambda T = 0.2898\ \mathrm{cm}-{}^\circ\mathrm{K}} = 0 \tag{4.35}$$

The numerical solution of Eq. (4.35), following the substitution of Eq. (4.34) for u_λ, gives the value of Planck's constant h as 6.625×10^{-27} erg-sec.

We can also evaluate the Stefan–Boltzmann constant σ using Eq. (4.34a),

$$E = \sigma T^4 = \int_0^\infty E_\lambda\, d\lambda = \int_0^\infty E_\nu\, d\nu \tag{4.36}$$

The transformation of the integrand of this expression is accomplished as follows:

$$E_\lambda\, d\lambda = \frac{2\pi h c_l^2}{(c_l/\nu)^5[\exp(h\nu/kT) - 1]}\left(-\frac{c_l}{\nu^2}\, d\nu\right)$$

or

$$E_\lambda\, d\lambda = -\frac{2\pi h}{c_l^2}\frac{\nu^3\, d\nu}{\exp(h\nu/kT) - 1} = -E_\nu\, d\nu$$

With the substitution of this value of $E_\nu\, d\nu$ and with the introduction of the change of variable $x = h\nu/kT$, Eq. (4.36) becomes

$$\sigma T^4 = \frac{2\pi k^4 T^4}{c_l^2 h^3}\int_0^\infty \frac{x^3 e^{-x}}{1 - e^{-x}}\, dx \tag{4.37}$$

The term $(1 - e^{-x})^{-1}$ can be expanded in a power series and the series of terms can subsequently be integrated by parts. Then the introduction of the Riemann zeta function[3] of 4,

$$\zeta(4) = \sum_{i=1}^{\infty} \frac{1}{i^4} = \frac{\pi^4}{90} \tag{4.38}$$

leads to the result

$$\sigma = \frac{2\pi^5 k^4}{15 c_l^2 h^3} \tag{4.39}$$

Planck's theory thus accomplishes a complete and accurate description of blackbody radiation with the help of a single experimental fact, Eq. (4.8). When we take up a discussion of quantum statistics we find that his success (like that of many other great theorists) was the result of sound intuition embedded in a slightly incorrect exposition. Today's quantum model for radiation is more complex than an oscillator with discrete allowable energy levels, as we see in Secs. 4.4 through 4.6. Furthermore, the oscillators are not conserved $\left(\sum_i N_i \neq \text{constant}\right)$, because radiation is absorbed and reemitted at the walls.

These objections are removed by treating radiation as a collection of degenerate particles that obey Bose–Einstein statistics. We discuss these particles in chapter 6 and give them the name "photons." The results obtained in this way are the same as those of Planck, but it is not surprising that a full acceptance of quantum theory had to wait for further clarification.

It is interesting, in this connection, to read Jeans's discussion[4] of his perplexity over the meaning of Planck's correction of his own prediction. The quest for the true nature of radiant energy during the early 1900s is reminiscent of the fable of the blind beggars and the elephant.[5] The difficulties raised by the dual view (wave versus

[3]The Riemann zeta function $\zeta(a)$ is defined as $\sum_{i=1}^{\infty} i^{-a}$. It has important applications in the subject of analysis. [See, e.g., E. T. Whitaker and G. N. Watson, *A Course in Modern Analysis*, 4th ed., Cambridge University Press, New York, 1963, chap. 13. See also Appendix E, 2(b).]

[4]Sir James Jeans, *The Dynamical Theory of Gases*, 4th ed., Dover Publications, Inc., New York, 1954, secs. 484–493, 525–528.

[5]Seven blind beggars approached an elephant. The first bumped into its side and cried, "The elephant is very like a wall." "No" said the second, seizing its tail, "The elephant is like a rope." The third, feeling an ear, said "He is like a leaf." "He is like a serpent," said the one who laid hold of its trunk. "A tree," said the one by its foreleg. The sixth struck a tusk and shrieked, "He is like a spear and is going to impale me." The seventh beggar heard all this and thought, "Whatever he is, he is bewitched," and he fled.

corpuscular) of radiation were not to be resolved until about 1927. Meanwhile Albert Einstein provided the early quantum theory with its second great success.

In 1906 Einstein, who was later to oppose certain features of the modern quantum theory, helped to put the new theory on its feet by using the quantization of energy to predict the specific heat of crystalline solids. We study this application next.

EXAMPLE 4.1 Derive Planck's law for a two-dimensional space. In this case, the solutions of Maxwell's equations are

$$E_x = A \sin(2\pi\nu t) \sin\left(\frac{\pi n_x x}{a}\right) \sin\left(\frac{\pi n_y y}{b}\right)$$

and

$$E_y = B \sin(2\pi\nu t) \sin\left(\frac{\pi n_x x}{a}\right) \sin\left(\frac{\pi n_y y}{b}\right)$$

where the n_x and n_y must satisfy

$$\left(\frac{2\nu}{c_l}\right)^2 = \left(\frac{n_x}{a}\right)^2 + \left(\frac{n_y}{b}\right)^2$$

The number of standing waves with frequency ν is given by analogy with Eq. (4.22) as

$$n_0 = \frac{\pi}{4}\frac{2a\nu}{c_l}\frac{2b\nu}{c_l} = \frac{\pi A \nu^2}{c_l^2}$$

where A is the area. Then

$$u_\nu = \frac{\bar{\epsilon}}{A}\frac{dn_0}{d\nu} = \frac{2\pi\nu}{c_l^2}\bar{\epsilon}$$

and $\bar{\epsilon}$ is one-half the value given by Eq. (4.32), because only single polarization is meaningful in two dimensions. Thus

$$u_\nu = \frac{2\pi h \nu^2}{c_l^2}\left[\exp\left(\frac{h\nu}{kT}\right) - 1\right]^{-1}$$

It can be shown (Problem 2.19) that the two-dimensional analogue of Eq. (2.52) is

$$J_{\text{particles}} = \frac{nc_l}{\pi}$$

So the energy "flux" in this case is

$$E = \frac{c_l}{\pi}\int_0^\infty u_\nu \, d\nu$$

Under the substitution of the preceding expression for u_ν we obtain from this

$$E = \sigma_2 T^3$$

where the two-dimensional "Stefan–Boltzmann" constant σ_2 is

$$\sigma_2 \equiv \frac{2k^3}{c_t h^2} \int_0^\infty \frac{x^2}{e^x - 1}\, dx$$

The evaluation of σ_2 is left as an exercise (Problem 4.6).

Thus two-dimensional radiation would be less temperature-sensitive than three-dimensional radiation, because E is proportional only to T^3.

4.3 SPECIFIC HEATS OF SOLIDS

LAW OF DULONG AND PETIT

DuLong and Petit noted in 1819 that *the molar specific heat of all elementary solids is very nearly 6*. Neumann extended this law in 1831 to say that each atom in the molecule of a solid contributed 6 cal/g mole-°C to the specific heat of the solid. Typical experimental specific heats for solids are given in Table 4.1.

TABLE 4.1 Application of the Laws of DuLong and Petit and of Neumann

Molar c_v for Monatomic Solids			Molar c_v for Diatomic and Triatomic Solids		
Substance	T, °C	c_v, cal/g mole–°C	Substance	T, °C	c_v/2 or 3, cal/g mole–°C
Ag	0	6.00	AgCl	28	6.27
Au	0	6.07	CuO	22	5.20
Cr	0	5.35	KCl	23	6.20
Fe	0	5.85	CuS	25	5.95
Ni	0	6.05	PbO_2	24	5.17
Sb	0	6.00	CaF_2	15–99	5.61
Graphite	0	1.82	ZnO	16–99	5.08
Diamond	0	1.25	$PbCl_2$	0–20	6.08

Although these laws are by no means exact, they strongly suggest that some underlying physical principle might be responsible for the degree of success they do enjoy.

CLASSICAL EQUIPARTITION THEORY FOR THE SPECIFIC HEATS OF SOLIDS

The equipartition theory provides the following simple explanation for the laws of DuLong and Petit and of Neumann: An atom in a solid is considered to be a localized harmonic oscillator, with three modes of vibration. There are two degrees of freedom associated with each mode of vibration, or a total of six degrees of freedom. Then Eq. (3.31) gives, for a monatomic solid,

$$c_v = \tfrac{1}{2}(6)R^0 = 5.972 \text{ cal/g mole-°C} \tag{4.40}$$

Equation (4.40) verifies the empirical laws beautifully, but it fails to explain why certain substances deviate very strongly from these laws. The specific heat of diamond is only 1.25 at 0°C, for example, but if it is heated to relatively high temperatures, its specific heat also approaches 5.972 cal/g mole-°C. Indeed, if the specific heats of a variety of simple solids are plotted against temperature,[6] the results will be of the form shown in Fig. 4.5. Equation (4.40) thus appears to provide a limiting value of c_v as the temperature is increased. However, it fails at low temperatures just as the classical Rayleigh–Jeans radiation expression failed at low values of λT.

EINSTEIN'S QUANTUM-MECHANICAL SPECIFIC-HEAT LAW

Einstein recognized that a quantum explanation might resolve the failure of the classical specific-heat theory. He argued that Planck's equation for the average energy of an oscillator should be used for atoms in a solid as well as for electromagnetic waves. Accordingly he wrote the molar internal energy for monatomic solids

$$u = N_A \bar{\epsilon}$$

Fig. 4.5 Temperature dependence of c_v for many solids. $\Theta_{Pb} = 88$, $\Theta_{Ag} = 215$, $\Theta_{KCl} = 218$, $\Theta_{Zn} = 235$, $\Theta_{NaCl} = 287$, $\Theta_{Cu} = 315$, $\Theta_{Al} = 392$, $\Theta_{CaF_2} = 499$, $\Theta_C = 1860$.

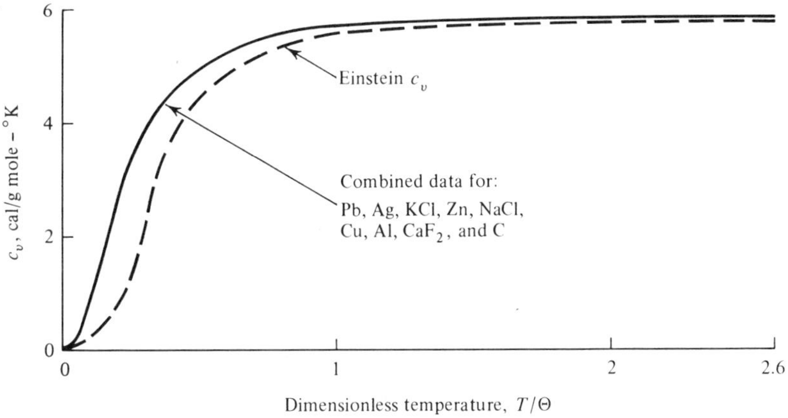

[6] Figure 4.5 actually correlates data for nine different substances onto a single curve, with the aid of a dimensionless temperature, T/Θ. This correlation is explained fully in chapter 5.

The substitution of Eq. (4.32) for $\bar{\epsilon}$, and multiplication of the result by $\frac{3}{2}$, since there are now three instead of two modes of vibration, gives

$$u = \frac{3N_A h\nu}{\exp(h\nu/kT) - 1}$$

Einstein's specific-heat expression is obtained by differentiating this expression,

$$c_v \equiv \left.\frac{\partial u}{\partial T}\right|_v = 3N_A k \frac{(h\nu/kT)^2 \exp(h\nu/kT)}{[\exp(h\nu/kT) - 1]^2} \tag{4.41}$$

or

$$c_v = 3R^0 \frac{x^2 e^x}{(e^x - 1)^2} \tag{4.41a}$$

where $x \equiv h\nu/kT$. The characteristic temperature Θ is $h\nu/k$, however; thus $x = \Theta/T$. It is easily shown (Problem 4.3) that

$$\lim_{T/\Theta \to \infty} (c_v) = 3R^0 \tag{4.42}$$

Hence both Einstein's specific-heat equation (4.41) and the Planck radiation expression approach their classical limits as $kT/h\nu$ or $\lambda kT/hc_l$ becomes large.

Einstein's specific-heat relationship gives the proper qualitative form for $c_v(T)$. It is exact in the limits as $T \to 0$ and $T \to \infty$, but it is a little inaccurate in the intermediate range (see Fig. 4.5). The reason lies in Einstein's assumption that the atoms are independent and all have the same classical frequency of oscillation.[7] Debye and others subsequently improved upon Einstein's prediction. We take up this work in chapter 10.

4.4 WAVE CHARACTERISTICS OF MATTER

By the end of the nineteenth century, general acceptance was given to the view that matter was corpuscular in character and that light (as well as other kinds of radiation) was some sort of wave action. Planck's quantum hypothesis then suggested that radiation might, after all, be endowed with certain corpuscular properties. By 1920 the dual nature of radiant energy was (as we have observed) generally recognized and radiant energy was thought to be carried in the form of corpuscular elements called *photons*. A

[7]This assumption was implied when Eq. (4.32) was used for $\bar{\epsilon}$. Equation (4.32) was derived by considering waves of only one classical frequency ν.

photon had an energy equal to $h\nu$ and was endowed with certain wavelike properties.

There was no suggestion up to this time that material particles might conversely exhibit certain wavelike characteristics. However, a formal analogy between geometric optics and classical mechanics had long been recognized.

ANALOGY BETWEEN CLASSICAL MECHANICS AND GEOMETRIC OPTICS

The laws of geometric optics are derivable from Fermat's *principle of least time* (1650). This principle states that of all the possible paths light might take between two points A and B, the actual path will be that for which the time of travel is minimal. Since the velocity of light is constant we can express the principle in variational form, calling S/λ the distance in wavelengths,

$$\delta\left(\int_A^B \frac{dS}{\lambda}\right) = 0 \tag{4.43}$$

The variation δ compares neighboring values of distances along paths between fixed end points A and B. That it vanishes implies that the path is of minimal length.

Maupertuis's *principle of least action* (1740) is analogous to Fermat's principle. It says that[8]

$$\delta\left(\int_A^B p\, dS\right) = 0 \tag{4.44}$$

where p is the momentum of a *particle*. The product $p\, dS$ is called *action*. It has the same units as Planck's constant, h, which is sometimes called *quantum action*.

De Broglie examined this analogy in 1924 in connection with his doctoral thesis. The result of his work was a bold suggestion as to the character of material particles.

MATTER WAVES

De Broglie argued that if the corpuscular and wave concepts of light were inseparable, then perhaps the corpuscular and wave concepts of matter might also be inseparable. Thus he postulated that there would be a frequency ν and a wavelength λ "associ-

[8]This is a special case of Hamilton's principle, which says that for conservative systems, $\delta\left(\int_{t_1}^{t_2} L\, dt\right) = 0$, where t is time and L, the "Lagrangian," is the difference between the kinetic and potential energies of particles.

ated'' with a particle of total energy[9] ϵ and momentum p such that

$$\epsilon = h\nu \qquad \text{and} \qquad p = \frac{h}{\lambda} \tag{4.45}$$

The de Broglie wavelength λ can be evaluated, for example, for a simple particle of mass m translating at speed v as

$$\lambda = \frac{h}{mv} \tag{4.46}$$

Let us consider an important experimental result and some aspects of wave behavior before we complete our interpretation of the notion of matter waves.

DAVISSON–GERMER EXPERIMENT

Soon after de Broglie's dissertation appeared, it was pointed out that, if de Broglie's ideas were correct, particles such as electrons should exhibit diffraction effects. In 1927 Davisson and Germer succeeded in observing a diffraction pattern in low-energy electrons reflected from a nickel surface (see Fig. 4.6). This diffraction effect could only be explained by attributing wavelike properties to the electrons.

Since the low-energy electrons used in the experiment do not penetrate the crystal appreciably, interference depends primarily on the scattering by the surface layer of the crystal. Figure 4.7 illustrates this interference for a case in which $\lambda = d/2$. It shows a row of atoms in the surface of the crystal, each of which acts, according

Fig. 4.6 Intensity of electron current as a function of θ.

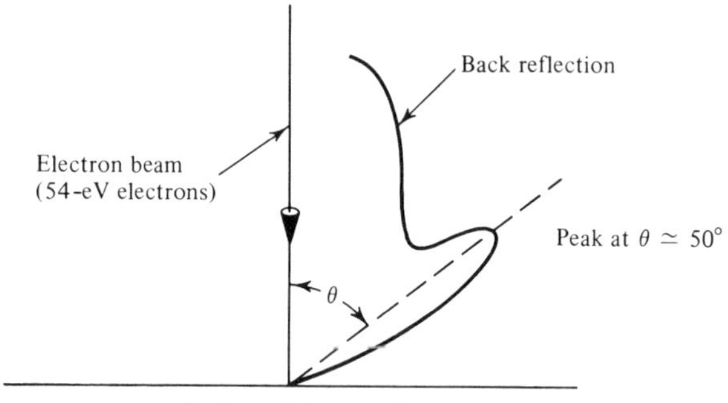

[9]The total energy includes the relativistic energy $m_o c_l^2$, where m_o is the rest mass. For a particle in translational motion: $m = m_o(1 - v^2/c_l^2)^{-1/2}$. When $v \ll c_l$ as in most actual situations, $m \simeq m_o(1 + v^2/2c_l^2)$, or $mc_l^2 = m_o c_l^2 + \frac{1}{2} m_o v^2$.

Fig. 4.7 Explanation of diffraction maximum in terms of wave interference, for the special case $\lambda = d/2$.

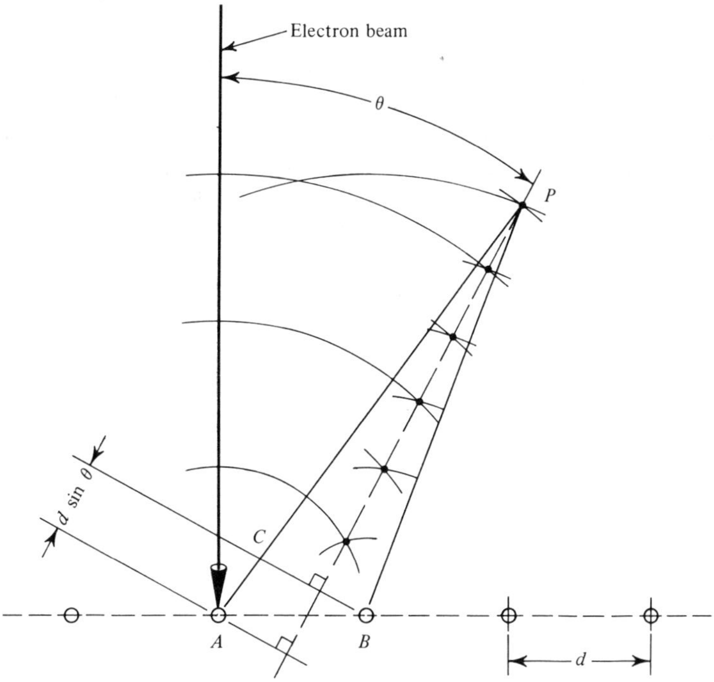

to Huygen's principle, as a secondary source of waves. Waves emanating from atoms A and B interfere constructively at P if the path difference $AP - BP = n\lambda$, where λ is the wavelength of the waves associated with the electrons and n is an integer. Hence for a first-order maximum ($n = 1$) the angle θ should be given by

$$AP - BP = d \sin \theta = \lambda \tag{4.47}$$

as long as $PC \gg d$. For nickel, the line spacing d is known from x-ray data to be 2.15 Å and we find that for 54-eV electrons, $\lambda = 1.67$ Å. Therefore, θ should be 51 deg in Fig. 4.6—in excellent agreement with the experimental results.

SOME QUANTITATIVE ASPECTS OF THE WAVELIKE CHARACTER OF MATTER

Combining Eqs. (4.45) and (4.46) gives us a wave speed w for "matter waves,"

$$w = \nu\lambda = \frac{\epsilon}{p} \tag{4.48}$$

With the Einstein relation converting mass to energy $\epsilon = mc_l^2$ and $p = mv$, we have

$$wv = c_l^2 \tag{4.49}$$

Thus, the velocity of a particle will equal the velocity of the waves it comprises, only if the particle velocity can approach the speed of light. For all actual moving particles, v is less than c_l and Eq. (4.61) requires that the corresponding phase velocity exceed c_l. But we have no way of relating a velocity greater than c_l to any physical process. It is therefore impossible to devise any means for measuring or observing the phase motion.

What kind of wave is this that moves at a speed that differs so greatly from the speed of the particle but which should in some sense be associated with the particle? We can gain some insight into the answer to this question by considering the general behavior of waves. The "displacement" of a wave can be represented in the following convenient form:

$$\Psi = C \cos (kx - \omega t) \tag{4.50}$$

where C is the amplitude; the *wave number*, $k \equiv \dfrac{2\pi}{\lambda}$; and the *angular frequency*, $\omega \equiv 2\pi\nu$. Thus

$$w = \nu\lambda = \frac{\omega}{k} \tag{4.51}$$

We also define a quantity called the phase, $\phi \equiv kx - \omega t$, so that

$$\left(\frac{\partial x}{\partial t}\right)_{\phi=\text{const}} = \frac{\omega}{k} \equiv \textit{phase speed, } w \tag{4.52}$$

The phase speed is thus the speed of a point on the wave at a specified phase angle.

Now let two waves with different angular frequencies and wave numbers be superposed as shown in Fig. 4.8. The equation of the combined waves is

$$\Psi_p = C[\cos (k_1 x - \omega_1 t) + \cos (k_2 x - \omega_2 t)] \tag{4.53}$$

Using the trigonometric identity

$$\cos A + \cos B = 2 \cos \left(\frac{A - B}{2}\right) \cos \left(\frac{A + B}{2}\right)$$

we can write Ψ_p in the form

$$\Psi_p = 2C \cos (k_g x - \omega_g t) \cos (k_w x - \omega_w t) \tag{4.54}$$

Fig. 4.8 Superposition of two waves.

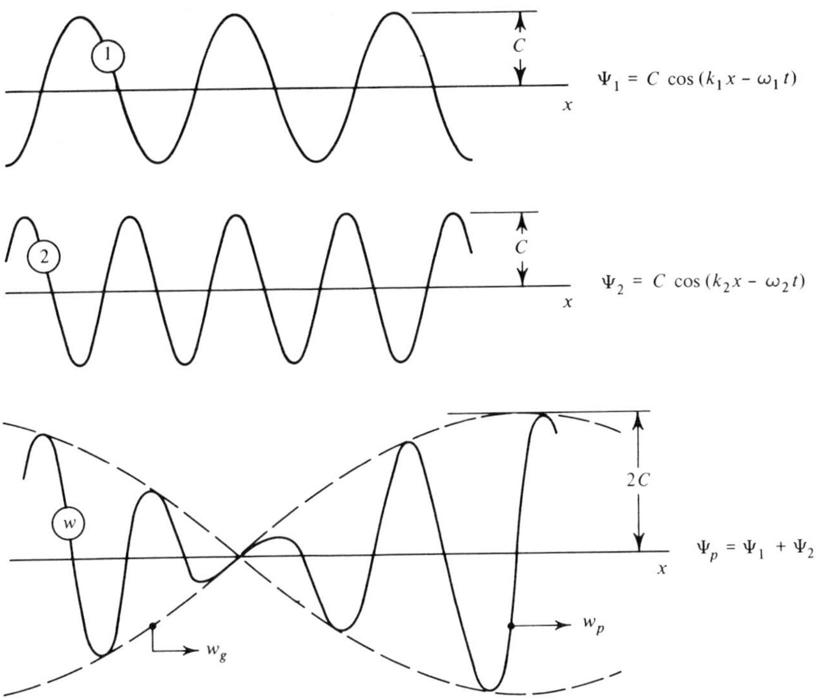

where

$$k_g \equiv \frac{k_2 - k_1}{2} \qquad \omega_g \equiv \frac{\omega_2 - \omega_1}{2}$$

$$k_w \equiv \frac{k_2 + k_1}{2} \qquad \omega_w \equiv \frac{\omega_2 + \omega_1}{2}$$

Obviously $\omega_g < \omega_w$. Therefore, $\cos(k_g x - \omega_g t)$ is the low-frequency term that determines the amplitude of the envelope of the high-frequency curve $\cos(k_w x - \omega_w t)$.

The speed with which the envelope propagates to the right is called the *group velocity* w_g. It is the phase velocity of the envelope

$$w_g = \frac{\omega_g}{k_g} = \frac{\omega_2 - \omega_1}{k_2 - k_1} \tag{4.55}$$

The phase speed w_p of the high frequency waves inside the envelope is

$$w_p = \frac{\omega_w}{k_w} = \frac{\omega_1 + \omega_2}{k_1 + k_2} \tag{4.56}$$

Suppose that these two frequencies become very close to one another. Then, for a given phase velocity, $\omega_2 \rightarrow \omega \leftarrow \omega_1$ and $k_2 \rightarrow k \leftarrow k_1$.

This would give a very low frequency "group" for the two waves. In the limit Eqs. (4.55) and (4.56) become[10]

$$\text{group velocity, } w_g = \frac{d\omega}{dk} \tag{4.57}$$

$$\text{phase speed, } w_p = \frac{\omega}{k} \tag{4.58}$$

Thus the wave (or phase) speed will differ from the group (or envelope phase) speed in a way that depends upon a *dispersion relation* $\omega = \omega(k)$ appropriate to the particular process. It turns out that an actual particle is the manifestation of the beating of a packet of many adjacent waves — not of just two waves — and Eq. (4.55) is not the appropriate dispersion relation for a particle. The actual particle velocity v is thus the group velocity for the packet. The phase velocity w of the matter wave is given by Eq. (4.52) as being faster than light and not subject to easy physical interpretation.

4.5 UNCERTAINTY PRINCIPLE

The idea that matter must be explained in terms of wave action should rightly imbue us with a certain uneasiness. Allied with this idea is the suggestion that matter is a more ephemeral thing than our intuition first tells us. Equations (4.45) show that the wavelengths of fairly substantial material particles are unthinkably small and frequencies are inconceivably high. But what about very small particles? Clearly they are somehow smeared about in time and space. Heisenberg's Uncertainty Principle expresses the indefiniteness of matter in a quantitative way.

The principle can be developed by consideration of a problem of *measurement*. Suppose a cosine wave passes a point and we wish to count the number of wave crests going by in unit time. We have a standard clock consisting of an oscillator that produces waves whose frequency we wish to compare with that of the incoming wave.

A beat may arise because the frequencies of the two waves

[10]Musical instruments provide vivid demonstrations of group velocity. The lower notes on a piano, for example, use two strings tuned to the same frequency (or "pitch"). When the piano is out of tune, different frequencies — fairly close together — are sounded when the note is played. These produce an evident group frequency or "beat," as it is called. If ω_g is faster than about 1 cps the beat will be offensive to the ear.

differ. If the waves are precisely the same, we will detect no beats. The beats (if they occur) will be of the frequency

$$\Delta\nu = \frac{|\nu_1 - \nu_2|}{2} \tag{4.59}$$

where ν_1 is the frequency of the incoming wave and ν_2 the frequency of the wave we are measuring.

The time required to observe one beat is $1/\Delta\nu$ — the period of the group frequency. To be confident of observing at least one beat we must make a measurement over a time Δt equal to at least the time for the occurrence of one beat,

$$|\Delta t| \geq \frac{1}{|\Delta\nu|} \tag{4.60}$$

Thus the longer we wait to observe a beat, the smaller is the uncertainty in the frequency. Therefore,

$$|\Delta t\, \Delta\nu| \geq 1$$

In time Δt the wave will have to move through a distance Δx. Therefore, Δx is the uncertainty in the location of the wave. If w is the velocity of the wave, then $\Delta t = \Delta x/w$, so

$$|\Delta x\, \Delta\nu| \geq w$$

But $\nu = w/\lambda$, so for small $\Delta\nu$ and $\Delta\lambda$,

$$|\Delta\nu| \simeq \frac{w|\Delta\lambda|}{\lambda^2}$$

Combining the preceding two equations we obtain

$$|\Delta x\, \Delta\lambda| \geq \lambda^2$$

Finally, we substitute de Broglie's relation $p = h/\lambda$ to get the uncertainty relation,

$$|\Delta x\, \Delta p| \geq h \tag{4.61}$$

Equation (4.61) is only an estimate, of course. Other methods of making this estimate lead to more accurate relations. One of these is

$$|\Delta x\, \Delta p| \geq \frac{\hbar}{2} = \frac{h}{4\pi} \tag{4.62}$$

where $\hbar$ is a modified Planck's constant equal to $h/2\pi$. However, the precise value of the uncertainty is not important in subsequent work.

At this point we have derived an analytical result that sets a limit on the certainty of *any* kind of observation. It implies that if we increase our certainty of the position of a particle we must also increase our ignorance of its momentum. This required ignorance is very small — so small, in fact, that it is generally swamped by experimental inaccuracy. Its value to us lies not in what it says about the accuracy of measurements but what it suggests about the nature of things.

The uncertainty relation tells us that *we can, under no circumstances, devise operations to specify position and momentum more accurately than Eq.* (4.62) *allows*. Our intuitive feeling that definite positions and momenta exist at any instant for particles is thus false, because we have no way of giving meaning to this feeling. Heisenberg has provided much more than a conclusion as to the measurability of matter. He has established the basis for a *principle*, which states, in effect, that questions about measurability become meaningless below the level established by Eq. (4.61) or (4.62). The uncertainty principle allows us to take the view that de Broglie waves characterize the fact that particles really are smeared in phase space. They *have no definite position and momentum!* By thus removing the intuitive concept of definiteness, Heisenberg removed that last obstacle to a fully quantum-mechanical view of the world.[11]

EXAMPLE 4.2 How small would a particle moving at typical molecular speeds have to be before the uncertainty of its location reached molecular dimensions?

In accordance with Eq. (4.62) we find that

$$|\Delta x\,\Delta p| \geq \frac{h}{4\pi} = 5.25 \times 10^{-28}\ \text{erg-sec}$$

Typical molecular dimensions given in Table 7.3 are on the order of 1 Å or 10^{-8} cm; and the molecular speed of air molecules at normal conditions is generally within $\pm\bar{C}$ of $\bar{C}$, where $\bar{C}$ is about 50,000 cm/sec. Thus

$$(10^{-8}\ \text{cm})(m \times 50{,}000\ \text{cm/sec}) \geq 5.25 \times 10^{-28}\ \text{g-cm}^2/\text{sec}$$

so the mass, m, must be at least as small as 10^{-24} g. From Appendix D we find that this is a little less than the mass of a proton or a neutron.

[11] If the reader finds this statement a little strong, he is in good company. There are others (notably Albert Einstein) who have objected to the concept that the world is ultimately indeterminate.

4.6 SCHRÖDINGER EQUATION

That matter can be interpreted as wave action suggests that a complete description of matter should include a wave equation for matter. The conventional wave equation for the displacement, Ψ, is [recall Eqs. (4.9) and (4.10)]

$$\nabla^2\Psi = \frac{1}{w^2}\frac{\partial^2\Psi}{\partial t^2} \tag{4.63}$$

Combination of the de Broglie relations (4.45) gives $\epsilon = p(\lambda\nu)$, but $\lambda\nu = w$. Thus

$$p_x^2 + p_y^2 + p_z^2 = p^2 = \frac{1}{w^2}\epsilon^2 \tag{4.64}$$

The similarity between the forms of Eqs. (4.63) and (4.64) is striking. This similarity suggested to Schrödinger the means for developing a wave equation for matter.

MOMENTUM AND ENERGY OPERATORS

The strategy for capitalizing upon the analogy between Eqs. (4.63) and (4.64) is as follows. If the displacement Ψ of a unidimensional wave is expressed in conventional complex form,

$$\Psi = C \exp[i(kx - \omega t)]$$

and then differentiated, it turns out that

$$-i\hbar\frac{\partial\Psi}{\partial x} = \frac{h\nu}{w}\Psi = p_x\Psi \tag{4.65}$$

and

$$i\hbar\frac{\partial\Psi}{\partial t} = h\nu\Psi = \epsilon\Psi \tag{4.66}$$

Equations (4.65) and (4.66) are faithful to the analogy between Eqs. (4.63) and (4.64). We are thus tempted to make an identification of the operators $-i\hbar(\partial/\partial x)$ and $i\hbar(\partial/\partial t)$ with p_x and ϵ, respectively.

The total mechanical energy, ϵ, of a nonrelativistic system (called the "Hamiltonian," H) can be written

$$\epsilon = \text{KE} + \text{PE} = \frac{p^2}{2m} + U(\mathbf{r})$$

or

$$\epsilon\Psi = \frac{p^2}{2m}\Psi + U(\mathbf{r})\Psi \tag{4.67}$$

Replacing p and ϵ in Eq. (4.67) with the operators $-i\hbar\nabla$ and $i\hbar[\partial/\partial t)$, we obtain for three-dimensional motion ($p^2 = p_x^2 + p_y^2 + p_z^2$)

$$\nabla^2\Psi - \frac{2mU\Psi}{\hbar^2} = \frac{2m}{i\hbar}\frac{\partial\Psi}{\partial t} \tag{4.68}$$

which is the *Schrödinger equation* for this system. The function Ψ is called the *wave function*. It is interesting that the Schrödinger equation is not a true wave equation. It is a diffusion equation, which possesses certain wave properties because of the imaginary coefficient of the first-order term. The fact that the Schrödinger equation is not a proper wave equation is of no importance, because it is really a mathematical postulate. Its correctness and usefulness can only be judged by comparing its results with experimental observations.

INTERPRETATION OF THE WAVE FUNCTION

It is postulated in the Schrödinger equation that the wave function $\Psi(\mathbf{r}, t)$ provides a complete quantum-mechanical description of the motion of a particle. But the physical significance of $\Psi(\mathbf{r}, t)$ is rather unclear. The wave-packet concept suggests that the wave function should be large where the particle is likely to be, and small elsewhere. The phrase "likely to be" is a consequence of the uncertainty principle, and it emphasizes the need for interpreting Ψ in statistical terms. It is, therefore, natural to regard Ψ as a measure of the probability of finding a particle at a particular position. However, a probability must be real and nonnegative, and Ψ is generally complex. It turns out that the probability density function for the existence of a particle at $\mathbf{r}$ and t is the product of Ψ and its complex conjugate, Ψ^*.[12] Accordingly (recall Sec. 2.2)

$$\int_V \Psi\Psi^*\,dV = \int_V |\Psi|^2\,dV = 1 \tag{4.69}$$

which is only to say that a particle will *certainly* "be" within whatever contains it.

By the same token, the spatial average of molecular properties can be evaluated as

$$\langle\phi\rangle \equiv \int_V \Psi^*\phi\Psi\,dV \tag{4.70}$$

The momentum of a particle in unidimensional motion, for example, is

$$\langle p\rangle \equiv \int_V \Psi^*\left(-i\hbar\frac{\partial}{\partial x}\right)\Psi\,dV \tag{4.71}$$

[12]The complex conjugate z^* of a complex number $z = x + iy$, is $x - iy$. The product $zz^* = x^2 + y^2$ is always real and equal to $|z|^2$.

4.7 QUANTUM STATE AS AN EIGENVALUE PROBLEM

The Schrödinger equation dictates the wave function Ψ which in turn tells us how a particle (if we can still use the word) occupies phase space. Before trying to solve the Schrödinger equation, it is well to reconsider what happens when we solve the simpler unidimensional wave equation for a vibrating string. The solution consists of a Fourier series of discrete solutions — each one of which describes a discrete standing wave in the string. The Schrödinger equation also dictates a *discrete* sequence of standing wave functions — each one of which corresponds with one of the meaningful energy levels of the particle.

Equation (4.68) will generally yield to a separation-of-variables solution of the form

$$\Psi(\mathbf{r}, t) = f(t)\psi(\mathbf{r})$$

Thus the equation becomes

$$\frac{\hbar i}{f}\frac{df}{dt} = \frac{1}{\psi}\left[U(\mathbf{r})\psi - \frac{\hbar^2}{2m}\nabla^2\psi\right]$$

Since the left-hand side depends upon t alone and the right-hand side depends upon $\mathbf{r}$ alone, each side must equal the same constant, which we shall call ϵ. Thus

$$U(\mathbf{r})\psi - \frac{\hbar^2}{2m}\nabla^2\psi = \epsilon\psi \tag{4.72}$$

and

$$f(t) = c\exp\left(-i\frac{\epsilon}{\hbar}t\right) \tag{4.73}$$

The solution to the equation is then

$$\Psi(\mathbf{r}, t) = c\psi(\mathbf{r})\exp\left(-i\frac{\epsilon}{\hbar}t\right)$$

Differentiating this result gives an expression identical to Eq. (4.66)

$$\hbar i\frac{\partial\Psi}{\partial t} = \epsilon\Psi \tag{4.66}$$

Thus if we are to satisfy Eq. (4.66), the separation constant ϵ must be the energy.

Our chief concern at the present will be with Eq. (4.72), the *stationary wave equation*, or *standing wave equation*. This equation and the boundary conditions that we generally impose upon it are a Sturm–Liouville system[13] for which the separation constant ϵ is the *eigenvalue. A very important feature of this system is that there exist solutions (in this case, standing waves) only for certain discrete eigenvalues (in this case, energy levels).* Thus solution of the stationary wave equation for a system will yield the allowable quantum energy levels for the system.

In some cases there is more than one eigenvalue for each solution. The physical meaning of this is that a system can occupy an energy level in more than one way. Such a system is then called *degenerate*, and the number of eigenvalues per eigenfunction is called the *degeneracy*, g_i.

Very often Eq. (4.72) is called the Schrödinger equation because it is used a great deal. Actually it is only an equation for the wave amplitude ψ.

Problems 4.1 Express the limits of Planck's radiation law for $\lambda kT/hc_l$ very large and very small. Are these the results that you would expect?

4.2 Fill in the missing mathematics and arithmetic in the numerical evaluations of h and σ.

4.3 Verify Eq. (4.42).

4.4 Prove that Einstein's specific-heat law approaches the limits required of it by (a) the fourth postulate, and (b) the classical theory of specific heats.

4.5 Solve $\nabla^2 E_x = (1/c_l^2)(\partial^2 E_x/\partial t^2)$ subject to the boundary conditions given in connection with the development of the Rayleigh–Jeans law.

4.6 Complete all the steps in the derivation of the Stefan–Boltzmann law for a two-dimensional space (Example 4.1) and evaluate σ_2.

4.7 Calculate the de Broglie wavelength of a mercury atom at 200°C and of a 1-oz rifle bullet traveling 1000 m/sec.

4.8 Rifle bullets weighing 20 g and moving at 45,000 cm/sec are fired from a distance of 100 meters at a target. If the only errors in accuracy arose from the Uncertainty Principle, roughly how much spread would there be around the center of the bull's eye?

4.9 A perfectly elastic tennis ball is dropped onto another perfectly elastic tennis ball from a height of ten meters. If the only imperfections in the system arise from the Uncertainty Principle, roughly how many bounces can the tennis ball make before it misses the lower ball?

[13]Some important properties of the Sturm–Liouville system are outlined in Appendix C.

5 the application of quantum mechanics

The development of quantum mechanics was described from a basically chronological viewpoint in chapter 4. We found that the energy of all systems must be viewed as varying in a discrete way rather than continuously. The partition of energy is thus far more fundamental than we were led to believe in chapter 3. In chapter 3 we divided energy into discrete levels simply as a convenience to make possible the enumeration of microstates. Now we have discovered that such a subdivision is, in fact, *specified* by the wavelike character of all matter and energy.

Next we explore the quantum partitioning of energy more fully, using the Schrödinger equation to compute the energy levels that are intrinsic in a variety of physical systems.

5.1 SOLUTIONS OF THE SCHRÖDINGER EQUATION FOR THREE IMPORTANT CASES

FREE PARTICLE IN A BOX

Suppose that a molecule translates freely in a box. The motion is *free* in that no external force fields or intermolecular forces act upon it except at the walls. The simplest such case would be unidimensional motion in the interval (0, L) with collisions at the "walls" ($x = 0$ and L). Thus the potential energy $U(\mathbf{r})$ is zero in (0, L) and infinite at $x = 0$ and L. This potential-energy function is illustrated in Fig. 5.1(a). Since no wave function can exist where $U \to \infty$, $\psi(0)$ and $\psi(L)$ must both vanish. And since the amplitude ψ should be continuous, it should still be close to zero at $x = 0^+$ and L^-, where U is zero. We can therefore drop U from consideration and write Eq. (4.72) in the form

$$-\frac{\hbar^2}{2m}\frac{d^2\psi}{dx^2} = \epsilon_x\psi \tag{5.1}$$

with the boundary conditions $\psi(0^+) = \psi(L^-) = 0$.

The general solution of Eq. (5.1) is

$$\psi(x) = A \sin\sqrt{\frac{2m\epsilon_x}{\hbar^2}}\,x + B\cos\sqrt{\frac{2m\epsilon_x}{\hbar^2}}\,x \tag{5.2}$$

Substitution of the first boundary condition gives $B = 0$. Substitution of the second one specifies the eigenvalues as

$$\epsilon_x = \frac{\hbar^2\pi^2}{2mL^2}n_x^2 \quad \text{or} \quad \frac{h^2}{8mL^2}n_x^2 \qquad n_x = 0, 1, 2, \ldots \tag{5.3}$$

We call n_x the *quantum number*, because it specifies the quantum state. Finally, Eq. (5.3) can be substituted into Eq. (5.2) and the result substituted into the normalization condition, Eq. (4.69).

$$\int_0^L |\psi|^2\,dx = 1 \tag{5.4}$$

or, in this case,

$$\int_0^L A^2\left|\sin^2\left(\frac{\pi n_x x}{L}\right)\right|dx = \frac{A^2L}{2} = 1 \tag{5.4a}$$

Thus $A = \sqrt{2/L}$ and

$$\psi = \sqrt{\frac{2}{L}}\sin\left(\frac{\pi n_x x}{L}\right) \qquad n_x = 0, 1, 2, \ldots \tag{5.5}$$

The energy levels of translating particles are shown by Eq. (5.3) to be very closely spaced. Typically $\Delta\epsilon_x \sim h^2/m = \mathcal{O}(10^{-30}\text{ erg})$. We also find that for a free particle in a three-dimensional cubic box of volume,[1] $V = L^3$,

$$\psi(x, y, z) = \left(\frac{2}{L}\right)^{3/2}\sin\left(\frac{\pi n_x x}{L}\right)\sin\left(\frac{\pi n_y y}{L}\right)\sin\left(\frac{\pi n_z z}{L}\right) \tag{5.6}$$

and

$$\epsilon = \epsilon_x + \epsilon_y + \epsilon_z = \frac{h^2}{8mV^{2/3}}(n_x^2 + n_y^2 + n_z^2) \tag{5.7}$$

[1] Actually these results are not restricted to a cubic shape (see, e.g., Example 5.5).

and the quantum state is now given by a set of three independent quantum numbers (n_x, n_y, n_z).

NOTION OF DEGENERACY

A particle in a unidimensional box has one standing wave function for each energy level. But in three-dimensional translation a particle's energy is specified by *three* quantum numbers. For each possible combination of n_x, n_y, n_z consistent with a given value of $n_x^2 + n_y^2 + n_z^2$, there is a different wave function. There are, in other words, a very large number of ways in which a particle can occupy a given energy level.

The number of modes of occupancy of the *i*th energy level is called the *degeneracy*, g_i, of that level. Particles are said to be degenerate when there is more than one mode of energy occupancy. If there is only one way in which a particle can occupy each of the energy levels of a particle (i.e., if $g_i = 1$), then we call that particle *nondegenerate*.

The fact that higher energy levels might (for some particles) be occupied in more ways than the lower levels will require important alterations of the simple statistics developed in chapter 3. We defer these considerations until chapter 6. However, we make note of g_i in each of the cases that we treat in this section for future reference.

HARMONIC OSCILLATOR

The vibrational motion of a molecule can often be treated as the motion of a harmonic oscillator. A unidimensional harmonic oscillator is a particle moving about an equilibrium position ($x = 0$) subject to a restoring force F that is linearly dependent upon x. Thus $F = -Kx$, where K is the "spring constant." Actually molecular oscillators are usually subject to nonlinear forces. The linear force law must be regarded as the first term in a series expansion for the actual law and should be applied only for small displacements. The potential energy [see Fig. 5.1(b)] for such a particle is defined for any conservative force field such that

$$\mathbf{F} \equiv -\nabla U \tag{5.8}$$

Thus, we have, in this case, the scalar relation

$$U(x) = -\int_0^x F\,dx = \frac{Kx^2}{2}$$

and the corresponding Schrödinger equation is

$$-\frac{\hbar^2}{2m}\frac{d^2\psi}{dx^2} + \left(\frac{Kx^2}{2} - \epsilon\right)\psi = 0 \tag{5.9}$$

The boundary conditions are $\psi(+\infty) = \psi(-\infty) = 0$.

Fig. 5.1 Potential-energy functions for three kinds of unidimensional motion.

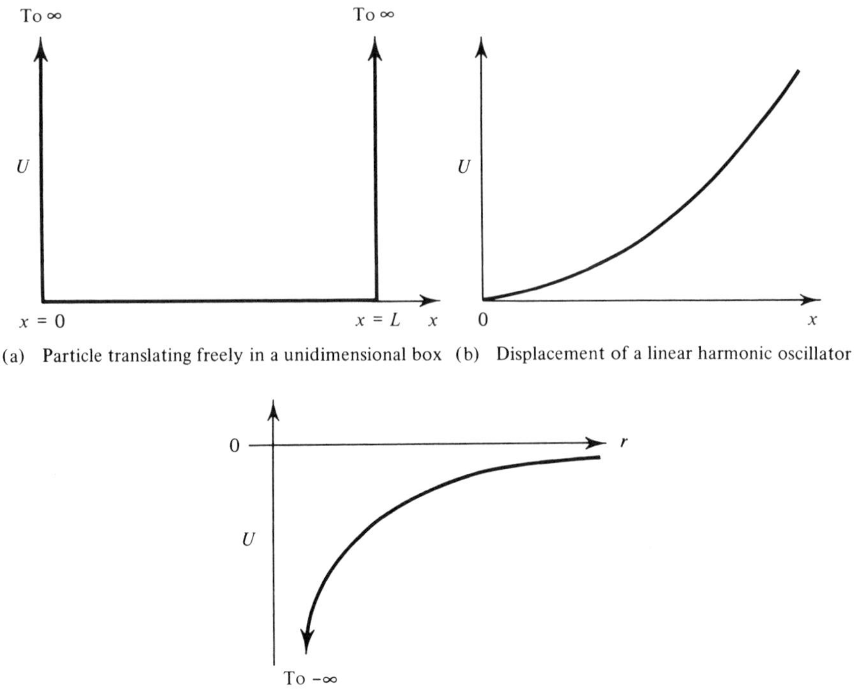

(a) Particle translating freely in a unidimensional box (b) Displacement of a linear harmonic oscillator

(c) Particle in a central field subject to a Coulomb interaction

Under the substitutions

$$\beta \equiv \frac{2m}{\hbar^2}\epsilon \qquad \text{and} \qquad \alpha^2 \equiv \frac{Km}{\hbar^2}$$

Eq. (5.9) becomes

$$\frac{d^2\psi}{dx^2} + [\beta - (\alpha x)^2]\psi = 0$$

This can be further transformed using

$$\xi \equiv \alpha^{1/2}x \qquad \text{and} \qquad \psi \equiv Q(\xi)\exp\left(\frac{-\xi^2}{2}\right)$$

where the exponential factor will satisfy the boundary conditions. The result is

$$\frac{d^2Q}{d\xi^2} - 2\xi\frac{dQ}{d\xi} + \left(\frac{\beta}{\alpha} - 1\right)Q = 0 \tag{5.9a}$$

Equation (5.9a) is the *Hermite equation*. Its solutions can be shown to *exist* only when the eigenvalues are

$$\left(\frac{\beta}{\alpha} - 1\right) = 2n \qquad n = 0, 1, 2, \ldots \tag{5.10}$$

The solutions are the *Hermite polynomials*,

$$H_n(\xi) = (-1)^n e^{\xi^2} \frac{d^n}{d\xi^n}(e^{-\xi^2})$$

Hence the wave-amplitude functions are

$$\psi_n(x) = \left(\sqrt{\frac{\alpha}{\pi}}\frac{1}{2^n n!}\right)^{1/2} H_n(\alpha^{1/2}x) \exp\left(-\frac{\alpha x^2}{2}\right)$$

where the lead coefficient has again been determined from the normalization condition.

The energy levels can be obtained by expressing Eq. (5.10) in terms of the original variables, whence

$$\epsilon = (n + \tfrac{1}{2})\,\hbar\sqrt{\frac{K}{m}} = (n + \tfrac{1}{2})\,h\left(\frac{1}{2\pi}\sqrt{\frac{K}{m}}\right) \tag{5.10a}$$

The group, $(1/2\pi)\sqrt{K/m}$, is the angular frequency ν of the oscillator. Thus

$$\epsilon = (n + \tfrac{1}{2})\,h\nu \tag{5.10b}$$

This result shows that Planck's original hypothesis, given in Eqs. (4.25) and (4.26), is in error to the extent that it omits one-half quantum of vibrational energy in the "ground" or "zero" state. This residual energy serves to satisfy the uncertainty principle by hedging a possible statement that the energy at any $x = 0$ is "known" to be zero.

RIGID ROTOR

Consider a molecule of arbitrary shape rotating about its own center of gravity. If the molecule is diatomic with atoms of masses m_1 and m_2 rigidly located a distance r_0 from one another, then its moment of inertia I will be

$$I = m_1 m_2 r_0^2/(m_1 + m_2)$$

In this case no potential field exists and the Schrödinger equation for the system is obtained from Eq. (4.72) as

$$r_0^2(\nabla^2\psi) = \frac{1}{\sin\theta}\frac{\partial}{\partial\theta}\left(\sin\theta\,\frac{\partial\psi}{\partial\theta}\right) + \frac{1}{\sin^2\theta}\frac{\partial^2\psi}{\partial\phi^2} = -\frac{2I}{\hbar^2}\epsilon\psi \tag{5.11}$$

The rigidity of the rotor prevents any radial variation of ψ. Motion can occur in both of the angular directions, however. The energy (or Hamiltonian) operator [recall Eq. (4.67)] takes the form $p^2/2m$, or $-(\hbar^2/2I)\nabla^2$, in this case, where m_r is the reduced mass, $m_1m_2/(m_1 + m_2)$.

Equation (5.11) is a partial differential equation; therefore, another separation of variables solution, $\psi(\theta, \phi) = P(\theta)\Phi(\phi)$, must be used and a new eigenvalue m introduced. The separation gives the ordinary equations

$$\frac{d^2\Phi}{d\phi^2} = -m^2\Phi \tag{5.12}$$

with the periodic boundary conditions $\Phi(0) = \Phi(2\pi)$ and $(d\Phi/d\phi)_0 = (d\Phi/d\phi)_{2\pi}$, and

$$\frac{1}{\sin\theta}\frac{d}{d\theta}\left(\sin\theta\frac{dP}{d\theta}\right) + \left(\frac{2I\epsilon}{\hbar^2} - \frac{m^2}{\sin^2\theta}\right)P = 0 \tag{5.13}$$

with the same boundary conditions on P.

The solution for Eq. (5.12) is

$$\Phi = A\sin m\phi + B\cos m\phi$$

The first boundary condition fixes the eigenvalues as $m = 0, \pm 1, \pm 2, \ldots$. Then, under the transformation $\cos\theta \equiv \zeta$, Eq. (5.13) takes the form

$$(1-\zeta^2)\frac{d^2P}{d\zeta^2} - 2\zeta\frac{dP}{d\zeta} + \left(\frac{2I\epsilon}{\hbar^2} - \frac{m^2}{1-\zeta^2}\right)P = 0 \tag{5.13a}$$

Equation (5.13a) is exactly Legendre's differential equation if m is integral — and we have just shown that it is. The corresponding eigenvalues, the energy levels, then have to be

$$\epsilon = \frac{\hbar^2}{2I}l(l+1) \qquad l = \text{integer} \geq |m| \tag{5.14}$$

The solutions of Eq. (5.13) are the associated Legendre polynomials, $P_l^m(\zeta)$. They are of the form

$$P_l^m(\zeta) = \frac{(1-\zeta^2)^{m/2}}{2^l l!}\frac{d^{l+m}}{d\zeta^{l+m}}(\zeta^2-1)^l$$

Finally one can show that the standing wave functions are given by

$$\psi = \frac{1}{\sqrt{2\pi}}\left[\frac{(2l+1)(l-m)!}{2(m+l)!}\right]^{1/2}P_l^m(\cos\theta)e^{im\phi} \tag{5.15}$$

(This problem is also discussed in relation to the Sturm–Liouville theory in Example C. 1, Appendix C.)

Our main interest in the preceding standing-wave equation lies in l, the *principal* or *rotational* quantum number. For each value of l, m can take on any value such that $|m| \leq l$. This means that there are $2l+1$ values (including $m = 0$) of m for each l, and $2l+1$

possible wave functions for each l. But there is only one energy ϵ for each l. The system is $(2l + 1)$-*fold degenerate*, or $g_l = 2l + 1$. This means that as the energy of rotation increases, an increasing number of orientations of rotation are meaningful and consistent with the uncertainty principle.

We should note in this connection that the two degenerate configurations that we have thus far encountered have been two- and three-dimensional. This is because degeneracy generally arises out of an increased number of geometrical alternatives for storing energy in space. Unidimensional motion generally provides only one alternative, however.

EXAMPLE 5.1 Compare the energy levels and degeneracies of a rigid rotor in a two dimensional space with those of one in a three-dimensional space.

In this case the Schrödinger equation is

$$r_0^2\left[\frac{1}{r_0^2 \sin\theta}\frac{\partial}{\partial\theta}\left(\sin\theta\frac{\partial\psi}{\partial\theta}\right)\right] = -\frac{2I}{\hbar^2}\epsilon\psi$$

Under the transformation $\cos\theta \equiv \zeta$, this becomes

$$(1 - \zeta^2)\frac{d^2\psi}{d\zeta^2} - 2\zeta\frac{d\psi}{d\zeta} + \frac{2I}{\hbar^2}\epsilon\psi = 0$$

which is Legendre's differential equation with $m = 0$. The resulting energy eigenvalues are

$$\epsilon = \frac{\hbar^2}{2I}l(l+1); \quad l = \text{integer} \geq 0$$

This differs from the three dimensional result only in that l can take on all values greater than zero. Since there is only one value of m in this case the wave function is uniquely specified by l alone. Thus there is only one way the particle can possess a given energy and it is therefore *non-degenerate*.

EXAMPLE 5.2 Obtain the degeneracy of a particle translating in n-dimensional space.

The Schrödinger equation for this situation is

$$\frac{\partial^2\psi}{\partial x_1^2} + \frac{\partial^2\psi}{\partial x_2^2} + \cdots + \frac{\partial^2\psi}{\partial x_n^2} = -\frac{2m\epsilon}{\hbar^2}\psi$$

Separating variables with $\psi = \prod_{i=1}^{n} X_i(x_i)$ we obtain

$$\frac{X''_1}{X_1} + \frac{X''_2}{X_2} + \cdots + \frac{X''_n}{X_n} = -\frac{2m\epsilon}{\hbar^2}$$

where we have used primes to denote differentiation. This can be written as

$$\frac{X''_1}{X_1} = -\left[\frac{X''_2}{X_2} + \frac{X''_3}{X_3} + \cdots + \frac{X''_n}{X_n} + \frac{2m\epsilon}{\hbar^2}\right] = -\alpha_1^2$$

so that the left hand side gives $\alpha_1 = \pi m_1$ where $m_1 = 0, 1, 2, \ldots$. Separating the right-hand equation we find

$$\frac{X''_2}{X_2} = -\left[\frac{X''_3}{X_3} + \frac{X''_4}{X_4} + \cdots + \frac{X''_n}{X_n} + \frac{2m\epsilon}{\hbar^2} - \pi^2 m_1^2\right] = -\alpha_2^2$$

Thus $\alpha_2 = \pi m_2$, and by induction we see that, $\alpha_3 = \pi m_3$, $\alpha_4 = \pi m_4$, $\ldots$, $\alpha_{n-1} = \pi m_{n-1}$. The final eigenvalue is given by

$$\frac{2m\epsilon}{\hbar^2} - \pi^2 \sum_{i=1}^{n-1} m_i^2 = \pi^2 m_n^2$$

so

$$\underline{\epsilon = \frac{\hbar^2 \pi^2}{2m} \sum_{i=1}^{n} m_i^2}$$

By paraphrasing the counting arguments that gave Eq. (4.22) we obtain the total number of wave functions for all energies up to a given level as

$$\frac{\pi^{n/2}}{2^{n-1}\Gamma\left(\frac{n}{2}+1\right)} m^n$$

For large m, the degeneracy can be approximated by the derivative of this with respect to m, or the rate of change of number with the principle quantum number:

$$g_m = \frac{\pi^{n/2}}{2\Gamma\left(\frac{n}{2}\right)} m^{n-1}$$

Thus for unidimensional space, g_m = unity; for three dimensional space, $g_m = \pi m^2/2$; etc. It is true in general that the degeneracy increases as the $n - 1$ power of the principal quantum number in any configuration since each additional degree of freedom adds additional ways in which energy can be stored.

Before returning to the problems of statistical mechanics it will help to complete our historical description of quantum mechanics with a brief look at the quantum description of the hydrogen atom.

5.2 HYDROGEN-ATOM PROBLEM

ENERGY LEVELS OF THE ELECTRON IN A CENTRAL FIELD

The third major application of the evolving science of quantum mechanics was made by Niels Bohr in 1913. Bohr employed the early quantum mechanics of Planck to explain the emission and absorption of radiation by the hydrogen atom. His explanation provided us with a remarkably understandable and compelling model for atomic structure. Crude as it was in some respects, it gave almost exact results. In this section we work the hydrogen-atom problem in reverse, looking first at the more correct description given by the Schrödinger equation and then showing how Bohr's great insight led him through the problem.

The solutions of the Schrödinger equation developed in Sec. 5.1 give quantum-mechanical descriptions of the basic modes of motion of a particle, the particle might be an electron, an atom, or a molecule *as a whole*. In an atom, however, there exist two kinds of particles: the nucleus, and electrons that are attracted to the nucleus by a force of electrostatic origin. The Schrödinger equation can also be used to describe a structured particle such as this.

The simple hydrogen atom consists of a single electron moving around the nucleus, and the attractive force is given by Coulomb's law as $F(r) \sim e^2/r^2$, where e is the charge of one electron, -1.60210×10^{-19} Coulombs and r is the distance between the electron and the nucleus. This Coulomb interaction can be represented by a potential-energy function $U(r)$,

$$\frac{dU(r)}{dr} \equiv -F(r) \tag{5.8a}$$

In this case the potential energy can most conveniently be called zero when it is at algebraic maximum (when $r = \infty$). Thus, if we denote the product of e^2 and a proportionality constant as C_1, we get [see Fig. 5.1(c)]

$$U = -\int_{\infty}^{r} F\,dr = -\frac{C_1}{r} \tag{5.16}$$

The Schrödinger equation (4.72) can then be written for the electron as

$$\frac{\hbar^2}{2m_e}\nabla^2\psi + \left(\epsilon + \frac{C_1}{r}\right)\psi = 0 \tag{5.17}$$

where m_e is the mass of the electron and the nucleus is considered to be approximately stationary.[2]

Equation (5.17) will yield to a separation-of-variables solution, $\psi(\theta, \phi, r) = Y(\theta, \phi)R(r)$. The result is a pair of equations for Y and R. The equation for Y is of the same form as Eq. (5.11), and its solutions are of the form of Eqs. (5.14) and (5.15). The equation for R is

$$\frac{1}{r^2}\frac{d}{dr}\left(r^2\frac{dR}{dr}\right) + \frac{2m_e}{\hbar^2}\left[\epsilon + \frac{C_1}{r} - \frac{l(l+1)\hbar^2}{2I}\right]R = 0$$

The solution of this equation[3] gives

$$R_n^{\,l}(r) = -\left[\left(\frac{2}{na_0}\right)^3\frac{(n-l-1)!}{2n[(n+l)!]^3}\right]^{1/2}\exp\left(-\frac{W}{2}\right)W^l L_{n+l}^{2l+1}(W)$$

where the $L_{n+l}^{2l+1}(W)$ are the *associated Laguerre polynomials*,

$$L_{n+l}^{2l+1}(W) = \sum_{k=0}^{n-l-1}(-1)^{k+1}\frac{[(n+l)!]^2W^k}{(n-l-1-k)!(2l+1+k)!k!} \tag{5.18}$$

and

$$a_0 \equiv \frac{\hbar^2}{m_eC_1} \qquad \text{and} \qquad W \equiv \frac{2}{na_0}r$$

The energy eigenvalues are given by

$$\epsilon_n = -\frac{m_eC_1^2}{2n^2\hbar^2} \tag{5.19}$$

where the negative sign of the energy has been included to signify that work must be supplied to pull the electron completely away from the nucleus.

The positive integer n is called the *principal quantum number.* For each n, l can vary from zero to $n-1$ as indicated in the summation limits in Eq. (5.18). For each l, the quantum number, m (which enters through the Y contribution), can have any one of $2l+1$ values, as we noted in the arguments following Eq. (5.15). The total degeneracy g_n of the wave function corresponding to a given energy level ϵ_n is then

$$\sum_{l=0}^{n-1}(2l+1) = \left[2\sum_{l=0}^{n-1}l\right] + n = 2\left[(n-1)\left(\frac{n-1}{2}\right) + \frac{n-1}{2}\right] + n$$

or

$$g_n = n^2 \tag{5.20}$$

[2]If the slight motion of the nucleus is to be considered, the Schrödinger equation will be unchanged, but m_e should be replaced with the reduced mass, $m_r = m_em_n/(m_e + m_n)$, where m_n is the mass of the nucleus.

[3]Details of solution are given in, for example, L. I. Schiff, *Quantum Mechanics,* 2nd ed., McGraw-Hill, Inc., New York, 1955, sec. 16.

Since emission or absorption of radiation results from a change of energy levels, the frequency of light emitted when such a change occurs between the two principal quantum states, $n = a$ and $n = b$, is given by the de Broglie relation,

$$\nu_{ab} = \frac{\epsilon_a - \epsilon_b}{h} = \frac{2\pi^2 m_e C_1^2}{h^3}\left(\frac{1}{b^2} - \frac{1}{a^2}\right) \tag{5.21}$$

The wave number[4] k_{ab} of such emitted light is

$$k_{ab} = \frac{\nu_{ab}}{c_l} = \frac{2\pi^2 m_e C_1^2}{c_l h^3}\left(\frac{1}{b^2} - \frac{1}{a^2}\right) \tag{5.21a}$$

The above expression is in agreement with the experimentally established formula,

$$k_{mn} = R\left(\frac{1}{n^2} - \frac{1}{m^2}\right) \tag{5.21b}$$

where R is the *Rydberg constant.* The original value of R given by Rydberg in 1890 was 109,720 cm^{-1}. Rydberg obtained this value and Eq. (5.21b) from the earlier experimental investigation of Balmer, who observed the series of spectral lines for atomic hydrogen. This series is now known as the *Balmer series.* Rydberg's original value of R is in remarkably good agreement with the value (109,737.3 cm^{-1}) obtained from the preceding quantum-mechanical expressions.

BOHR'S THEORY OF THE HYDROGEN ATOM

In keeping with Planck's elementary concept of energy quantization, Bohr assumed that the electron in the field of a hydrogen nucleus was allowed to move only in certain *discrete* circular orbits around the nucleus. The electron was assumed to radiate no energy while it was in these allowed orbits.

The circular orbit of the electron in the Bohr atom is depicted in Fig. 5.2. Bohr made two assumptions relative to the orbits. The first said that energies of neighboring orbits were spaced in accordance with Planck's hypothesis,

$$\Delta\epsilon = h\nu$$

The second said that the least increment of "action" in one orbit should be Planck's constant. Thus

$$p(2\pi r) = nh \quad \text{or} \quad m_e v r = n\hbar \qquad n = 1, 2, \ldots \tag{5.22}$$

[4]Here we are defining k as (frequency in cycles/second)/speed. Elsewhere we have defined it as (frequency in rad/sec)/speed. The two k's differ by a factor of 2π and both are in conventional use. With the exception of this application, k is expressed in (rad/sec)/speed in this book.

Fig. 5.2 Bohr hydrogen atom.

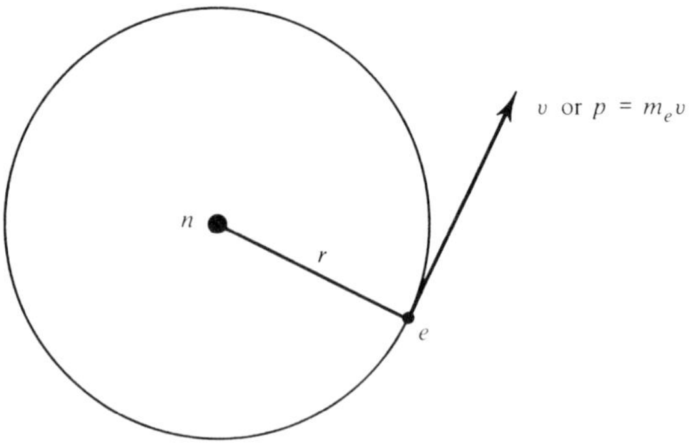

The latter condition can be used with the classical force balance

$$\frac{C_1}{r^2} = m_e\left(\frac{v^2}{r}\right) \tag{5.23}$$

to eliminate the velocity v. The radius is then

$$r = \frac{(n\hbar)^2}{m_e C_1} \tag{5.24}$$

The sum of kinetic and potential energies of the electron is given by

$$\epsilon = \text{KE} + \text{PE} = \frac{C_1}{2r} + \left(-\frac{C_1}{r}\right) = -\frac{C_1}{2r}$$

where Eq. (5.23) has been used to express kinetic energy in terms of C_1. The substitution of Eq. (5.24) for r then gives

$$\epsilon_n = -\frac{m_e C_1^2}{2n^2\hbar^2} \tag{5.19}$$

which is exactly the result obtained from the Schrödinger equation.

The frequency of the radiation emitted is given by Eq. (5.19) and $\Delta\epsilon = h\nu$,

$$\nu_{ab} = \frac{\epsilon_a - \epsilon_b}{h} = \frac{2\pi^2 m_e C_1^2}{h^3}\left(\frac{1}{b^2} - \frac{1}{a^2}\right) \tag{5.21}$$

which is also the result obtained from Schrödinger's equation. For frequencies that are close to one another [e.g., n and $(n + \Delta n)$, where $n \gg \Delta n$] this becomes, to a very close approximation,

$$\nu_n = \frac{4\pi^2 m_e C_1^2\,\Delta n}{(nh)^3} \tag{5.21c}$$

The simple Bohr theory indicates the existence of an infinite number of discrete electron orbits characterized by the quantum number, n. As the quantum number is increased, the radius of the orbit as well as the energy associated with the electron is increased, as indicated by Eqs. (5.24) and (5.19). Since for a free electron $r \rightarrow \infty$ and $\epsilon = 0$, the energy given by Eq. (5.19) can be regarded as the work needed to remove the electron from the state n to the state infinity, or the *ionization energy* from the state n.

For a heavier atom of nuclear charge Ze^+ with Z orbital electrons, the electrons are distributed among the various orbits according to Bohr's theory. The electrons will occupy the states of lowest allowed energy or the orbits of least n and least radius. It is found from a more advanced consideration that there exists an upper limit for the number of electrons in a given orbit. For the first four orbits ($n = 1, 2, 3,$ and 4), the limits on the number of electrons are 2, 8, 18, and 32, respectively.

CORRESPONDENCE PRINCIPLE

The above quantum-mechanical results from the Schrödinger equation and the hybrid theory of Bohr can be used to demonstrate an important general principle called the correspondence principle, put forward by Bohr in 1923. The principle states that *a quantum-mechanical result must reduce to the result of the corresponding classical calculation in the appropriate limit.*

For a hydrogen atom the difference between energy levels at large quantum numbers is small enough that the discrete quantum energy distribution can be approximated by a continuous energy distribution that is classical in concept. Thus at large quantum numbers the frequency of light emitted, according to the correspondence principle, should be in agreement with the classical result for the frequency of revolutions of the electron in orbit. This may be illustrated by using Eq. (5.22) to eliminate h from Eq. (5.21c):

$$\nu_n = \frac{m_e C_1^2}{2\pi(m_e v r)^3}\,\Delta n$$

But Eq. (5.23) can be used in turn to eliminate C_1,

$$\nu_n = \frac{v/r}{2\pi}\,\Delta n$$

where v/r is the angular velocity. When Δn is chosen as unity the frequency of emitted light becomes exactly the angular frequency of rotation of the electron. This is what a purely classical view would lead us to expect.

We use the correspondence principle implicitly in the following two sections. Each time we compute a partition function we then show how, in the appropriate limit of high temperature, the partition function approaches its classical limit.

EXAMPLE 5.3 Calculate the radius of a Bohr electron orbit for which $n = 100$ and check the validity of the correspondence principle at this quantum level.

The radius is given by Eq. (5.24) in terms of C_1, but C_1 is given by Coulomb's law:

$$\text{force} = \frac{C_1}{r^2} = \frac{k_c e \cdot e}{r^2} \qquad C_1 = k_c e^2$$

where k_c is identically unity in cgs units, and the charge e, on an electron is 4.8030×10^{-10} esu. Then from Eq. (5.24) we have

$$r = \frac{(n\hbar)^2}{m_e e^2} = \frac{(100)^2(1.0544 \times 10^{-27})^2}{(9.1085 \times 10^{-28})(4.8030 \times 10^{-10})^2} = 5.29 \times 10^{-5}\ \text{cm}$$

which is on the order of the mean free path at standard conditions. This is quite a large orbit.

Now to make a comparison with Bohr's correspondence principle, let us first compute λ_{mn} for two cases: one in which the electron drops from level 101 to 100 and one in which it drops from level 103 to 100. From Eq. (5.21b) and $k = 1/\lambda$, we obtain

$$\lambda_{101;100} = \frac{1}{109{,}737.3} \Big/ \left(\frac{1}{100^2} - \frac{1}{101^2}\right) = 4.56\ \text{cm}$$

and

$$\lambda_{103;100} = \frac{1}{109{,}737.3} \Big/ \left(\frac{1}{100^2} - \frac{1}{103^2}\right) = 1.63\ \text{cm}$$

From the correspondence principle we can note that $\lambda = c_l/\nu$ and obtain

$$\lambda_{\Delta n} = c_l \frac{2\pi m_e^2 v^3 r^3}{e^4\, \Delta n}$$

but from Eq. (5.22), $vr = n\hbar/m_e$, so

$$\lambda_{\Delta n} = c_l \frac{2\pi\hbar^3}{m_e e^4} \frac{n^3}{\Delta n}$$

Thus

$$\lambda_{100;101} = (2.9979 \times 10^{10}) \frac{2\pi(1.0544 \times 10^{-27})^3}{(9.1085 \times 10^{-28})(4.8029 \times 10^{-10})^4} \frac{100^3}{1} = 4.57\ \text{cm}$$

and

$$\lambda_{103;100} = \frac{1}{3}\lambda_{100;101} = 1.52\ \text{cm}$$

The quantum level 100 is therefore sufficiently high that the correspondence principle gives good accuracy.

Actually we should note that at this quantum level the radius of orbit is exceedingly large, and the additional energy required for complete removal of the electron,

$$\int_r^\infty F\,dr = -e^2\left(\frac{1}{r}\right)_{n^2\hbar^2/m_e e^2}^{\infty} = \frac{m_e e^4}{n^2\hbar^2}$$

is negligibly small by virtue of its dependence on n^{-2}. Thus such large orbits are highly unstable with respect to complete removal of the electron, or "ionization," and we would expect them to occur very seldom.

5.3 EVALUATION OF THE PARTITION FUNCTION

QUANTUM-MECHANICAL PARTITION FUNCTION

In chapter 3 the classical partition function was shown to be the generating function for thermodynamic properties. Now we have the tools with which we can return to its evaluation.

The partition function is the sum of exp $(-\epsilon_i/kT)$ over all possible states that a particle might assume. Within the framework of quantum mechanics, there are g_i possible states that have energy ϵ_i. Thus we must sum over g_i exp $(-\epsilon_i/kT)$ instead of just exp $(-\epsilon_i/kT)$, as we did in Eq. (3.19):

$$Z = \sum_{i=0}^{\infty} g_i \exp\left(-\frac{\epsilon_i}{kT}\right) \tag{5.25}$$

We obtain Eq. (5.25) in a formal way in Sec. 6.1 from quantum-statistical considerations.

The partition function shown here can also be decomposed into factors on the basis of the Schrödinger equation. When particles are independent, as they are in an ideal gas, it is possible to show[5] that the Schrödinger wave equation for one particle can be separated into independent wave equations. One characterizes the translational motion of the particles; the other characterizes the internal motions of the particles.

One consequence of this is a direct result of the multiplicative law of probabilities, which states that the probability of a set of independent events occurring is equal to the product of the prob-

[5]See, for example, J. L. Powell and B. Crasemann, *Quantum Mechanics*, Addison-Wesley Publishing Company, Inc., Reading, Mass., 1961.

abilities of each event in the set. In accordance with Eq. (4.69), the wave functions are multiplicative,

$$\psi = \psi_t \psi_{\text{int}}$$

because the $|\psi|^2$'s are to be interpreted as probabilities and the wave functions for translation ψ_t and for internal motion ψ_{int} are independent.

The other consequence of this independence relates to the energy of the particles. We have

$$\epsilon = \epsilon_{ijk\ldots} = \epsilon_i + \epsilon_j + \cdots$$

each of which might be degenerate. Accordingly, there are g_i ways for the particle to possess energy ϵ_i and g_j ways for it to possess energy ϵ_j. As long as ϵ_i and ϵ_j are independent,[6] for each of the g_i ways in which ϵ_i can be stored, there are g_j ways in which ϵ_j can be stored. There are therefore $g_i g_j$ ways in which both ϵ_i and ϵ_j can be specified simultaneously. The sum over states, or partition function, then becomes

$$Z = \sum_{ij\cdots} g_i g_j \cdots \exp\left(-\frac{\epsilon_i}{kT}\right) \exp\left(-\frac{\epsilon_j}{kT}\right) \cdots$$

And by virtue of the independence of g_i and ϵ_i upon any other g and ϵ,

$$Z = \sum_i g_i \exp\left(-\frac{\epsilon_i}{kT}\right) \sum_j g_j \exp\left(-\frac{\epsilon_j}{kT}\right) \cdots$$

so that

$$Z = Z_i Z_j \cdots \tag{5.26}$$

Equation (5.26) will make it possible for us to compute the partition functions for the component modes of energy storage and to multiply them together to get the partition function for a structured particle. With this in mind we proceed to calculate partition functions for some component modes of storage.

FREELY TRANSLATING PARTICLE

The partition function for a freely translating particle in a rectangular box of dimensions (a, b, c) is given by

$$Z_t = \sum_i g_i \exp\left(-\frac{\epsilon_i}{kT}\right) \tag{5.25a}$$

[6]Usually they are. For a translating particle, for example, ϵ_x and ϵ_y have to be independent to satisfy the principle of equal a priori probabilities. The rotational and vibrational modes of energy storage are also generally independent of translation and frequently independent of each other.

where g_i, the number of ways of choosing the numbers n_x, n_y, n_z, is simply unity. Using Eq. (5.3) for ϵ_i results in

$$Z_t = \sum_{n_x,n_y,n_z} \exp\left[-\frac{h^2}{8mkT}\left(\frac{n_x^2}{a^2}+\frac{n_y^2}{b^2}+\frac{n_z^2}{c^2}\right)\right] \tag{5.27}$$

Before trying to reduce this to a more compact form, let us look at the quantity $\Theta_t \equiv h^2/8mka^2$. For a hydrogen molecule in a 1-cm³ box, $a = b = c$,

$$\Theta_t \equiv \frac{h^2}{8mka^2} = \frac{(6.62 \times 10^{-27})^2}{8(3.3 \times 10^{-24})(1.38 \times 10^{-16})(1)^2} = 1.2 \times 10^{-14}\ ^\circ\mathrm{K}$$

The quantity Θ_t, the so-called "characteristic temperature for translation," provides an indication of the closeness of quantum spacings. Figure 5.3 shows how these extremely close spacings expedite the evaluation of the partition-function summations. When $\Theta_t \ll T$, Eq. (5.27) can be written

$$Z_t = \sum_{n_x,n_y,n_z} \exp\left(-\frac{\Theta_t}{T}n_x^2\right)\exp\left(-\frac{\Theta_t}{T}n_y^2\right)\exp\left(-\frac{\Theta_t}{T}n_z^2\right)$$

$$\simeq \int_0^\infty \exp\left(-\frac{\Theta_t}{T}n_x^2\right)dn_x \int_0^\infty \exp\left(-\frac{\Theta_t}{T}n_y^2\right)dn_y \int_0^\infty \exp\left(-\frac{\Theta_t}{T}n_z^2\right)dn_z$$

These integrals are easy to evaluate because $\int_0^\infty \exp(-x^2)\,dx$ is well known and equal to $\sqrt{\pi}/2$. The result for Z_t is

$$Z_t \simeq \left(\frac{\pi T}{4\Theta_t}\right)^{3/2} = \frac{V}{(h/m)^3}\left(\frac{2\pi kT}{m}\right)^{3/2} \tag{5.28}$$

Fig. 5.3 Partition function for a gas with a low characteristic temperature

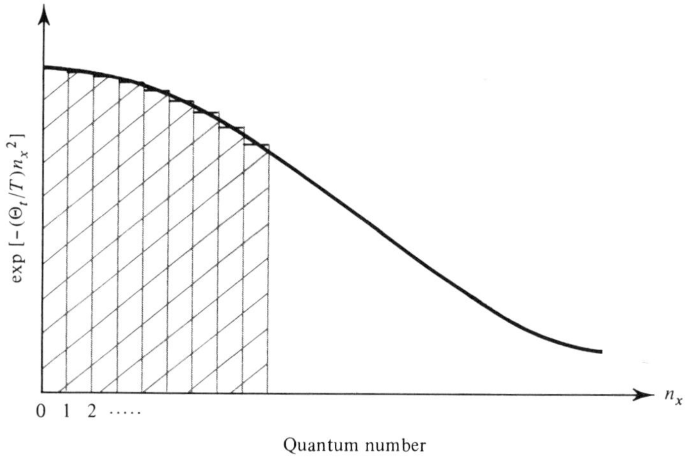

where the product (abc) has been characterized as the volume V of the box.

The approximation that has been made in writing this result will be valid as long as the temperature is greater than 10^{-14} °K. It will be valid for *still lower temperatures* in the case of heavier gases in smaller boxes. The quantum jumps in the energy of translation thus prove to be so small that we can forget quantum behavior in virtually all real situations. Equation (5.28) is almost the same as Eq. (3.33)—the result at which we had to stop for lack of ability to specify the velocity cell size $\Delta\Omega$. The only difference is that the cell space is actually specified in Eq. (5.28). Its value is

$$\Delta\Omega = \frac{(h/m)^3}{V} \qquad \text{or} \qquad \Delta\tau = (m^3\,\Delta\Omega)V = h^3 \tag{5.29}$$

Equation (5.28) illustrates the validity of Bohr's correspondence principle, in that it is the same as the classical partition function, Eq. (3.33), and it is the limit reached by a partition function, Eq. (5.27), at "high" temperature.

In the preceding case Z_t was obtained for translation in three dimensions by using, in essence, Eq. (5.26). The degeneracies of the component translatory energies were unity, so the particle was considered to be nondegenerate. It is also possible to evaluate Z_t by considering the translating particle to be degenerate and by using Eq. (5.25) directly. To do this we must describe ϵ with a single quantum number,

$$\epsilon_n = \frac{h^2}{8mL^2}n^2 \qquad n = 0, 1, 2, \ldots \tag{5.3a}$$

The degeneracy g_n can be shown to equal $\pi n^2/2$ by counting all of the quantum states consistent with any $n = (n_x^2 + n_y^2 + n_z^2)^{1/2}$. It is left as an exercise (Problem 5.11) to do this with the aid of a Rayleigh-Jeans counting scheme and to verify that Eq. (5.28) still results if we consider a translating particle to be degenerate.

EXAMPLE 5.4 In the study of dense gases, Sec. 9.1, we define a *thermal de Broglie wavelength* Λ as

$$\Lambda \equiv \left(\frac{h^2}{2\pi mkT}\right)^{1/2} = \left(\frac{V}{Z}\right)^{1/3}$$

Explain why this quantity is given this name.

From Eq. (5.7) we have

$$\epsilon = \frac{h^2n^2}{8mV^{2/3}}$$

But $\epsilon = mc^2/2 = p^2/2m$, and in accordance with the de Broglie relation, $p = h/\lambda$. Thus

$$\epsilon = \frac{h^2}{2m\lambda^2} = \frac{h^2n^2}{8mV^{2/3}}$$

so

$$\lambda = \frac{2}{n}V^{1/3}$$

Thus Λ is defined in such a way that

$$\Lambda = \left(\frac{n}{2Z^{1/3}}\right)\lambda$$

Thus Λ is proportional to λ, and $1/\Lambda^3$ is on the order of the partition function for a volume equal to λ^3.

LINEAR HARMONIC OSCILLATOR

The linear harmonic oscillator is the second of three typical modes of microscopic motion that we shall want to have at our disposal subsequently. In this case, Θ_v, the characteristic temperature of vibration, is far higher — on the order of 10^3 °K. One cannot pass from summation to integration at temperatures in the range of practical interest. It is not necessary to do so, however, because in this case the partition function can be summed exactly.

The partition function constructed from the energy given in Eq. (5.10b) is

$$Z_v = \sum_{n=0}^{\infty} \exp\left[-\frac{h\nu}{2kT}(2n+1)\right] \tag{5.30}$$

where $h\nu/k \equiv \Theta_v$. Equation (5.30) can be expanded,

$$Z_v = \exp\left(-\frac{\Theta_v}{2T}\right)\left[1 + \exp\left(-\frac{\Theta_v}{T}\right) + \exp\left(-\frac{2\Theta_v}{T}\right) + \cdots\right]$$

and summed. We recognize that the bracketed term is in the form of a binomial expansion. Thus the vibrational partition function becomes

$$Z_v = \frac{\exp(-\Theta_v/2T)}{1 - \exp(-\Theta_v/T)}$$

When the top and bottom have been divided by $\exp(-\Theta_v/2T)$ this becomes

$$Z_v = \left[2\sinh\left(\frac{\Theta_v}{2T}\right)\right]^{-1} \tag{5.31}$$

Although Eq. (5.31) is continuous, it is not the same partition function that would be obtained from a purely classical computation. Consider the value of Z_v as $x = \Theta_v/T$ becomes small:

$$Z_v = (e^{x/2} - e^{-x/2})^{-1} = \left\{\left[1 + \frac{x}{2} + \frac{1}{2!}\left(\frac{x}{2}\right)^2 + \cdots\right] - \left[1 - \frac{x}{2} + \frac{1}{2!}\left(\frac{x}{2}\right)^2 - \cdots\right]\right\}^{-1}$$

At high temperature only the linear terms in x contribute, and

$$Z_v \to x^{-1} = \frac{T}{\Theta_v}$$

or

$$Z_v \to \frac{kT}{h\nu} \tag{5.32}$$

In Sec. 5.4 we see how to obtain this using a strictly classical computation (see also Problem 5.14).

RIGID ROTOR

The energy levels for the rigid rotor were given by Eq. (5.14), and the degeneracies of the energy states were shown to be $(2l + 1)$. A tacit assumption was made but not mentioned when we did this: The ends of the rotor were assumed to be distinguishable from one another! If this were not so, the degeneracy would have been lower than we computed. The reason for this is shown in Fig. 5.4. Here a diatomic molecule whose ends can be distinguished from one another by virtue of nuclear spin occupies its orbit in two ways. Without nuclear spin these states would be indistinguishable and could only be counted as one state.

We delay a more detailed consideration of this problem until we take up the diatomic molecule in chapter 7. For the moment we regard the ends as distinguishable and write the partition function as

$$Z_r = \sum_{l=0}^{\infty} (2l + 1) \exp\left[-\frac{l(l + 1)h^2}{8\pi^2 IkT}\right] \tag{5.33}$$

Once again we wish to develop the limiting value of this sum.

Fig. 5.4 Two ways for a rigid rotor (whose ends are distinguishable by virtue of nuclear spin) to occupy the same energy level.

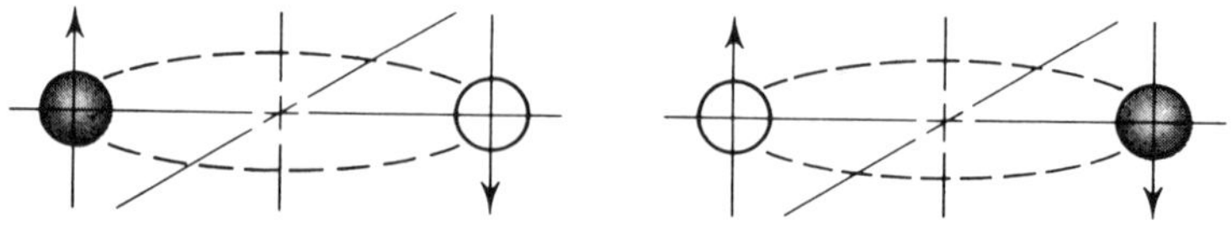

Changing variables with $x \equiv l(l+1)$, $dx \simeq (2l+1)\,\delta l$, where δl is taken as unity leads to

$$Z_r \simeq \int_0^\infty \exp\left[-x\left(\frac{h^2}{8\pi^2 IkT}\right)\right] dx$$

or

$$Z_r \simeq \frac{8\pi^2 IkT}{h^2} = \frac{2IkT}{\hbar^2} = \frac{T}{\Theta_r} \tag{5.34}$$

where the characteristic temperature, $\Theta_r \equiv \hbar^2/2Ik$. It can be shown that Eq. (5.33) can be summed and put in the following power-series form:

$$Z_r = \frac{T}{\Theta_r}\left[1 + \frac{1}{3}\frac{\Theta_r}{T} + \frac{1}{15}\left(\frac{\Theta_r}{T}\right)^2 + \frac{4}{315}\left(\frac{\Theta_r}{T}\right)^3 + \cdots\right]$$

This approaches Eq. (5.34) rapidly as T exceeds Θ_r. However, Θ_r generally assumes values between 2 to 100°K, so Eq. (5.34) is valid in much of the range of practical interest.

5.4 RELATION BETWEEN THE CLASSICAL AND QUANTUM PARTITION OF ENERGY

The replacement of summations with integrals in the preceding section had a significance that ran deeper than mere arithmetic convenience. The approximation is one that becomes legitimate only when the classical or nonquantum description of matter becomes accurate. Let us illustrate this in the following way:[7]

Consider a particle to translate in a one-dimensional box of breadth, a, as illustrated in Fig. 5.5. The energies of two adjacent states are

$$\epsilon_1 = \frac{h^2(n+1)^2}{8ma^2} \quad \text{and} \quad \epsilon_2 = \frac{h^2 n^2}{8ma^2}$$

so that

$$(n+1) - n = \frac{2a(\sqrt{2m\epsilon_1} - \sqrt{2m\epsilon_2})}{h}$$

The area of the shaded portions of Fig. 5.5, which represents the least possible uncertainty as to the location of the particle in phase space, is

$$2(a)(\sqrt{2m\epsilon_1} - \sqrt{2m\epsilon_2}) = h \tag{5.35}$$

[7]This explanation is similar to one offered by G. S. Rushbrooke in *Introduction to Statistical Mechanics*, Oxford University Press, New York, 1962, chap. IV.

Fig. 5.5 Particle translating in a one-dimensional box.

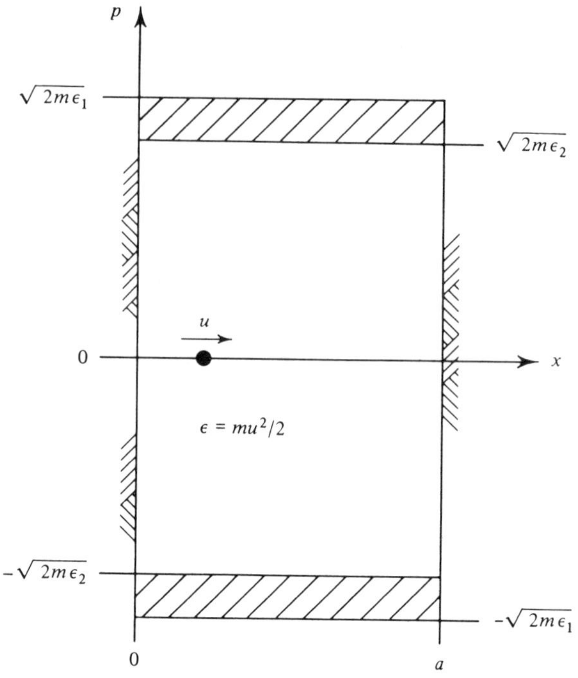

Equation (5.35) can be recast in terms of velocity-position space, using $\epsilon = mu^2/2$,

$$2a(u_1 - u_2) = \frac{h}{m} \tag{5.35a}$$

The result given by Eq. (5.35) can be generalized for more dimensions: *The classical surfaces of constant total energy in phase space, which correspond to the energies allowed by quantum theory, subdivide phase space into units of h^n, where n is the number of spatial coordinates.*

Thus the partition functions that were obtained from integrals in Sec. 5.3 could have been obtained in a more general way. For $\Theta_t \ll T$, we have, from the definition of integration,

$$h^n \sum_i \exp\left(-\frac{\epsilon_i}{kT}\right) \simeq \int \exp\left(-\frac{\epsilon}{kT}\right) d\tau \tag{5.36}$$

where the argument of the summation is now energy instead of number ($n = 1, 2, \ldots$). The spacing is therefore h^n instead of unity. Then

$$Z = \frac{1}{h^n} \int \exp\left(-\frac{\epsilon}{kT}\right) d\tau \tag{5.37}$$

but $d\tau = dV(m^n d\Omega)$, so

$$Z = \left(\frac{m}{h}\right)^n \int \exp\left(-\frac{\epsilon}{kT}\right) dV\, d\Omega \tag{5.37a}$$

The integration over τ is at least sixfold in three-dimensional space. For a translating and rotating "dumbbell" molecule it would consist of 10 coordinates: 3 Cartesian, 2 angular, 3 translational momentum, and 2 rotary momentum. In this case n would still be 5.

The total energy ϵ will be recognizable to the reader who has studied dynamics as the Hamiltonian. Equation (5.37) was first developed by J. W. Gibbs on the basis of purely dynamical considerations. The fact that the factor h^{-n} did not appear in his formulation only meant that he had to remain ignorant of an additive constant in the entropy and free-energy functions. Since Z appears under the derivative of a logarithm in the energy, pressure, and specific-heat functions [recall Eqs. (3.22) to (3.25)], this factor left these functions unaltered. Gibbs named the integral in Eq. (5.37) the "phase integral." With the factor of h^{-n} included, it is called the "modified phase integral."

It might be disturbing at first glance that the degeneracies do not appear in the phase integral. This omission is treated fully in more advanced texts. For our purposes it must suffice to point out that the degeneracy ceases to have meaning in a classical description. There is no walling-off of discrete energy levels. And, although we moved to the phase integral from a quantum-mechanical description, we have actually reached a result that would have been directly obtainable by strictly classical means.

Degeneracies do not properly belong in the phase integral. When the integral is formed directly from a degenerate partition function (as we have done) instead of from the Hamiltonian of the particle, the degeneracies are absorbed in the integration so as to give the same result as a direct formulation of the phase integral would. This happened in the derivation of Eq. (5.34) (compare with Problem 5.10), and it also happened when we derived Eq. (5.28).

The phase integral also possesses the important multiplicative quality of the partition function. As long as the Hamiltonian is a linear function of the energy we can write

$$Z = \frac{1}{h^n}\int \exp\left(-\frac{\epsilon_i}{kT}\right) dx_i\, dp_i \int \exp\left(-\frac{\epsilon_j}{kT}\right) dx_j\, dp_j \cdots$$

or

$$Z = Z_i Z_j \cdots \tag{5.26}$$

just as we did for the degenerate partition function.

EXAMPLE 5.5 Suppose we first express the energy of a translating particle in spherical coordinates. Then let us see how we could compute Z_t for the particles in a sphere of radius R using this expression.

The energy of a particle is $(p_x^2 + p_y^2 + p_z^2)/2m$, where p_x, p_y, and p_z are Cartesian components of momentum. Now, with reference to Fig. 2.3, suppose that we rotate an (x, y, z) coordinate frame until x aligns with $|\mathbf{r}|$, represented in a spherical coordinate frame. It would follow that

$$dx = dr \qquad dy = r\,d\theta \qquad dz = r \sin\theta\, d\phi$$

Thus if the components of any vector, say p, are to be equal in both frames,

$$p_x = p_r \qquad p_y = \frac{p_\theta}{r} \qquad p_z = \frac{p_\phi}{r \sin\theta}$$

so

$$\epsilon = \frac{p_x^2 + p_y^2 + p_z^2}{2m} = \frac{p_r^2 + (p_\theta^2/r^2) + (p_\phi^2/r^2 \sin^2\theta)}{2m} \tag{5.38}$$

To evaluate Z_t we now use Eq. (5.37) and replace ϵ with the right hand side of Eq. (5.38),

$$Z_t = \frac{1}{h^3} \underbrace{\int\!\!\int\!\!\int\!\!\int\!\!\int\!\!\int}_{\text{over } V, \text{ and all } p} \exp\left[\frac{\left(-\dfrac{p_r^2}{kT} - \dfrac{p_\theta^2}{r^2 kT} - \dfrac{p_\phi^2}{r^2 \sin^2\theta\, kT}\right)}{2m}\right] dx\,dy\,dz\,dp_x\,dp_y\,dp_z$$

but $dx\,dy\,dz = r^2 \sin\theta\, dr\,d\theta\,d\phi$ (recall Fig. 2.3 and context) and $dp_x\,dp_y\,dp_z = dp_r\,dp_\theta\,dp_\phi/r^2 \sin\theta$, so

$$Z_t = \frac{1}{h^3} \overbrace{\underbrace{\int\!\!\int\!\!\int}_{-\infty}}^{\infty} \int_0^{2\pi}\int_0^{\pi}\int_0^{R} \exp \left[\frac{\left(-\dfrac{p_r^2}{kT} - \dfrac{p_\theta^2}{r^2 kT} - \dfrac{p_\phi^2}{r^2 \sin^2\theta\, kT}\right)}{2m}\right] dr\,d\theta\,d\phi\,dp_r\,dp_\theta\,dp_\phi$$

$$= \frac{1}{h^3}\frac{4\pi R^3}{3}(2\pi mkT)^{3/2} = V\left(\frac{2\pi mkT}{h^2}\right)^{3/2} \tag{5.28}$$

It turns out that this result is independent of the shape of the volume that contains the particles for which we are computing Z_t.

Problems 5.1 Verify Eqs. (5.6) and (5.7).

5.2 Draw, in the two-dimensional phase space of an oscillator, the classical paths corresponding to the quantum-mechanically allowed energies. What is the "volume" of the cells into which these paths divide the space? What is the physical meaning of this result?

5.3 A string is tightly stretched between two supports. The string is struck, and the displacement travels along the string. Write the one-dimensional Schrödinger equation for this system.

5.4 Determine the quantum energy level for a linear harmonic oscillator for which there exists a constant potential energy at the equilibrium position.

5.5 Find the energy levels of a particle moving in a potential field of the shape $U(x < 0) = \infty$; $U(x \geq 0) = Ax^2$.

5.6 Complete the steps in the derivation of the expressions for $\psi_n(x)$ and ϵ_n for a harmonic oscillator.

5.7 Complete the steps leading to Eq. (5.15).

5.8 Complete the steps in the derivation of the standing-wave function for an electron in orbit.

5.9 Nothing was said in Sec. 5.2 about the possibility of elliptical electron orbits in the Bohr atom. Prepare a discussion of the quantum description of elliptical orbits and show how Eq. (5.21b) must be modified for more complicated atoms.

5.10 Use the modified phase integral to compute the partition function for a rigid rotor directly from the classical Hamiltonian. Compare your result with Eq. (5.34).

5.11 Compute Z_t for a translating particle using one quantum number to characterize the energy and regarding the particle as degenerate.

5.12 Determine the molecular distribution functions for translating molecules with both three and one quantum numbers. (Note that in one case you should obtain a velocity distribution, because three components are specified. In the other case you should obtain a speed distribution.)

5.13 Derive an expression for the pressure exerted by an isothermal atmosphere, as a function of elevation, using the modified phase integral with a Hamiltonian that accounts for both kinetic and potential energy of particles. Compare your result with Example 3.2.

5.14 Use the modified phase integral to obtain the partition function for a linear harmonic oscillator.

5.15 Calculate the internal energy of an ideal monatomic gas enclosed in a cylindrical vessel of radius R and length L rotating about its axis with angular velocity ω.

5.16 Using the anharmonic potential energy, $U(x) = cx^2 - gx^3 - fx^4$, show that the specific heat of a classical anharmonic oscillator is given, to first order in T, by the approximate relation

$$c_v \simeq k\left[1 + \left(\frac{3f}{2c^2} + \frac{15g^2}{8c^3}\right) kT\right]$$

Note that g and f characterize small corrections to otherwise harmonic motion.

6 quantum statistics

6.1 FERMI-DIRAC AND BOSE-EINSTEIN DISTRIBUTIONS AND THEIR CLASSICAL LIMIT

ROLE OF INDISTINGUISHABILITY AND DEGENERACY

We now return to the problem that we left in Sec. 3.5 — that of formulating distribution functions for particles. The problem is changed in two ways when we take a quantum viewpoint toward matter and energy.

First, the uncertainty principle requires that we abandon the concept of distinguishability except in certain restrictive cases. To determine whether or not particles are distinguishable we ask: Is it possible to conceive of any operation that would determine whether or not two particles have been interchanged? The answer to the question could only be yes under special circumstances. If, for example, the particles are *localized* by virtue of being located within a crystal lattice, then we must specify, a priori, that they cannot be interchanged. Localized particles accordingly have identity by virtue of position. Figure 5.4 also illustrates how opposing nuclear spins might serve to distinguish two nuclei in the same molecule from one another.

The second change in the statistical problem is introduced by the degeneracy of the energy levels of particles. If the higher energy levels are possessed of increasing degeneracy, then these states will also be possessed of increasing statistical weight. The lower energy levels will not be as highly favored as they are in nondegenerate distributions.

THERMODYNAMIC PROBABILITIES FOR THREE CASES

Three particular situations will be of subsequent interest to us. Each will have a slightly different statistical description.

1. *The particles are distinguishable* (e.g., they might be localized) and the degeneracies g_i, may or may not all be equal to one another. The number of ways of arranging particles for a given macrostate can be determined as follows. We have already seen that there are $N!\Big/\prod_{i=0}^{\infty} N_i!$ ways of arranging the particles into the various energy levels. But it is also possible to rearrange the N_i distinguishable particles within the g_i distinguishable quantum states at this level. This number of rearrangements is given by Problem 5, Sec. 3.2, as $g_i^{N_i}$. It follows that

$$W \equiv W_{\mathrm{B}} = N! \prod_{i=0}^{\infty} \frac{(g_i)^{N_i}}{N_i!} \tag{6.1}$$

where the subscript B evokes the similarity of this probability to Boltzmann's thermodynamic probability.

2. *The particles are indistinguishable.* The number of ways of arranging N_i indistinguishable particles in g_i distinguishable ways, at any energy level, is given by Problem 4, Sec. 3.2, as $(g_i + N_i - 1)!/(g_i - 1)!N_i!$. Thus the number of ways of arranging the corresponding macrostate is

$$W \equiv W_{\mathrm{BE}} = \prod_{i=0}^{\infty} \frac{(g_i + N_i - 1)!}{(g_i - 1)!N_i!} \tag{6.2}$$

where the subscript BE anticipates the relationship of this thermodynamic probability to Bose-Einstein statistics.

3. *Particles are indistinguishable, and no more than one of them can occupy the* i*th energy level in each of the* g_i *ways.*[1] There are

[1]This, we see shortly, would be true of particles that are bound to obey Pauli's exclusion principle.

$g_i!/N_i!(g_i - N_i)!$ ways in which the g_i quantum states can be divided into N_i that are "occupied" by a single particle and $g_i - N_i$ that are not. Accordingly,

$$W \equiv W_{\text{FD}} = \prod_{i=0}^{\infty} \frac{g_i!}{N_i!(g_i - N_i)!} \tag{6.3}$$

In this case the subscript FD anticipates the relevance of this result to Fermi–Dirac statistics.

It is a matter of great convenience that in dilute gases the degeneracies are largely unoccupied and $g_i \gg N_i$. When this is true,

$$\frac{(g_i + N_i - 1)!}{(g_i - 1)!N_i!} = \frac{(g_i + N_i - 1)\cdots(g_i + 1)(g_i)}{N_i!} \text{ is slightly} > \frac{g_i^{N_i}}{N_i!}$$

and

$$\frac{g_i!}{(g_i - N_i)!N_i!} = \frac{g_i(g_i - 1)\cdots(g_i - N_i + 1)}{N_i!} \text{ is slightly} < \frac{g_i^{N_i}}{N_i!}$$

And in the limit, as g_i becomes much much larger than N_i, Eqs. (6.1), (6.2), and (6.3) give

$$W_{\text{FD}} \simeq W_{\text{BE}} \simeq \frac{W_{\text{B}}}{N!} \tag{6.4}$$

DEVELOPMENT OF THE DISTRIBUTIONS

Distribution functions for particles of the preceding three types are obtained using the same basic strategy that underlay the derivation of the Boltzmann distribution. The particles must be distributed in such a way as to maximize the thermodynamic probability subject to the usual constraints

$$\sum_{i=0}^{\infty} N_i = N \tag{3.1}$$

and

$$\sum_{i=0}^{\infty} \epsilon_i N_i = N\bar{\epsilon} = U \tag{3.2a}$$

The procedure, using the method of Lagrangian multipliers, will be the same as outlined in Sec. 3.5. The details will be left as an exercise (Problem 6.1), and the resulting distributions will be listed below.

1. The idea of a hybrid classical-quantum particle — distinguishable, yet degenerate — is a useful device. Furthermore, it is not wholly unreasonable from a physical viewpoint. For example, particles that occupy fixed positions and have identity by virtue of their individual locations would properly be described in this way.

Such particles are called *boltzons*, or *degenerate boltzons*. The distribution of the N_i's for boltzons is

$$N_i = \frac{g_i}{e^{\alpha}e^{\beta\epsilon_i}} \tag{6.5}$$

or

$$\frac{N_i}{N} = \frac{g_i e^{-\beta\epsilon_i}}{Z} \tag{6.5a}$$

where

$$Z = \sum_{i=0}^{\infty} g_i e^{-\beta\epsilon_i} \tag{5.25}$$

Thus Eq. (5.25), which was written down heuristically in Sec. 5.3, is justified by a straightforward statistical development.

2. We next wish to obtain the distribution of the N_i's for the general case of indistinguishable particles. If no constraints are placed upon the occupancy of the energy levels, the thermodynamic probability is properly given by Eq. (6.2). This statistical method was advanced by S. N. Bose in 1924 and it was applied to the ideal gas by Einstein in 1924 and 1925. Particles that are governed by this statistical description are called *bosons*.

The resulting distribution of the N_i's for bosons is [2]

$$N_i = \frac{g_i}{e^{\alpha}e^{\beta\epsilon_i} - 1} \tag{6.6}$$

3. The Pauli exclusion principle, advanced by W. Pauli in 1925, requires the development of a third type of quantum-statistical description. The principle generalizes certain experimental evidence and can be stated in a variety of ways. For our purposes it will suffice to say that a quantum state cannot be occupied by more than one such primary indistinguishable particle as an electron or a proton unless the wave function for the system of particles is antisymmetric. Photons and phonons, for example, are *not* bound by this principle and will therefore be subject to Bose–Einstein statistics.

A year after Pauli stated this principle, Enrico Fermi and P. A. M. Dirac independently developed the statistical description that made use of it. The thermodynamic probability for a collection of indistinguishable particles was given by Eq. (6.3) for the case in which only

[2] Equation (6.6) is derived subject to the usual constraints — Eqs. (3.1) and (3.2). Later in this chapter we encounter applications of Bose–Einstein statistics in which Eq. (3.1) is not imposed. At that time we shall have to develop a form of Eq. (6.6) in which there appears only one Lagrangian multiplier.

one particle could occupy each energy state. Such particles are often called *fermions*. The distribution of the N_i's for fermions is

$$N_i = \frac{g_i}{e^{\alpha}e^{\beta\epsilon_i} + 1} \tag{6.7}$$

EXAMPLE 6.1 Consider the distribution of personal income in the United States. In a population of N people there are N_1 who earn t_1 dollars/yr, N_2 who earn t_2, and so on. The average income is known to be $\bar{t}$. In our present inflationary economy it has turned out that if a person can draw an income of t_i, then the number of ways in which he can do so is approximately proportional to t_i^2. We should like to predict the distribution of income under these circumstances.

In this case

$$\sum_{i=0}^{\infty} N_i = N \qquad \sum_{i=0}^{\infty} t_i N_i = \bar{t}N \qquad g_i = At_i^2$$

and the thermodynamic probability is given by Eq. (6.1), because the incomes (or the people that earn them) are distinguishable. With the help of Stirling's approximation, we can maximize W:

$$d \ln W = \sum_{i=0}^{\infty} (\ln At_i^2 - \ln N_i)\, dN_i = 0$$

Combining this with the differential form of the constraints gives

$$\sum_{i=0}^{\infty} \left(\ln \left[\frac{N_i}{At_i^2} \right] + \alpha + \beta t_i \right) dN_i = 0$$

Finally, we set the coefficients of the dN_i's equal to zero and obtain

$$N_i = Ae^{-\alpha} t_i^2 \exp(-\beta t_i)$$

The multiplier β can be eliminated by substituting this N_i into the two constraints:

$$N = Ae^{-\alpha} \sum_{i=0}^{\infty} t_i^2 \exp(-\beta t_i) \simeq \frac{Ae^{-\alpha}}{\Delta t} \int_0^{\infty} t^2 \exp(-\beta t)\, dt$$

or

$$N = \frac{2Ae^{-\alpha}}{\beta^3\, \Delta t}$$

Likewise,

$$N\bar{t} = Ae^{-\alpha} \sum_{i=0}^{\infty} t_i^3 \exp(-\beta t_i) = \frac{6Ae^{-\alpha}}{\beta^4\, \Delta t}$$

Solving between these expressions we obtain $\beta = 3/\bar{t}$.

The distribution is then given by

$$\frac{N_i}{N} = \frac{Ae^{-\alpha} t_i^2 \exp(-3t_i/\bar{t})}{2Ae^{-\alpha}/(3/\bar{t})^3 \,\Delta t}$$

or, according to Eqs. (2.11) and (2.14),

$$\bar{t}F(t) = \frac{27}{2}\left(\frac{t}{\bar{t}}\right)^2 \exp\left(\frac{-3t}{\bar{t}}\right)$$

This distribution characterizes income between 1946 and 1962 very accurately. In a depression economy there is less access to income and $g_i \sim t_i^{1/2}$. This gives a more skewed and less uniform distribution of the wealth.

BOLTZONS AS THE CLASSICAL LIMIT FOR FERMIONS AND BOSONS

Equation (6.4) showed us that if gases are dilute and few of the modes of energy occupancy are filled, then both the W_{BE} and W_{FD} approach[3] $W_{\mathrm{B}}/N!$

$$W_{\mathrm{BE}},\, W_{\mathrm{FD}} \rightarrow \prod_{i=0}^{\infty} \frac{g_i^{N_i}}{N_i!} \tag{6.8}$$

The distribution of N_i's derived from this thermodynamic probability (Problem 6.2) is identical to Eq. (6.5a). It therefore follows that, as $(N_i/g_i) \rightarrow 0$,

$$N_{i\mathrm{BE}},\, N_{i\mathrm{FD}} \rightarrow N_{i\mathrm{B}} \tag{6.9}$$

or

$$\frac{g_i}{e^{\alpha}e^{\beta\epsilon_i} - 1},\ \frac{g_i}{e^{\alpha}e^{\beta\epsilon_i} + 1} \rightarrow \frac{g_i}{e^{\alpha}e^{\beta\epsilon_i}} \tag{6.9a}$$

Thus, for $(N_i/g_i) \rightarrow 0$, we find that

$$e^{\alpha}e^{\beta\epsilon_i} \gg 1 \tag{6.10}$$

This condition can be reduced to

$$e^{\alpha} \gg 1 \tag{6.10a}$$

[3]Some authors argue that the thermodynamic probability for indistinguishable particles can be obtained by dividing W_{B} by the $N!$ rearrangements that become meaningless when particles are indistinguishable. Although this rationale *does* result in Eq. (6.8), it is in fact not correct, and it makes Eq. (6.8) appear to be exact. The same rationale carried out fully would result also in eliminating the $N_i!$'s from Eq. (6.8). Accordingly, W, the number of microstates per macrostate in a group of classical, nondegenerate, indistinguishable particles, would be *unity*. This result points up the folly of trying to discuss indistinguishability as a classical idea. Equation (6.8) *must* be regarded as the limiting approximation of the quantum-thermodynamic probabilities, W_{BE} and W_{FD}.

because the lowest value of ϵ_i/kT, ϵ_0/kT, is always quite small. Thus if particles are to behave as boltzons,

$$N = \sum_{i=0}^{\infty} N_i \simeq \sum_{i=0}^{\infty} g_i e^{-\alpha} e^{-\epsilon_i/kT}$$

or

$$N \simeq \frac{Z}{e^{\alpha}} \tag{6.11}$$

By way of illustration of the implications of this result, suppose that we introduce the partition function for a monatomic ideal gas,

$$Z_t = \left(\frac{\pi}{4}\frac{T}{\Theta_t}\right)^{3/2} \tag{5.28}$$

In this case Eq. (6.11) becomes

$$e^{\alpha} = \frac{1}{N}\left(\frac{\pi}{4}\frac{T}{\Theta_t}\right)^{3/2} \tag{6.12}$$

Equation (6.12) shows that condition (6.10a) will be satisfied for temperatures well in excess of $N^{2/3}\Theta_t$. Thus monatomic translating particles can be treated as boltzons only at temperatures vastly in excess of Θ_t — typically a few degrees Kelvin. Since $\Theta_t = h^2/8mka^2$, Eq. (6.12) can also be written in the form

$$e^{\alpha} = \left(\frac{2\pi mkT}{h^2}\right)^{3/2} \frac{V}{N} \tag{6.12a}$$

If we suppose that N/V corresponds roughly with the molecular density at standard conditions in a given configuration, then e^{α} will be less than unity below 10^4 °K for electrons, 1°K for H_2, and at much lower temperatures for other molecular gases. This means that quantum-statistical effects are very important at temperatures of practical importance in electron gases, but there is little application for molecular gases. The "degenerate ideal gas" occupies what can be a wide temperature range from Θ_t up to $e^{\alpha} = 0(1)$.

EVALUATION OF THE LAGRANGIAN MULTIPLIERS

In chapter 3, the multiplier β was calculated for nondegenerate boltzons and the multiplier α was ignored. Now we follow an argument by ter Haar[4] that gives both multipliers for all three statistics.

[4] D. ter Haar, *Elements of Statistical Mechanics*, Holt, Rinehart and Winston, Inc., New York, 1960, chap. 4, sec. 5.

Equation (6.1) can be rewritten in the form

$$d \ln W_{\mathrm{B}} = \sum_{i=0}^{\infty} \ln \left(\frac{g_i}{N_i}\right) dN_i \tag{6.1a}$$

but Eq. (6.5) gives $\ln (g_i/N_i) = \alpha + \beta\epsilon_i$, so we obtain

$$d \ln W_{\mathrm{B}} = \alpha \sum_{i=0}^{\infty} dN_i + \beta \sum_{i=0}^{\infty} \epsilon_i \, dN_i \tag{6.13}$$

It turns out (Problem 6.7) that this is also the case for W_{BE} and W_{FD},

$$d \ln W_{\mathrm{BE}} = d \ln W_{\mathrm{FD}} = \alpha \sum_{i=0}^{\infty} dN_i + \beta \sum_{i=0}^{\infty} \epsilon_i dN_i \tag{6.13a}$$

Consider next a change of energy, dU,

$$dU = \sum_{i=0}^{\infty} \epsilon_i dN_i + \sum_{i=0}^{\infty} N_i d\epsilon_i \tag{6.14}$$

The two summations in Eq. (6.14) represent two fundamentally different ways in which a system's energy can change. Consider, first, changes represented by the sum $\sum_{i=0}^{\infty} N_i \, d\epsilon_i$. The change in U is accomplished without altering the distribution numbers (the N_i's). Accordingly, W remains unaltered and the entropy, $S = k \ln W$, is also unchanged. But we already have a name for reversible interactions of a system that occur at constant entropy. Such interactions are called *work* interactions. Thus we can write

$$\delta Wk = \sum_{i=0}^{\infty} N_i \, d\epsilon_i \tag{6.15}$$

Equation (6.14) then takes the form

$$\sum_{i=0}^{\infty} \epsilon_i \, dN_i = dU - \delta Wk \tag{6.16}$$

We know from the first law of phenomenological thermodynamics that the sum of work and energy change is the heat interaction of the system. Thus $\sum_{i=0}^{\infty} \epsilon_i \, dN_i = \delta Q$, and Eq. (6.16) is simply a statement of the first law. Thus, if we write $W = W_{\mathrm{B}}$ or W_{BE} or W_{FD} and substitute Eq. (6.16) back into Eq. (6.13) or (6.13a), we get

$$d \ln W = \alpha \, dN + \beta(dU - \delta Wk) \tag{6.17}$$

or

$$d \ln W = \alpha \, dN + \beta \, \delta Q \tag{6.17a}$$

If the system is subjected to a change that alters W without changing the number of particles, we get, from Eq. (6.17a),

$$kd \ln W = dS = k\beta\, \delta Q \tag{6.18}$$

But $\delta Q = T\,\delta S$ because we are dealing with equilibrium particles [recall Eq. (1.13)]; thus

$$\beta = \frac{1}{kT} \tag{2.34}$$

in all three statistical formulations.

To evaluate α we begin by writing the Helmholtz function, $F = U - TS$, in differential form:

$$dF = (dU - T\,dS) - S\,dT$$

But, from Eq. (6.17),

$$T\,dS = kT[\alpha\, dN + \beta(dU - \delta Wk)]$$

Thus

$$dF = -\frac{\alpha}{\beta}\, dN + \delta Wk - S dT \tag{6.19}$$

The evaluation of α is completed by using Eq. (6.19) to form the partial of F with respect to N, at constant T, and with all extensive properties related to work modes held constant. For simple systems $\delta\, Wk = -p\, dV$, only, and we have

$$\left.\frac{\partial F}{\partial N}\right|_{T,V} = -\frac{\alpha}{\beta} \tag{6.20}$$

In accordance with Eqs. (1.38) and (1.41) we can write $dF = -S\,dT - p\,dV + \mu\, dN$ for single-component systems. It follows that $(\partial F/\partial N)_{T,V}$ is the chemical potential μ which in turn is equal to the molar Gibbs function [recall Eq. (1.48)]. It follows that

$$\alpha = -\frac{\mu}{kT} \quad \text{or} \quad -\frac{\mu}{R^0 T} \tag{6.21}$$

depending upon whether μ is expressed as a molecular or molar quantity. We defer consideration of the physical meaning of this result until the subsequent section.

6.2 IDEAL MONATOMIC BOSE–EINSTEIN AND FERMI–DIRAC GASES

We saw in Sec. 6.1 that the quantum-statistical description had to be used up to temperatures that vastly exceeded Θ_t. In some cases the temperature range $\Theta_t < T < N^{2/3}\Theta_t$ will be of practical importance. We must therefore consider the statistical description of these "degenerate ideal gases" in more detail.

GENERATING FUNCTION FOR DEGENERATE GAS PROPERTIES

We need to develop a generating function for the properties of Bose–Einstein and Fermi–Dirac gases. The concept of a generating function for properties is one that we have followed somewhat carefully up to this point. The Massieu function F/T was one of the classical generating functions that we considered in Sec. 1.5. In Sec. 3.5 we discovered a direct relationship [Eq. (3.27)] between F/T and the microscopic generating function Z.

For boltzons we can write, in accordance with Eqs. (6.5) and (6.11),

$$\sum_{i=0}^{\infty} g_i \exp\left[-(\alpha + \beta\epsilon_i)\right] = N = e^{-\alpha} Z$$

Equations (6.21) and (3.27) can be used to eliminate α and Z from this expression, so

$$\ln\left\{\sum_{i=0}^{\infty} g_i \exp\left[-(\alpha + \beta\epsilon_i)\right]\right\} = \frac{(\mu N - F)/kT}{N} \tag{6.22}$$

Equation (6.22) suggests the definition of a new generating function closely related to the partition function. We call it the *q potential:*

$$q \equiv \frac{\mu N - F}{kT} \tag{6.23}$$

Replacing $-F$ with $TS - U$, S with $k \ln W$, and μ/kT with $-\alpha$ in Eq. (6.23) permits us to express it in the form,

$$q = \ln W - \alpha N - \beta U \tag{6.23a}$$

The derivative of q,

$$dq = d \ln W - \alpha\, dN - N\, d\alpha - U\, d\beta - \beta\, dU$$

can be simplified by combining it with Eq. (6.17). The resulting differential q potential for a simple system, in which $\delta Wk = -p\, dV$, is

$$dq = -N\, d\alpha - U d\beta + \beta p dV \tag{6.24}$$

The q potential is a potential in the sense that the following derivatives can be generated from it:

$$N = -\left(\frac{\partial q}{\partial \alpha}\right)_{\beta,V} \tag{6.25}$$

$$U = -\left(\frac{\partial q}{\partial \beta}\right)_{\alpha,V} \tag{6.26}$$

$$\beta p = \left(\frac{\partial q}{\partial V}\right)_{\alpha,\beta} \tag{6.27}$$

And from Eqs. (6.23), (1.32), and (1.41) we have

$$q = \frac{pV}{kT} \tag{6.28}$$

We should note that the q potential is, in fact, the Legendre transform function with the independent parameters $1/T$, V, and μ/T. In other words,

$$\frac{pV}{kT} = q\left(\frac{1}{T}, V, \frac{\mu}{T}\right) \tag{6.28a}$$

as we saw in Example 1.3. The explicit expression for $q(1/T, V, \mu/T)$ constitutes the fundamental equation and thus contains complete thermodynamic information for a substance.

In addition to its importance in the evaluation of the properties of degenerate gases, the q potential is the fundamental macroscopic parameter in the grand canonical ensemble theory to be discussed in chapter 8.

EVALUATION OF THE PROPERTIES OF THE BOSE-EINSTEIN AND FERMI-DIRAC GASES

With the aid of Eqs. (6.2), (6.3), (6.6), and (6.7) it is easy to show (Problem 6.8) that[5]

$$\left.\begin{matrix}\ln W_{\mathrm{BE}} \\ \ln W_{\mathrm{FD}}\end{matrix}\right\} = \sum_{i=0}^{\infty} N_i(\alpha + \beta\epsilon_i) \mp \sum_{i=0}^{\infty} g_i \ln(1 \mp e^{-\alpha-\beta\epsilon_i}) \tag{6.29}$$

Using Eqs. (6.29) and (6.23a) for the q potential we find that

$$q_{\mathrm{BE\ or\ FD}} = \mp \sum_{i=0}^{\infty} g_i \ln(1 \mp e^{-\alpha-\beta\epsilon_i}) \tag{6.30}$$

Equation (6.30) looks something like a partition function. Like a partition function it can be put in integral form for $T \gg \Theta_t$. This

[5]In the rest of this section the upper symbol in a $\mp$ or $\pm$ sign will refer to the Bose–Einstein case, the lower symbol to the Fermi–Dirac case.

can be done by writing $g_i = g_i(\epsilon_i)$ and noting that in the continuous limit $dg = (dg/d\epsilon)\,d\epsilon$, where we view g_i as being an increment of the cumulative number of modes of occupancy, g, up to a given quantum level. Thus

$$q_{\text{BE or FD}} = \mp \int_0^\infty \ln\,(1 \mp e^{-\alpha-\beta\epsilon})\,\frac{dg}{d\epsilon}\,d\epsilon \tag{6.31}$$

The problem of getting a continuous functional relation between g and ϵ is solved by first determining the functional relation between g and n, then using Eq. (5.3) or (5.7) to express n in terms of ϵ. The Rayleigh–Jeans arguments that led to Eq. (4.22) can be paraphrased to show that if $n = \sqrt{n_x^2 + n_y^2 + n_z^2}$, then

$$g = \frac{1}{8}\frac{4\pi}{3}n^3$$

or

$$\frac{dg}{dn} = \frac{\pi}{2}n^2$$

and, with the help of Eq. (5.7),

$$\frac{dg}{d\epsilon} = \frac{dg}{dn}\frac{dn}{d\epsilon} = \frac{\pi}{2}\,\frac{8mV^{2/3}\epsilon}{h^2}\cdot\frac{1}{2}\sqrt{\frac{8m}{\epsilon}}\,\frac{V^{1/3}}{h}$$

Thus

$$\frac{dg}{d\epsilon} = 2\pi V\left(\frac{2m}{h^2}\right)^{3/2}\sqrt{\epsilon} \tag{6.32}$$

and Eq. (6.31) takes the form

$$q_{\text{BE or FD}} = \mp 2\pi\left(\frac{2m}{h^2}\right)^{3/2} V\int_0^\infty \sqrt{\epsilon}\,\ln\,(1 \mp e^{-\alpha-\beta\epsilon})\,d\epsilon \tag{6.33}$$

Changing the variable with $\xi \equiv \beta\epsilon = \epsilon/kT$ we obtain

$$q_{\text{BE or FD}} = \mp 2\pi\left(\frac{2mkT}{h^2}\right)^{3/2} V\int_0^\infty \sqrt{\xi}\,\ln\,(1 \mp e^{-\alpha}e^{-\xi})\,d\xi \tag{6.33a}$$

The integrand can be expanded, because $e^{-\alpha}e^{-\xi} < 1$ and the resultant terms can be integrated by parts. The result is

$$q_{\text{BE or FD}} = \pm\left(\frac{2\pi mkT}{h^2}\right)^{3/2} V\sum_{n=1}^{\infty}\frac{(\pm e^{-\alpha})^n}{n^{5/2}} \tag{6.34}$$

Equation (6.34) is the fundamental equation, $q = q(1/T, V, \mu/T)$, and the properties of the gas can then be obtained immediately from it using Eqs. (6.25) through (6.28). From either Eq. (6.27) or (6.28) and Eq. (6.25) we obtain, upon substitution of Eq. (6.34),

$$\left.\frac{pV}{kTN}\right|_{\text{BE or FD}} = \frac{\sum_{n=1}^{\infty}(\pm 1)^n e^{-\alpha n}/n^{5/2}}{\sum_{n=1}^{\infty}(\pm 1)^n e^{-\alpha n}/n^{3/2}} \tag{6.35}$$

or, after some algebra,

$$\left.\frac{pV}{kTN}\right|_{\text{BE or FD}} = 1 \mp \frac{e^{-\alpha}}{2^{5/2}} + \cdots \tag{6.35a}$$

The ideal-gas constant is therefore subject to a correction for indistinguishability that is on the order of $\mp e^{-\alpha}$ or more. Equation (6.35a) is consistent with our macroscopic definition for the ideal-gas law, even though the constant changes.

Combining Eqs. (6.26) and (6.28) in Eq. (6.34) gives

$$U = \tfrac{3}{2} NkT\left(1 \mp \frac{e^{-\alpha}}{2^{5/2}} + \cdots\right) \tag{6.36}$$

where we only need one term if $e^{\alpha} \geq 1$ [recall the context of Eq. (6.12)].

We have now seen that the question of distinguishability of particles — although it might have first appeared to be a very subjective question — embodies a very real effect upon ideal-gas properties once we undertake a quantum description of matter.[6] Before leaving the ideal gas we should take brief note of a particular kind of extreme deviation from classical (Boltzmann) ideal-gas behavior.

EINSTEIN CONDENSATION

Einstein pointed out a curious phenomenon related to the Bose-Einstein distribution in 1925. This subsequently received attention for its possible relevance to the λ transition of liquid helium.[7] Suppose that a Bose-Einstein gas is cooled to a very low temperature. Since $\beta \gg 1$, $\epsilon_0 < \epsilon_1 < \cdots$, and $e^{\beta\epsilon_0} \ll e^{\beta\epsilon_1} \ll \cdots$, it follows that

$$N = \sum_{i=0}^{\infty} N_i = \frac{g_0}{e^{\alpha+\beta\epsilon_0} - 1} + \frac{g_1}{e^{\alpha+\beta\epsilon_1} - 1} + \cdots = N_0 + N_1 + \cdots$$

approaches

$$N \simeq \frac{g_0}{e^{\alpha+\beta\epsilon_0} - 1} = N_0 \tag{6.37}$$

The preceding result indicates that there can be a pileup of particles in the ground state, ϵ_0, at low temperatures. This rapid increase in the population of the ground state for a Bose-Einstein

[6]The Bose-Einstein description is, of course, *correct* for conventional gases, while the Boltzmann description differs from it and is *incorrect* — even at high temperatures. Fortunately, the difference is negligible for almost any temperature of practical interest.

[7]This matter is discussed by ter Haar, reference 4, chap. 9, sec. 2; ter Haar also presents a broad bibliography on the subject.

gas is what we call the Einstein condensation. The transition temperature at which this takes place can be identified. This temperature is rather like a condensation point and, indeed, it is found to lie at $T = 3.2°K$, which is comparable with the observed λ transition for helium, $T = 2.19°K$.

6.3 PHOTON GAS AND PHONON GAS

The name *photon* is given to the "particle" (or particle-like qualities) associated with electromagnetic waves and it is possessed of energy $h\nu$. This energy can be occupied in as many ways as there are standing waves. A "gas" consisting of photons is thus a degenerate Bose–Einstein gas. If Planck's result for the distribution of radiant energy was correct we should now be in a position to rederive it by treating photons from this viewpoint. And we also should be able to evaluate all the other thermodynamic properties of a photon gas.

The discussion that follows can also be applied to a related "particle" called the *phonon*. As electromagnetic waves travel through a medium, "acoustic" or vibrational waves also travel through a crystal lattice. And in keeping with de Broglie's concepts, a particle can also be associated with these waves. Such a "particle" is called a phonon. It is also a boson and, in fact, its description does not differ fundamentally from that of a photon. The discussion that follows could easily be adapted to the phonon as well.

PLANCK DISTRIBUTION

We should recall that Planck envisioned a *nondegenerate* assembly of distinguishable boltzon oscillators. These oscillators were permitted to assume rising energy levels, $h\nu$, $2h\nu$, $3h\nu$, ... He also assumed that these oscillators were conserved, when in fact they are absorbed and generated at the walls of their container. To rederive the distribution we must first correct Eq. (6.6) for the removal of the conservation constraint

$$\sum_{i=0}^{\infty} N_i = N \tag{3.1}$$

The result, obtained from Eq. (6.2) and the second constraint (3.2) (Problem 6.12), is

$$N_i = \frac{g_i}{e^{\epsilon_i/kT} - 1} \tag{6.38}$$

The degeneracy is again obtainable from Eq. (4.22) and again g_i can be viewed as the increment dg of a continuous variable g. Thus

$$dg = \frac{dn_0}{d\lambda}\,d\lambda = -2\left(\frac{4\pi V}{\lambda^4}\right)d\lambda$$

where the factor 2 is introduced to account for the two degrees of polarization.

Equation (6.38) can also be expressed in differential form, because *all* the ϵ_i's are equal to $h\nu$ or hc_l/λ and the degeneracy can be thought of as varying continuously with ν. Thus

$$dN = \frac{dg}{\exp(h\nu/kT) - 1}$$

or

$$\frac{dN}{d\lambda} = \frac{dN}{dg}\frac{dg}{d\lambda} = -\frac{8\pi V}{\lambda^4[\exp(h\nu/kT) - 1]} \tag{6.39}$$

This can finally be multiplied by ϵ/V or $h\nu/V$ to give the radiation density, $u_\lambda = -(hc_l/\lambda V)(dN/d\lambda)$, analogous to Eq. (4.21),

$$u_\lambda = \frac{8\pi c_l h}{\lambda^5[\exp(hc_l/\lambda kT) - 1]} \tag{4.34}$$

which was Planck's result. Although the result is the same, it has now been developed in a way that is consistent with both the Schrödinger equation and a fully quantum view of matter.

EXAMPLE 6.2 How many photons are there in a cavity of 1 cm³ volume at a temperature of 300°K?

For a photon gas we can sum Eq. (6.38) over all i and obtain

$$N = \sum_{i=0}^{\infty} \frac{g_i}{\exp(\epsilon_i/kT) - 1} \simeq \int_{\nu=0}^{\nu=\infty} \frac{dg(\nu)}{\exp[\epsilon(\nu)/kT] - 1}$$

but

$$dg = -2\left(\frac{4\pi V}{\lambda^4}\right)\frac{d\lambda}{d\nu}\,d\nu = \left(-\frac{8\pi V\nu^4}{c_l^4}\right)\left(-\frac{c_l}{\nu^2}\right)d\nu$$

so

$$N = \int_0^\infty \frac{8\pi V}{c_l^3}\frac{\nu^2 d\nu}{\exp(h\nu/kT) - 1} = \frac{8\pi V}{c_l^3}\left(\frac{kT}{h}\right)^3\int_0^\infty \frac{x^2 dx}{e^x - 1}$$

With the help of Appendix E-1-(e) we find that the integral is equal to $\Gamma(3) \cdot \zeta(3)$ or 2.404. Thus

$$N = 60.5V\left(\frac{kT}{hc_l}\right)^3$$

so

$$N = 60.5 \times 1\ \text{cm}^3\left(\frac{300^0\text{K}}{1.439\ \text{cm}-{}^0\text{K}}\right)^3 = \underline{5.45 \times 10^8\ \text{photons}}$$

THERMODYNAMIC RELATIONS FOR PHOTONS

The properties of degenerate Bose–Einstein gases that were formulated in Sec. 6.2 can now be reformulated for photons. The only difference lies in the relaxation of the conservation constraint (3.1) and the resultant loss of the Lagrangian multiplier, α [recall Eq. (6.33)]. Thus the q potential for photons can be obtained from Eq. (6.33) with the multiplier α removed, or

$$q_{\text{photon}} = -\frac{8\pi\sqrt{2}V}{h^3}\int_0^\infty \sqrt{\epsilon m}\ \ln\left(1 - e^{-\beta\epsilon}\right) d(\epsilon m) \qquad (6.33\text{b})$$

with the help of the following heuristic argument.

The quantity ϵm in Eq. (6.33b) is bothersome because the *rest mass* of a photon is zero. However, if we write $\epsilon m = p^2/2$ and then introduce the de Broglie relation $p = h/\lambda$ or $p = h\nu/c_l$, this suggests the equivalence

$$\epsilon m = \frac{1}{2}\left(\frac{h\nu}{c_l}\right)^2$$

Substituting this in Eq. (6.33b) and integrating the result with respect to ν, we can obtain from Eq. (6.33b) the q potential for a photon gas (Problem 6.13),

$$q_{\text{photon}} = \frac{8\pi^5}{45}\left(\frac{kT}{hc_l}\right)^3 V \qquad (6.33\text{c})$$

The remaining properties of a photon gas can now be obtained directly. Using Eqs. (6.26) and (6.28), for example, we can readily obtain a familiar result from Eq. (6.33c),

$$U = \frac{3}{\beta}q_{\text{photon}} = 3pV \qquad (4.2\text{a})$$

The same result can also be obtained easily from Eqs. (6.27) and (6.28). So, too, can expressions for the entropy and the constant-volume specific heat for a photon gas. These simple derivations are left as an exercise (Problem 6.14).

The preceding discussion has centered on photon gases. We have actually described a phonon "gas" problem already in Sec. 4.3—the Einstein solid problem, in which we treated the vibration of atoms in a solid much as we treat photons. This notion will be carried to greater refinement in the improvements of Einstein's prediction, given by Debye and others, which make more accurate and descriptive use of the phonon concept (see chapter 10).

6.4 "ELECTRON GAS" IN A METAL

ELECTRON GASES

The thermal and electrical conductivities of a metallic solid are the results of the translatory motion of a gas of free electrons within the solid. The magnitudes of these transport properties are very great in metals because the velocities of low-mass electrons are very high, and we find a linear dependence of transport properties upon $\bar{C}$ in Sec. 11.2. The law of Wiedemann and Franz shows that the relation between the thermal and electrical conductivities is linear. But if these properties are chiefly the result of transport in an electron gas, then they should both be proportional to the velocity of the electrons. They should then, in turn, be proportional to one another, and the empirical law of Wiedemann and Franz shows that they are.

Since this law was subsequently derived by applying the methods of kinetic theory to free electrons (see Sec. 11.5), there appears to be little doubt that such "gases" exist. Furthermore, there is other experimental evidence which shows that the number of free electrons per atom in a metal is on the order of unity.

The presence of such a gas should rightly add to the specific heat of the solid. At least to a first order of approximation, such electrons should move independently of the lattice atoms as a free gas. Because there are three degrees of translational freedom for these electrons as compared with six degrees of vibrational freedom for the lattice atoms, and because there are about as many electrons as atoms, the classical equipartition theorem would lead us to expect the molar specific heat to be around $4.5R^\circ$ instead of $3R^\circ$.

Table 4.1 reveals that c_v exceeds the classical value by far less than 50 percent at high temperatures. Since free electrons have to obey the Pauli exclusion principle, a proper description of their behavior requires the use of Fermi–Dirac statistics. This important example shows the way in which a quantum-statistical description becomes important at much higher temperatures than those discussed in Sec. 6.2.

ANALYTICAL DESCRIPTION OF THE ELECTRON

It is necessary to learn the energy levels and the degeneracies of the electrons, and to establish some properties of the electrons themselves, before trying to predict their behavior. Experiments show that at room temperatures electrons do not escape or "emit" from the surfaces of metals. They are bound in some way to remain within the solid — at least as long as the temperature remains low.

The following model for the energy of a free electron within a solid was developed by Sommerfeld in 1928. An electron within the solid is viewed as having uniform energy throughout the box. The energy required to remove an electron at rest from the solid is ϵ_s. This is similar to our concept of a free particle in a box [recall the potential-energy function described for the free particle in Fig. 5.1(a)]. The cases are different in that we now grant that a finite energy, ϵ_s, will get the electron out of its container — the metallic solid. Figure 6.1 shows the potential-energy function for these two related cases. In both Figs. 6.1(a) and (b) are shown lines representing the electron energy levels given by the Schrödinger equation for translating particles, Eqs. (5.3) and (5.7).

The Fermi–Dirac distribution for these particles can be written in the form

$$N_i = \frac{g_i}{\exp[(\epsilon_i - \mu)/kT] + 1} \tag{6.40}$$

where Eq. (6.21) has been used to replace α in Eq. (6.7). The electron gas is degenerate just as the photon gas was. By arguments similar to those employed in Sec. 6.2 we would find that in its continuous limit

$$g_i \to 2\pi V\left(\frac{2m}{h^2}\right)^{3/2}\sqrt{\epsilon}\,d\epsilon \tag{6.32a}$$

Equation (6.32a) is actually low by a factor of 2. Electrons have a *spin degeneracy* of 2 because they can be spinning in either of two directions. Each mode of storage of translational energy can occur in combination with either mode of rotation. Equation (6.32) should accordingly be replaced, in this case, by

$$dg = 4\pi V\left(\frac{2m}{h^2}\right)^{3/2}\sqrt{\epsilon}\,d\epsilon \tag{6.41}$$

Fig. 6.1 Potential-energy functions for a molecule in a box and a free electron in a solid.

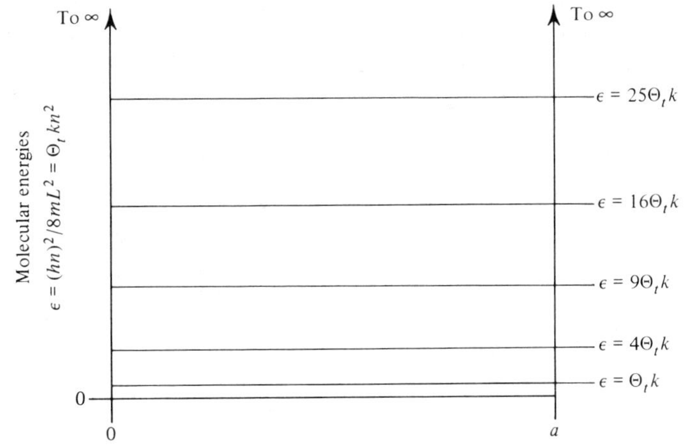

(a) Idealization of a translating molecule in a unidimensional box of length a. (recall Fig. 5.1a)

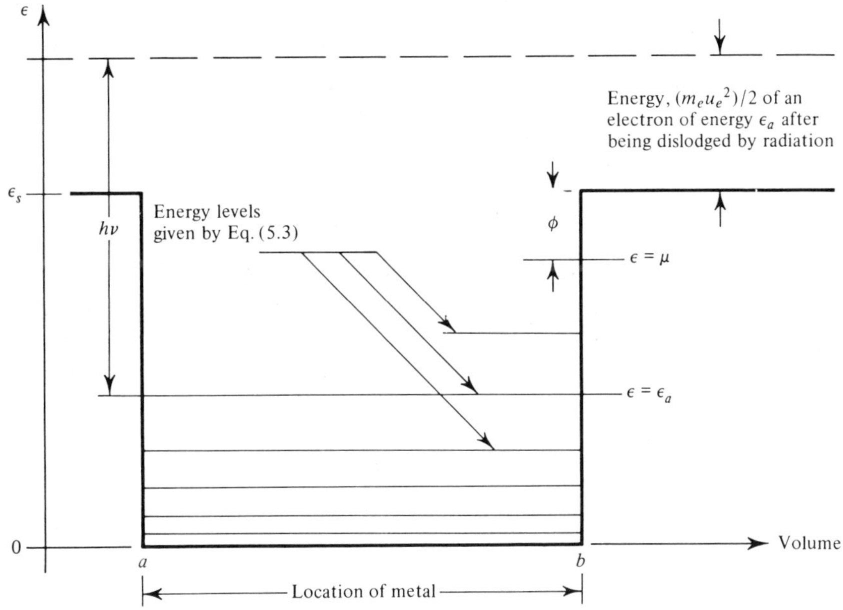

(b) Idealization of a free electron in a solid.

The distribution function for energy, $f(\epsilon)$, is then

$$f(\epsilon) = \frac{dN}{d\epsilon} = \left[4\pi V\left(\frac{2m}{h^2}\right)^{3/2}\right]\frac{\sqrt{\epsilon}}{\exp[(\epsilon-\mu)/kT]+1} \tag{6.42}$$

PROPERTIES OF AN ELECTRON GAS

With the distribution function, Eq. (6.42), in hand, the properties of an electron gas can be developed. For example, the number of particles with energies greater than zero is

$$N(\epsilon) = 4\pi V\left(\frac{2m}{h^2}\right)^{3/2} \int_0^\epsilon \frac{\sqrt{\xi}\, d\xi}{\exp[(\xi - \mu)/kT] + 1} \tag{6.43}$$

This expression will reveal a very interesting quality of this (or any other) Fermi–Dirac distribution, if we allow the temperature to go to zero. First, let us look at the integral

$$\lim_{T\to 0} \int_0^\epsilon \frac{\sqrt{\xi}\, d\xi}{\exp[(\xi - \mu)/kT] + 1} = \lim_{T\to 0} \left[\underbrace{\int_0^\mu \frac{\sqrt{\xi}\, d\xi}{\exp[(\xi - \mu)/kT] + 1}}_{=\frac{2}{3}\mu_0^{3/2}} + \underbrace{\int_\mu^\epsilon \frac{\sqrt{\xi}\, d\xi}{\exp[(\xi - \mu)/kT] + 1}}_{=0} \right]$$

where the $\lim_{T\to 0} \mu = \mu_0$, and $\epsilon > \mu_0$. We can see immediately from this integral that

$$\lim_{T\to 0} N(\epsilon) = \frac{8\pi}{3} V\left(\frac{2m}{h^2}\right)^{3/2} \mu_0^{3/2} \qquad \text{for } \epsilon \geq \mu_0 \tag{6.44}$$

and

$$\lim_{T\to 0} N(\epsilon) = \frac{8\pi}{3} V\left(\frac{2m}{h^2}\right)^{3/2} \epsilon^{3/2} \qquad \text{for } \epsilon < \mu_0 \tag{6.44a}$$

Equation (6.43) provides a basis for computing the chemical potential (or the molar Gibbs free energy) at zero temperature. When this has been done we can return to Eq. (6.42) with Eqs. (6.44) and (6.44a). The result for platinum is shown in Fig. 6.2, although the scaling of the coordinates has been left as an exercise (Problem 6.19).

The quantity μ_0—called the *Fermi level* of the gas—is an important quantity. All energy levels above it remain unoccupied at absolute zero. However, translational energy is still allowed at absolute zero as long as it does not exceed μ_0. The energy of classical particles would vanish at 0°K, on the other hand. In Table 6.1 Fermi levels are listed for free electrons in various common metals, along with other parameters.

At higher temperatures we can still approximate the free energy, μ, fairly easily. To do so we must observe that metals *melt* while kT is still far below typical electronic energies. At room temperature kT is only 0.025 eV while μ is typically in the range

TABLE 6.1 Electronic Properties of Selected Metals

Metal	Concentration of free electrons at 0°C, $10^{22}/cm^3$	Fermi level μ_0, eV	Linear c_v term: A experimental, (cal/g mole-°C²) × 10^4	Linear c_v term: $(\pi^2/2)(kR°/\mu_0)$, (cal/g mole-°C²) × 10^4	Work function ϕ, eV
Li	4.6	4.72	4.3	1.79	—
Na	2.5	3.12	1.73–1.80	2.81	2.3
K	1.3	2.14	—	—	2.26
Cr	—	—	—	—	4.37–4.60
Cu	8.5	7.04	—	—	—
Ag	5.8	5.51	1.54–1.60	1.53	—
Au	5.9	5.51	1.8	1.53	—
W	—	—	3.5	—	4.49–4.50
Pt	6.6	5.30	1.60–1.65	—	5.3–6.2
Zn	—	—	1.25–1.42	—	4.24

of 1 to 10 eV. Inspection of Eq. (6.42) reveals that when $(\epsilon - \mu)/kT$ is small, $f(\mu) \simeq f_{\max}/2$ (see Fig. 6.2). It turns out that the value of μ that satisfies this relationship is very nearly equal to μ_0. Sommerfeld did a more precise computation of μ and found that

$$\mu = \mu_0\left[1 - \frac{\pi^2}{12}\left(\frac{kT}{\mu_0}\right)^2 + \cdots\right] \tag{6.45}$$

The distribution curves for higher temperatures can then be calculated, once μ is known. This has been done and the results

Fig. 6.2 Energy distribution for the electron gas in platinum.

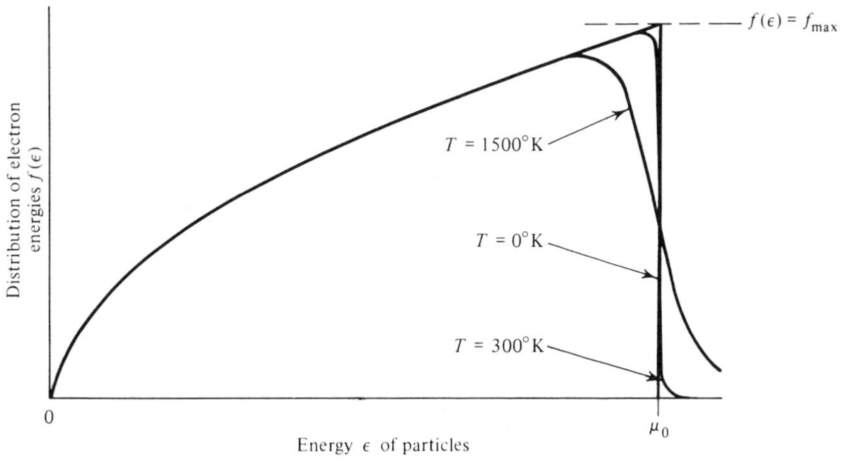

plotted in Fig. 6.2. It is most interesting to note in this figure that the distribution curves are not terribly sensitive to temperature.

The specific heat c_v can be calculated using

$$c_v \equiv \frac{\partial U}{\partial T}\bigg|_V = \frac{\partial(N_A\bar{\epsilon})}{\partial T}\bigg|_V$$

where

$$\bar{\epsilon} \equiv \frac{\int_0^\infty \epsilon f(\epsilon)\, d\epsilon}{\int_0^\infty f(\epsilon)\, d\epsilon} = \frac{\int_0^\infty \frac{\epsilon^{3/2} d\epsilon}{\exp[(\epsilon-\mu)/kT]+1}}{\int_0^\infty \frac{\epsilon^{1/2} d\epsilon}{\exp[(\epsilon-\mu)/kT]+1}} \tag{6.46}$$

At $T = 0°\text{K}$ this gives

$$\bar{\epsilon} = \frac{2\mu_0^{5/2}}{5} \Big/ \frac{2}{3}\mu_0^{3/2} = \frac{3\mu_0}{5}$$

Sommerfeld refined the computation for higher temperatures and obtained

$$\bar{\epsilon} = \frac{3}{5}\mu_0\left[1 + \frac{5\pi^2}{12}\left(\frac{kT}{\mu_0}\right)^2 + \cdots\right] \tag{6.47}$$

In all cases of practical importance $kT \ll \mu_0$, so we obtain, to a very close order of approximation,

$$c_v = \frac{\pi^2}{2}\left(\frac{kT}{\mu_0}\right)R^0 \tag{6.48}$$

which is far less than the classical value, $\frac{3}{2}R°$.

It is easy to show that when $T \to 0$, Debye's formula for the specific heat of a solid[8] gives $c_v \to BT^3$, where B is a constant. If we add to this the electronic c_v we obtain

$$c_v\,|_{T\to 0°\text{K}} = AT + BT^3 \tag{6.49}$$

where B is obtainable from Debye's theory and we would expect A to be $(\pi^2/2)(kR^0/\mu_0)$. The expression for BT^3 is given in Eq. (10.22).

Experiments reveal that our predicted value of A does not agree very closely with experimental values. The reason is that the free-electron model is too crude for great accuracy. Observed and calculated values of A are compared for various metals in Table 6.1.

We therefore see (if only qualitatively) that, as a direct consequence of the Pauli exclusion principle, electrons have appreciable kinetic energy at very low temperatures. But the excitation energy

[8]Debye's formula is an improvement on Einstein's early quantum result. We take it up in chapter 10.

required to jar electrons into the unoccupied higher energy states is too high, except near the Fermi level. As a consequence, only very few of the electrons actually contribute to the specific heat of a solid.

EXAMPLE 6.3 Evaluate the maximum energy of the free electrons in tungsten at $T = 0°\text{K}$, and calculate the contribution of the electrons to c_v at $T = 3000°\text{K}$. There are two free electrons per tungsten atom.

The maximum energy of electrons at 0°K is μ_0. From Eq. (6.44),

$$\mu_0 = \frac{h^2}{8m_e}\left(\frac{3N}{\pi V}\right)^{2/3}$$

where $N = 2 \times N_A = 1.205 \times 10^{24}/\text{g mole}$, $\rho = 19.3\ \text{g/cm}^3$, $V = M/\rho = (184/19.3)\ \text{cm}^3/\text{g mole}$, and $m_e = 9.1 \times 10^{-28}$ g. Then

$$\mu_0 = \frac{(6.62 \times 10^{-27})^2}{16 \times 9.1 \times 10^{-28}}\left(\frac{3}{\pi}\,\frac{1.205 \times 10^{24}}{184/19.3}\right)^{2/3} = 14.7 \times 10^{-12}\ \text{erg}$$

or

$$\mu_0 = 9.18\ \text{eV}$$

which compares favorably with the values given in Table 6.1.

The specific-heat contribution is given by Eq. (6.48) as

$$c_v = \frac{\pi^2}{2}\left(\frac{kT}{\mu_0}\right)R^0 = \frac{\pi^2 \times 1.38 \times 10^{-16} \times 3000}{2 \times 14.7 \times 10^{-12}}R^0$$

for $T = 3000°\text{K}$. The result is

$$c_v = 0.139R^0$$

as compared with $3R^0$ for the tungsten itself. The electronic contribution is only about 5 percent in this case.

Figure 6.1(b) illustrates another property of interest to solid-state physicists, the work function, ϕ:

$$\phi \equiv \epsilon_s - \mu \tag{6.50}$$

This is the free energy that must be supplied before electrons will flow from the surface of the solid in any number. It is typically in the range 2 to 6 eV (see Table 6.1).

PHOTOELECTRIC EFFECT

The Sommerfeld model for an electron gas made possible the explanation of still another phenomenon that had caused a great deal of concern. This was the *photoelectric effect*, discovered by Hertz in 1887. Hertz discovered (by accident) that ultraviolet light falling on a negatively charged plate would reduce the charge.

Since the identification of the electron was not made until a decade later, by Thompson, Hertz could not know that electromagnetic radiation was knocking negatively charged electrons out of the metal cathode.

The concept of an electron gas made it possible to attribute the phenomenon to the removal of electrons by the radiant energy $h\nu$. This effect is illustrated in Fig. 6.1.

Problems 6.1 Derive Eqs. (6.5a), (6.6), and (6.7).

6.2 Derive the distribution of N_i's based upon $W_B/N!$ and compare it with Eq. (6.5a).

6.3 Rain falls on a watershed. Evaluate the distribution in time of raindrops passing the gaging station, if the number of ways that a drop can find its way to the gaging station after time t_i is proportional to t_i^2. There are two constraints on the distribution. The total number of raindrops that run off is N. The root-mean-square time for the distribution is a known value, t_{rms}.

6.4 In Problem 6.3 replace t_i^2 with t_i^{n-1} and replace t_{rms} with $t_{rm\beta} \equiv [\sum (N_i/N)t_i^\beta]^{1/\beta}$. Show that the result now takes the form of a *generalized gamma distribution*,

$$F(t) = \frac{\beta}{a^n\Gamma(n/\beta)} t^{n-1} \exp\left[-\left(\frac{t}{a}\right)^\beta\right] \qquad a \equiv \left(\frac{\beta}{n}\right)^{1/\beta} t_{rm\beta}$$

6.5 Plot e^α against temperature for 1 cm^3 of atomic hydrogen and suggest a minimum temperature for which the atoms can be treated as boltzons, in this case.

6.6 Show that $e^\alpha \simeq (2\pi/3)^{3/2}\ V/N\lambda^3$ (where λ is the de Broglie wavelength) when particles can first be treated as boltzons. Discuss the meaning of this criterion.

6.7 Verify Eq. (6.13a).

6.8 Verify Eqs. (6.29) and (6.30).

6.9 Do a derivation, similar to the derivation of Eq. (6.33), for q_B. Under what conditions are the q potentials the same in each case?

6.10 Verify Eqs. (6.35a) and (6.36), adding some additional terms to the expansions. Derive similar expressions for S and c_p.

6.11 Show that a two-dimensional ideal Bose–Einstein gas does not condense.

6.12 Derive Eq. (6.38). Include a verification that $\beta = 1/kT$.

6.13 Complete the steps in the derivation of Eq. (6.33c).

6.14 Use Eq. (6.27) to show that $pV = U/3$ for a photon gas. Use the potential, q_{photon}, to show that

$$S = 4k\frac{pV}{kT}$$

and

$$c_v = \frac{32\pi^5}{15}\left(\frac{kT}{hc_l}\right)^3 kV$$

for a photon gas. Compare this c_v with the value given in Example 1.2.

6.15 In earlier sections we took great care to justify passing from finite increments to differential representations of quantities. We were fairly casual about doing it when we discussed the photon and electron gases, however. Explain why we were safe in ignoring the question in these cases.

6.16 Show that for the electron gas in a metal, $pV = 2U/3$. Compare this result with the corresponding result for a molecular gas. Calculate the pressure exerted by the electron gas in silver at 300°K.

6.17 Calculate the mean speed and the root-mean-square speed of an electron in a metal at 0°K.

6.18 Calculate the current density in Ampere/cm^2 across a surface in silver at 0°K in either direction.

6.19 Develop Fig. 6.2 to scale. Plot the ratio $(c_v)_{\mathrm{FD}}/(c_v)_{\mathrm{classical}}$ to scale as a function of temperature for platinum.

7 thermostatic properties of ideal gases

In chapter 6 we undertook a fully quantum-statistical description of ideal gases and found that the results resembled those of Boltzmann statistics, as long as the gas was dilute. Diluteness, in this case, meant that the degeneracies were far from filled up — that $N_i/g_i \ll 1$. We found that this would be true for most atoms and molecules at temperatures in excess of a few degrees Kelvin. But at these low temperatures only such gases as helium and hydrogen, with very lightweight molecules, continue to exist as gases. Our interest in the degenerate behavior of ideal gases is motivated more by the problems of describing gases of subatomic particles (such as electrons) and nonmaterial "particles" (such as photons) than it is by a few low-temperature molecular problems.

We now wish to describe nondegenerate ideal gases in greater detail. While the microscopic qualities of ideal-gas behavior were treated in chapter 2, the methods of statistical mechanics will give us complete macroscopic information.

7.1 THERMODYNAMIC PROBABILITY AND PARTITION FUNCTION FOR IDEAL GASES

THERMODYNAMIC PROBABILITY

In chapter 3 it was shown that for classical (distinguishable) particles

$$S_{\text{dist}} = k \ln W_{\text{dist}} = kN \ln Z_{\text{dist}} + \frac{U}{T} \tag{3.22}$$

from which it follows that

$$W_{\text{dist}} = Z^N \exp\left(\frac{U}{kT}\right) \tag{7.1}$$

This is related, in accordance with Eq. (6.4), to the thermodynamic probability for quantum (indistinguishable) particles by

$$W_{\text{dist}} = N!\, W_{\text{indist}} \tag{7.2}$$

Therefore,

$$W_{\text{indist}} = \frac{Z^N}{N!} \exp\left(\frac{U}{kT}\right) \tag{7.3}$$

With the help of Stirling's approximation we can write Eq. (7.3) as

$$W_{\text{indist}} = \left(\frac{Ze}{N}\right)^N \exp\left(\frac{U}{kT}\right) \tag{7.3a}$$

where e is the natural number 2.71828. . . . Thus

$$S_{\text{indist}} = k \ln W_{\text{indist}} = kN \ln\left(\frac{Z}{N}\right) + kN + \frac{U}{T} \tag{7.4}$$

which reduces to[1]

$$S_{\text{indist}} = S_{\text{dist}} - k \ln N! \tag{7.5}$$

Thus the fact that real atoms are not distinguishable results in a loss of entropy equal to $k \ln N!$. There are fewer microstates from among which the given macrostate can be chosen. Hence that macrostate is less improbable than it was when we could tell the atoms apart. Within the parlance of an information-theory formula-

[1]The missing algebra in this paragraph is left as an exercise (Problem 7.1).

tion[2] of statistical thermodynamics we would say that our a priori "uncertainty" of the state of the gas has been reduced.

PARTITION FUNCTION

The partition function depends upon two basic types of energy storage, translational and internal:

$$\epsilon = \epsilon_t + \epsilon_{\mathrm{int}}$$

Thus, in accordance with Sec. 5.3,

$$Z = Z_t Z_{\mathrm{int}} \tag{5.26a}$$

The internal partition function Z_{int}, can be obtained when g_{int} and ϵ_{int} have been evaluated either from quantum mechanics or by certain experimental methods. The energy ϵ_{int} might represent rotational, vibrational, electronic, and nuclear contributions, as well as energies of rotation–vibration interaction, vibration–electronic interaction, and so on. Thus

$$\epsilon_{\mathrm{int}} = \epsilon_r + \epsilon_v + \epsilon_e + \epsilon_n + \epsilon_{r,v} + \epsilon_{v,e} + \cdots \tag{7.6}$$

and the corresponding partition function, Z_{int}, is

$$Z_{\mathrm{int}} = Z_r Z_v Z_e Z_n Z_{r,v} Z_{v,e} \cdots \tag{7.7}$$

These energy modes need not all exist in a given particle. If they do exist, they need not all be excited, and might be negligible. The nuclear rotation and electronic modes of energy storage are usually available to a monatomic particle, for example, but they are only excitable at extremely high temperatures. They are therefore often neglected.

THERMODYNAMIC PROPERTIES

Any two thermodynamic properties of a single-component, single-phase system can be used to specify the state; the rest depend upon these two. We arbitrarily take T and V as independent, because the partition function is usually expressed in terms of these variables. We also introduce N as a variable by considering the *size* of the system to be variable. The fundamental equation that uses T, V, and N as independent variables is the equation for the

[2]Entropy can be viewed as a measure of our uncertainty as to the microstate of a thermodynamic system. For an advanced discussion, see E. T. Jaynes, "Information Theory and Statistical Mechanics," *Phys. Rev.*, **106,** 620; and **108,** 171 (1957).

Helmholtz function,[3] $F = F(T, V, N)$. If we combine $F \equiv U - TS$ with Eq. (7.4) we find that

$$F_{\text{indist}} = -kTN \ln\left(\frac{Ze}{N}\right) \tag{7.8}$$

This is almost identical with Eq. (3.24),

$$F_{\text{dist}} = -kTN \ln Z \tag{3.24}$$

The only difference lies in the additional term $-kTN \ln (e/N)$, which is equal to $kT \ln N!$ for large N.

Many of the properties that are derivable from Eq. (7.8) or (3.24) will not be altered by this term. For example, the pressure for indistinguishable particles

$$p = -\left.\frac{\partial F}{\partial V}\right|_{T,N} = kTN \left.\frac{\partial \ln (Ze/N)}{\partial V}\right|_{T,N}$$

or

$$p = kTN \left.\frac{\partial \ln Z}{\partial V}\right|_{T,N} \tag{3.25}$$

is identical to the result we developed for distinguishable particles. Likewise,

$$U = F + TS = F - T\left.\frac{\partial F}{\partial T}\right|_{V,N}$$

from which we again obtain the same result that we obtained for distinguishable particles

$$U = kT^2N \left.\frac{\partial \ln (Ze/N)}{\partial T}\right|_{V,N} = kT^2N \left.\frac{\partial \ln Z}{\partial T}\right|_{V,N} \tag{3.23}$$

The entropy and Helmholtz functions [Eqs. (3.22) and (3.24)] depend directly upon $\ln Z$ rather than upon its derivatives. They therefore retain the influence of the $\ln (e/N)$ contribution and are examples of properties that differ by an additive constant for distinguishable and indistinguishable properties.

It is often convenient to talk about the contributions of the various individual molecular energies to the extensive properties of molecules. Thus

$$\ln\left(\frac{Ze}{N}\right) = \ln\left(\frac{Z_t e}{N}\right) + \ln Z_r + \ln Z_v + \cdots \tag{7.9}$$

[3] We should recall that the Massieu function $-F/T = -F/T(1/T, V, N)$ is the Legendre transform of the entropy function $S(U, V, N)$. This is discussed in the context of Eq. (1.48). We should also remember that the symbol N is used indiscriminately to designate the number of particles in microscopic equations and the number of moles in macroscopic equations. The distinction is not really important because the two quantities differ only by a conversion factor N_A.

where we have assigned the indistinguishability contribution, $\ln(e/N)$, to the ever-present translational term. This is the proper place to put it because translation is required before indistinguishability can have meaning. If particles did not translate they would have *identity* by virtue of location and the $\ln(e/N)$ term would not arise. Any extensive property, X ($X = F$, or S, or U, or etc.), can now be calculated as the sum of a series of additive terms:

$$X = X_t + X_r + X_v + \cdots \tag{7.10}$$

each corresponding to one term of Eq. (7.9).

7.2 IDEAL MONATOMIC GASES

PARTITION FUNCTION

The monatomic "molecule" stores no energy in vibration. The rotational contribution *is* the nuclear-spin contribution in this case, and it has a characteristic temperature on the order of 10^{10} °K. This means that no rotation of the atom (or of the nucleus) can be excited beyond its zero ground state at reasonable temperatures. Thus the only internal mode of energy storage is electronic, and

$$\epsilon = \epsilon_t + \epsilon_{\text{int}} = \epsilon_t + \epsilon_e \tag{7.11}$$

and

$$g = g_t g_e \tag{7.12}$$

Then, with the aid of Eq. (5.28), we can write the partition function for an ideal monatomic gas as

$$Z = \frac{V}{h^3}(2\pi mkT)^{3/2} \sum_e g_e \exp\left(-\frac{\epsilon_e}{kT}\right) \tag{7.13}$$

The energy levels of electrons moving about the nucleus are fairly widely spaced; they have characteristic temperatures on the order of 10^4 °K. The matter of determining these levels becomes fairly complicated, owing to the three-dimensionality of motion.[4] The ground energy ϵ_0 and its degeneracy g_0 are zero and unity, respectively, so the summation term in Eq. (7.13) can be deleted at moderate temperatures.

[4]The interested reader will find values of electron and nuclear energy levels in C. E. Moore, *Atomic Energy Levels*, Natl. Bur. Std. Circ. 467, vols. I–III, 1949, 1952, 1958.

THERMODYNAMIC PROPERTIES

Combining Eqs. (7.8) and (7.13) yields the fundamental equation for an ideal monatomic gas:

$$F = -kTN \ln\left[\frac{Ve}{h^3N}(2\pi mkT)^{3/2}\sum_e g_e \exp\left(-\frac{\epsilon_e}{kT}\right)\right] \tag{7.14}$$

The computation of pressure will reveal no new information, but it will give us confidence in the use of this fundamental equation. Substituting F into Eq. (3.25) gives

$$p = \frac{NkT}{V} = nkT \tag{2.36}$$

If V is expressed on a molar basis, this becomes

$$p = \frac{R^0T}{v} \tag{2.5}$$

which is more comforting than edifying.

The substitution of Eq. (7.13) in Eq. (3.23) also leads to previously known results if we take the electrons to be in their ground state,

$$U = \tfrac{3}{2}R^0T \tag{1.27}$$

and

$$c_v = \left.\frac{\partial U}{\partial T}\right|_{V,N} = \tfrac{3}{2}R^0 \qquad \text{[recall (1.27) and (3.31)]}$$

Finally, the entropy is

$$S = -\left.\frac{\partial F}{\partial T}\right|_{V,N}$$

or

$$S = Nk\left\{\ln\left[\left(\frac{2\pi mkT}{h^2}\right)^{3/2}\frac{V}{N}\right] + \frac{5}{2} + \ln\sum_e g_e \exp\left(-\frac{\epsilon_e}{kT}\right) + \left[\frac{\sum_e g_e(\epsilon_e/kT)\exp(-\epsilon_e/kT)}{\sum_e g_e \exp(-\epsilon_e/kT)}\right]\right\} \tag{7.15}$$

SACKUR-TETRODE EQUATION

At most temperatures of practical interest, the electron terms vanish and we have

$$S = kN\left\{\ln\left[\left(\frac{2m\pi kT}{h^2}\right)^{3/2}\frac{kT}{p}\right] + \frac{5}{2}\right\} \tag{7.15a}$$

Equation (7.15a) is an exceedingly valuable result. It is called the Sackur–Tetrode equation and it expresses the absolute entropy of an ideal monatomic gas. It is the fundamental equation for such a gas and it can easily be written in the form of Eq. (1.23).

EXAMPLE 7.1 To show how difficult it can be to evaluate absolute entropy with the methods of classical thermodynamics and how helpful the Sackur–Tetrode equation can be, we consider an example. Suppose we wanted to know the entropy of mercury vapor at its normal boiling point, 630°K.

The classical expression for the absolute entropy of an ideal gas is

$$S = \int_0^T c_p{}^0(T)\, d(\ln T) - R^0 \ln\left(\frac{p}{p_{\text{ref}}}\right) + \sum \frac{\text{latent heats}}{T_{\text{phase changes}}} \qquad (7.16)$$

The integration is done at low pressure from absolute zero of temperature up to the temperature of interest — 630°K in this case. The term $c_p{}^0(T)$ designates the specific heat measured at a low-enough pressure — usually 1 atm — that it will not change with further reductions of pressure. It is the specific heat that is most commonly reported for a substance.

The path to be used here is as follows. Graphically integrate experimental values of $c_p{}^0\, d(\ln T)$ for the solid, up to the melting point, 234.2°K. Then graphically integrate $c_p{}^0\, d(\ln T)$ for the liquid, from 234.2 to 630°K. In the present example we can choose p_{ref} as 1 atm so that the second term in Eq. (7.16) is zero. The latent heats of both melting and vaporization must be obtained for use in the last term. After considerable labor, which can only be undertaken if data are available, Eq. (7.16) yields

$$S_{\text{exptl}} = (14.3 + 6.3)\,\frac{\text{cal}}{\text{g mole-°K}} - 0 + \left(\frac{558}{234.2} + \frac{14180}{630}\right)\frac{\text{cal}}{\text{g mole-°K}}$$

$$= 20.6 + (2.4 + 22.5) = 45.5\,\frac{\text{cal}}{\text{g mole-°K}}$$

The Sackur–Tetrode equation [Eq. (7.15a)], on the other hand, predicts that

$$S = 45.7\,\frac{\text{cal}}{\text{g mole-°K}}$$

The two computations differ by only about 0.4 percent. This discrepancy should probably be credited to experimental error.

It is interesting that the simple statistical considerations upon which the Sackur–Tetrode equation is based circumvent the difficulties introduced by phase transitions and the thermodynamics of solid bodies. The reason is that the entropy is an absolute characterization of the disorder of a given ensemble of particles. In classical thermodynamics, entropy (although it is a state property) is often expressed in terms of reversible heat transfer during a path from a point of known entropy. Equation (7.16) is such an expression. Statistical mechanics shows how the absolute entropy at a point can be found without reference to any such fictitious paths.

7.3 IDEAL DIATOMIC GASES

POTENTIAL-ENERGY FUNCTION FOR A DIATOMIC MOLECULE

The evaluation of the thermodynamic properties of diatomic molecules is made considerably more complicated by the existence of vibrational and rotational modes of energy storage. Of course, we have already noted that these modes are unexcited at low temperatures and diatomic behavior eventually reduces to monatomic behavior. But, by the same token, the behavior of diatomic molecules becomes a good deal more complicated than we indicated in chapter 3, as the temperature becomes very high.

The internal potential energy of a diatomic molecule provides the means for understanding much of the complication inherent in the diatomic molecule. The electrons of the component atoms interact and exert forces that hold the atoms in place. The intermolecular force F can be written in accordance with Eq. (5.8) as the negative gradient of the potential energy of electronic bonding $U(\xi)$,

$$F = -\frac{d}{d\xi} U(\xi) \tag{7.17}$$

where ξ is the distance between atomic nuclei.

Figure 7.1 shows a typical potential-energy function for a particular electronic state. For spacings smaller than an equilibrium value ξ_e the electronic forces repel and for larger spacings they attract. The zero-energy state is taken as the completely dissociated condition in which the atoms have been moved far from one another. The energy at equilibrium is accordingly the negative "dissociation energy." The actual energy that can actually be realized in dissociation will not be D (see Fig. 7.1) but a slightly smaller value, D_0. This is the case because an oscillator cannot exist in a zero-energy state. In accordance with Eq. (5.10b), there must be a ground level of vibrational energy ϵ_{v_0} equal to $\frac{1}{2} h\nu$, so D_0 is equal to $(D - \epsilon_{v_0})$.

Fig. 7.1 Potential-energy function for a diatomic molecule in a particular electronic state.

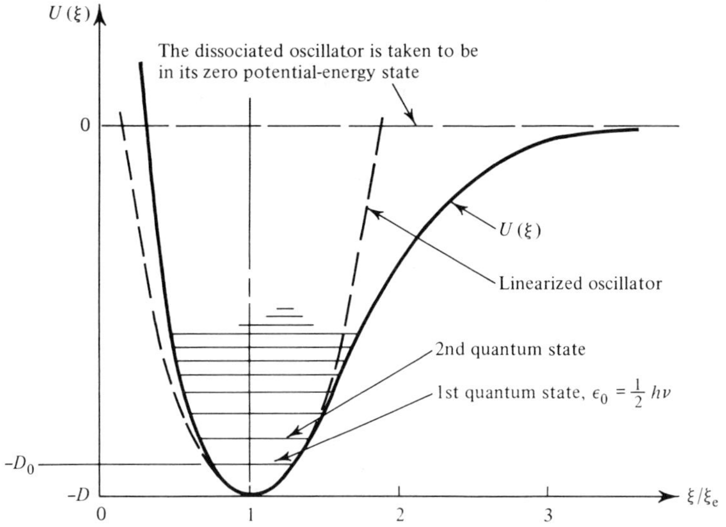

We should observe in this context that ξ_e is not a real position and it is reachable only by extrapolation. The molecule is never really in equilibrium because ϵ_{v_0} is always finite.

The analytical description of the diatomic molecule is rendered complicated by a variety of factors at higher temperatures.

1. The potential-energy function changes when the electronic states are excited. Fortunately, the electrons remain in their ground state up to pretty high temperatures; but when they are excited, ξ_e increases and the dissociation energy changes.

2. At the higher levels of vibrational excitation, the vibration becomes increasingly nonlinear, or *anharmonic*.

3. As the amplitude of vibration increases, the moment of inertia of the molecule varies during rotation, thus serving to couple rotation with vibration.

4. Centrifugal stretching also serves to influence vibration when the rotational mode becomes highly excited.

5. When vibration has been excited to a sufficiently high level, the atoms will dissociate and absorb the dissociation energy.

The internal energy of the diatomic molecule [recall Eq. (7.6)] then takes the form

$$\begin{aligned}\epsilon_{\text{int}} &= \epsilon_r + \epsilon_e + (\epsilon_v - \epsilon_{v_0}) + \epsilon_{v_0} - D + \cdots \\ &= \epsilon_r + \epsilon_e + (\epsilon_v - \epsilon_{v_0}) - D_0 + \cdots \end{aligned} \tag{7.18}$$

where we have not written down any of the coupling terms. Equation (7.7) also becomes

$$Z_{\text{int}} = Z_r Z_e Z_v Z_D \cdots \tag{7.19}$$

where

$$Z_v = \sum_{vj} \exp\left(-\frac{\epsilon_{vj} - \epsilon_{v0}}{kT}\right) \tag{7.20}$$

This can be put in nicer form since

$$\exp\left(-\frac{\epsilon_{v0}}{kT}\right) = \exp\left(-\frac{h\nu}{2kT}\right) = \exp\left(-\frac{\Theta_v}{2T}\right)$$

Thus, just as Eq. (5.30) summed to Eq. (5.31), Eq. (7.20) sums to

$$Z_v = \frac{\exp(\Theta_v/2T)}{2\sinh(\Theta_v/2T)}$$

so Eq. (5.31) should be replaced with the corrected expression,

$$Z_v = \frac{1}{1 - \exp(-\Theta_v/T)} \tag{7.21}$$

Finally, the partition function for dissociation is

$$Z_D = \exp\left(\frac{D_0}{kT}\right) \tag{7.22}$$

Some typical values of D_0 for diatomic molecules have been included in Table 7.3, Sec. 7.4.

RIGID-ROTOR HARMONIC-OSCILLATOR APPROXIMATION

To a first approximation valid for moderate temperature, the interactions among various internal energy modes can be neglected, and the molecule can simultaneously be characterized both as a rigid rotor and as a harmonic oscillator. Then the fundamental equation, Eq. (7.8), can be written in terms of a partition function,

$$Z = Z_t Z_{\text{int}} = Z_t Z_r Z_v Z_e Z_D \tag{7.19a}$$

The partition functions Z_t, Z_r, and Z_v were all given in chapter 5 and eq. (7.21), and the following results are obtained easily (Problem 7.5) for the electronic and dissociation contributions to properties:

$$\left.\begin{aligned} Z_e &= g_{e0} \\ U_e &= H_e = c_{v_e} = c_{p_e} = 0 \\ S_e &= R^0 \ln g_{e0} \\ F_e &= G_e = -R^0 T \ln g_{e0} \end{aligned}\right\} \text{for moderate temperatures} \tag{7.23}$$

and

$$\begin{aligned} U_D &= H_D = F_D = -N_A D_0 \\ S_D &= c_{pD} = c_{vD} = 0 \end{aligned} \tag{7.24}$$

Now we are in a position to explain, in more precise terms, the variation of c_v for diatomic gases that was exhibited in Fig. 3.3. This behavior can be computed directly for temperatures below the point of molecular dissociation or electronic excitation. We simply combine Eqs. (5.28), (5.33), and (5.31) with (7.19a) in

$$c_v = R^0 \frac{\partial}{\partial T}\left(T^2 \frac{\partial \ln Z}{\partial T}\right) \tag{7.25}$$

or

$$\frac{c_v}{R^0} = \frac{\partial}{\partial T}\left[T^2\left(\frac{\partial \ln Z_t}{\partial T} + \frac{\partial \ln Z_r}{\partial T} + \frac{\partial \ln Z_v}{\partial T}\right)\right] \tag{7.26}$$

Two provisions that were made in the statement of the kinetic hypothesis for ideal gases in Sec. 2.1 can profitably be re-emphasized at this point: (1) The gas was assumed dilute, but particles were *not shrunk to point masses*. This permitted us to talk about structured ideal gases. (2) Energy dissipation was disallowed in collisions, but we *did not go further to claim that collisions were elastic*. Internal modes of energy were thus permitted to exchange energy. The result was consistent with Boyle's and Charles's laws, but it did not require the additional assumption (which some authors make) that the definition of a macroscopic ideal gas must include constant specific heats.

Before leaving the diatomic molecule we should look at two complications in somewhat greater detail.

ON THE DISTINGUISHABILITY OF ATOMS IN THE HOMONUCLEAR DIATOMIC MOLECULE

A homonuclear diatomic molecule such as H_2, N_2, and O_2 is made up of two identical atoms. We restrict our detailed discussion of the indistinguishability of the atoms in a homonuclear diatomic molecule to hydrogen. This case, in which the ends are simple protons, presents a simple-enough structure to expose the difficulties of indistinguishability without undue complication.

The question as to whether or not the atoms are distinguishable in a hydrogen molecule is more difficult to answer than it might first seem. We must first write a proper wave function for the rotor—one that includes attributes of the ends. For the hydrogen molecule (neglecting electronic states and considering translation and vibration separately), we have in accordance with Sec. 5.3,

$$\psi_{\text{rigid rotor}} = \psi_r \psi_{\text{spin}} \tag{7.27}$$

The wave function is the product of the wave functions for the independent rotational and spin modes of energy storage.

The wave function ψ_{spin} is for nuclear spin. For the hydrogen molecule it describes the combined spin energy of both hydrogen nuclei. We have ignored the contribution of nuclear spin to the partition function because spin cannot be excited at physically plausible temperatures. The importance of nuclear spin lies in that it can give identity to the ends of the molecule, even in the ground state. If the wave function ψ_{spin} is *symmetric* in the sense that the nuclei could be interchanged without altering it, then the nuclei can be called *indistinguishable*. If ψ_{spin} changes when the particles are interchanged, then we say that it is *antisymmetric*. The nuclei can be called *distinguishable* in this case.

Consider first the *symmetric case*. The spin degeneracy of the pair of nuclei $g_n^{(1,2)}$ must be written in terms of the spin degeneracy of the individual nuclei g_n. There are several ways in which the ground-level degeneracy $g_0^{(1,2)}$ of the two indistinguishable particles can be selected from among g_0 distinguishable occupancy modes. Using Problem 4 in Sec. 3.2, we obtain

$$g_0^{(1,2)} = \frac{(g_0 + 2 - 1)!}{(g_0 - 1)!\,2!} = \frac{g_0(g_0 + 1)}{2} \tag{7.28}$$

Consider next the *antisymmetric case*. In this case $g_0^{(1,2)}$ is the number of ways that two distinguishable particles can be selected from among the g_0 occupancy modes. Using Problem 3 in Sec. 3.2 we find

$$g_0^{(1,2)} = \frac{g_0!}{(g_0 - 2)!\,2!} = \frac{g_0(g_0 - 1)}{2} \tag{7.29}$$

The total degeneracy of both cases will also be of interest subsequently [compare with Eq. (5.20)],

$$g_0^{(1,2)} = \frac{g_0(g_0 + 1)}{2} + \frac{g_0(g_0 - 1)}{2} = g_0^2 \tag{7.30}$$

The symmetry of the rotational wave function must also be questioned. If the nuclei are interchanged, it will be as though θ and ϕ were changed to $\pi - \theta$ and $\phi + \pi$. This substitution in the expression for ψ_r, Eq. (5.15), leads to

$$\psi_r(\pi - \theta, \phi + \pi) = (-1)^l \psi_r(\theta, \phi) \tag{7.31}$$

where l is the rotational quantum number. Thus ψ_r is symmetrical for even l and antisymmetrical for odd l.

The Pauli exclusion principle provides the basis for selecting the correct wave function for the molecule. As we noted in Sec. 6.1, it precludes a quantum state from being occupied by more than one proton unless the wave function for the system of protons is antisymmetric. Now with reference to Eq. (7.27) we note that the wave function for a rigid rotor must be antisymmetric:

$$\psi_{\text{rigid rotor}} = \psi_r \psi_{\text{spin}} = \text{antisymmetric} \tag{7.32}$$

so that two possibilities can take place,

$$\underbrace{\psi_{\text{rigid rotor}}}_{\text{antisym}} = \underbrace{\psi_r}_{\text{antisym}} \underbrace{\psi_{\text{spin}}}_{\text{sym}} \qquad \text{case 1} \tag{7.32a}$$

$$\underbrace{\psi_{\text{rigid rotor}}}_{\text{antisym}} = \underbrace{\psi_r}_{\text{sym}} \underbrace{\psi_{\text{spin}}}_{\text{antisym}} \qquad \text{case 2} \tag{7.32b}$$

Consider first, case 1: If ψ_{spin} is symmetric, the rotational wave function must be antisymmetrical. This requires that

$$l = \text{odd} \qquad \text{and} \qquad g_0^{(1,2)} = \frac{g_0(g_0+1)}{2} \tag{7.33}$$

Conversely, case 2 is defined by ψ_{spin} = antisymmetric, ψ_r = symmetric, and

$$l = \text{even} \qquad \text{and} \qquad g_0^{(1,2)} = \frac{g_0(g_0-1)}{2} \tag{7.34}$$

These two cases actually represent two identifiable types of hydrogen gas. The first is *ortho-hydrogen*, and its rotational partition function is[5]

$$Z_r^{\text{ortho}} = \frac{g_0(g_0+1)}{2} \sum_{l\ \text{odd}} (2l+1) \exp\left[-l(l+1)\frac{\Theta_r}{T}\right] \tag{7.35}$$

For hydrogen nuclei the lowest value of the quantum number for nuclear spin turns out to be $\frac{1}{2}$, and the ground-level degeneracy is given by $g_0 = 2(\frac{1}{2}) + 1 = 2$ [note the similarity to $g_l = (2l+1)$ in the rigid-rotor case in Sec. 5.1]. Thus

$$Z_r^{\text{ortho}} = 3 \sum_{l\ \text{odd}} (2l+1) \exp\left[-l(l+1)\frac{\Theta_r}{T}\right] \tag{7.35a}$$

For *para-hydrogen* it follows that

$$Z_r^{\text{para}} = \frac{g_0(g_0-1)}{2} \sum_{l\ \text{even}} (2l+1) \exp\left[-l(l+1)\frac{\Theta_r}{T}\right] \tag{7.36}$$

or

$$Z_r^{\text{para}} = \sum_{l\ \text{even}} (2l+1) \exp\left[-l(l+1)\frac{\Theta_r}{T}\right] \tag{7.36a}$$

[5]The reader will note that we treat H_2 molecules as boltzons even though the Pauli exclusion principle applies. The reason is that $e^{\alpha} \gg 1$ for almost any configuration of gaseous hydrogen [recall Eq. (6.12*a*)], and Fermi–Dirac statistics can be replaced with Boltzmann statistics.

If we ignore the distinction and write Z_r only for the sum, the result will be

$$Z_r = 3 \sum_{l \text{ odd}} (2l+1) \exp\left[-l(l+1)\frac{\Theta_r}{T}\right] + \sum_{l \text{ even}} (2l+1) \exp\left[-l(l+1)\frac{\Theta_r}{T}\right] \tag{7.37}$$

from which we can immediately conclude that there will be three ortho-H_2 molecules for each para-H_2 molecule. The specific heats computed from Eqs. (7.35a), (7.36a), and (7.37) are shown in Fig. 7.2.

The separation of the two kinds of hydrogen can be accomplished in the laboratory. Once it has been done, the 3 : 1 equilibrium will not reestablish itself in the separated component for months; for practical purposes the two gases retain their independence when they are mixed together, although they are not separate chemical species.

At higher temperatures, where the terms of the summations are significant up to high l's, we have

$$2 \sum_{\substack{l \text{ odd or} \\ l \text{ even}}} (2l+1) \exp\left[-l(l+1)\frac{\Theta_r}{T}\right] \simeq \sum_{\text{all } l} (2l+1) \exp\left[-l(l+1)\frac{\Theta_r}{T}\right]$$

It follows that as $\Theta_r/T \to$ small, Eq. (7.37) gives

$$Z_r \to 2 \sum_{\text{all } l} (2l+1) \exp\left[-l(l+1)\frac{\Theta_r}{T}\right] \tag{7.38}$$

Fig. 7.2 Specific heats of diatomic hydrogen.

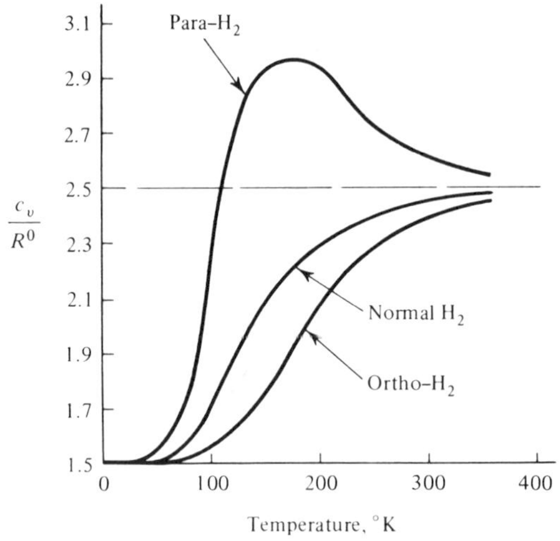

We saw in Eq. (5.34) that this series sums to T/Θ_r. The factor 2 is actually $g_n^2/2$ (Problem 7.10) evaluated for $g_n = g_0 = 2$. The constant g_n^2 arises from nuclear spin, which is ordinarily neglected in the classical evaluation of partition functions and the 2 in the denominator is called the *symmetry number* σ because it arises from molecular symmetry. Thus

$$Z_r = \frac{g_n^2}{\sigma}\frac{T}{\Theta_r} \tag{7.38a}$$

Although the present discussion of hydrogen is only illustrative of homonuclear molecule behavior and is not generally applicable, Eq. (7.38a) is a general result for $T \gg \Theta_r$. The use of the symmetry number is exploited much further in Sec. 7.4, for polyatomic molecules.

ANHARMONICITY, ROTATION-VIBRATION COUPLING, AND CENTRIFUGAL STRETCHING

Let us return to a consideration of the potential-energy function for a diatomic molecule as shown in Fig. 7.1. As more energy is stored in the vibrating system, the displacement from the equilibrium position $\xi/\xi_e = 1$ increases. The allowable quantum energy levels, designated by horizontal lines, are initially spaced equally, and as the nonlinear behavior becomes more pronounced, they begin to bunch more closely together.

Although the function $U(\xi)$ is usually complicated, it can be represented with a Taylor series as

$$U(\xi) = U(\xi_e) + \left.\frac{dU}{d\xi}\right|_{\xi_e}(\xi - \xi_e) + \frac{1}{2!}\left.\frac{d^2U}{d\xi^2}\right|_{\xi_e}(\xi - \xi_e)^2 + \cdots \tag{7.39}$$

But $U(\xi_e) = -D$ and $(dU/d\xi)_{\xi_e} = 0$. Thus $U(\xi)$ can be approximated as

$$U(\xi) \simeq -D + \frac{(\xi - \xi_e)^2}{2}\left.\frac{d^2U}{d\xi^2}\right|_{\xi_e} \tag{7.40}$$

whence $(d^2U/d\xi^2)_{\xi_e}$ is the "spring constant" for the equivalent linear vibration. Figure 7.1 also includes the approximation (7.40) to $U(\xi)$.

Now let us consider the vibrational energy of the molecule as it is influenced by anharmonicity. First, we return to the result for harmonic vibration, Eq. (5.10b),

$$\epsilon_v = h\nu(j + \tfrac{1}{2}) \tag{5.10c}$$

This expression can be extended approximately for use with the unevenly spaced energy levels in Fig. 7.1 if it is written as a power series in the quantum number, $(j + \frac{1}{2})$. Thus, comparing Eq. (5.10c) with Eq. (5.10a), we have

$$\frac{\epsilon_v}{hc_l} = k_e[(j + \tfrac{1}{2}) - x_e(j + \tfrac{1}{2})^2 + \cdots] \tag{7.41}$$

where k_e is the wave number of the vibration at the equilibrium point and x_e is a constant, much less than unity. k_e (or $1/\lambda_e$) is based on the wavelength λ_e that light oscillating at a frequency ν would have. It is equal to $[(d^2U/d\xi^2)_{\xi_e}/m]^{1/2}/2\pi c_l$.

The rotational energy also deviates from its rigid-rotor value,

$$\frac{\epsilon_r}{hc_l} = \frac{\hbar}{4\pi c_l I} l(l+1) = B_e l(l+1) \tag{5.14a}$$

by virtue of centrifugal stretching. This can also be corrected approximately to

$$\frac{\epsilon_r}{hc_l} = B_v l(l+1) - D_v l^2(l+1)^2 \tag{7.42}$$

where the constants B and D have been corrected from their equilibrium values B_e and D_e by

$$\begin{aligned} B_v &= B_e - \alpha(j + \tfrac{1}{2}) + \cdots \\ D_v &= D_e + \beta(j + \tfrac{1}{2}) + \cdots \end{aligned} \tag{7.43}$$

where $D_e = (4B_e^3/k_e^2)$

It is now possible to write a single expression for the coupled rotational–vibrational energy of an anharmonic oscillator,

$$\frac{\epsilon_{r,v}}{hc_l} \simeq k_e[(j + \tfrac{1}{2}) - x_e(j + \tfrac{1}{2})^2] + B_e l(l+1) - D_e l^2(l+1)^2$$
$$- \alpha(j + \tfrac{1}{2})l(l+1) \tag{7.44}$$

This expression depends upon correction constants x_e, D_e, and α, as well as k_e. The constant β has been dropped because it almost always proves to be far smaller than D_e. These constants are developed either by experiments or by semirational analysis based upon the Schrödinger equation. Some typical values are given in Table 7.1.

The partition function can be corrected to take account of the combined energy $\epsilon_{r,v}$. The process is involved[6] and here we can only summarize what is done. By the definition of the partition function we have

$$Z_r Z_v = \sum_{j=0} \sum_{l=0} g_l \exp\left(-\frac{\epsilon_{r,v} - \epsilon_{v0}}{kT}\right) \tag{7.45}$$

[6]See K. S. Pitzer, *Quantum Chemistry*, Prentice-Hall, Inc., Englewood Cliffs, N.J., 1953, app. 14; or N. Davidson, *Statistical Mechanics*, McGraw-Hill, Inc., New York, 1962, sec. 8.8.

TABLE 7.1 Rotational and Vibrational Constants for Typical Diatomic Molecules

Substance	k_e, cm^{-1}	B_e, cm^{-1}	D_e, cm^{-1}	$k_e x_e$, cm^{-1}	α, cm^{-1}
HB_r	2649.65	8.4665	3.53×10^{-4}	45.57	0.2325
Br_2	323.22	0.08092	2.035×10^{-8}	1.070	0.000275
CN	2068.705	1.8991	6.401×10^{-6}	13.144	0.01735
CO	2169.52	1.9302	6.43×10^{-6}	13.453	0.01746
Cl_2	561.1	0.2408	1.8×10^{-7}	4.0	0.0017
H_2	4405.3	60.848	0.04644	125.325	3.0664
I_2	214.52	0.037364	0.455×10^{-8}	0.6133	0.0001206
NO	1903.60	1.7042	5×10^{-6}	13.97	0.0178
N_2	2357.55	1.99825	6×10^{-6}	14.059	0.0179
O_2	1580.246	1.445	4.956×10^{-6}	12.071	0.0158

where $\epsilon_{v_0} = \frac{1}{2} h\nu$, the ground-level vibrational energy and $g_l = 2l + 1$ the rotational degeneracy. After some judicious approximations and considerable algebra, the double summation can be approximated as

$$Z_r Z_v = \left(\frac{1}{1 - (\alpha/2B_e)} \frac{T}{\sigma \Theta_r} \left\{ 1 - \exp\left[-\frac{(1 - 2x_e)\Theta_v}{T} \right] \right\}^{-1} \right) Z_{\text{corr}} \quad (7.45a)$$

where the term Z_{corr} is the following complicated function of the constants and temperature:

$$\begin{aligned} Z_{\text{corr}} \equiv 1 &+ \frac{1 - \alpha/2B_e}{3} \frac{\Theta_r}{T} + \frac{2D_e}{B_e - \alpha/2} \frac{T}{\Theta_r} \\ &+ \frac{1}{(B_e/\alpha - 1/2)\{\exp[(1 - 2x_e)\Theta_v/T] - 1\}} \\ &+ \frac{2x_e \Theta_v/T}{\{\exp[(1 - 2x_e)\Theta_v/T] - 1\}^2} \end{aligned} \quad (7.46)$$

Additional methods exist for simplifying the derivatives of Z so as to facilitate evaluations of the thermodynamic functions based on Z. Although the preceding methods account for the major nonlinear effects, they only do so in a range fairly near to equilibrium. It should be clear at this point that a great deal can be done toward evaluating properties using methods that, although semiempirical, remain reasonably close to the quantum-mechanical realities of the gases.

A knowledge of characteristic temperatures has proved to be of importance in what we have done and it will continue to be important in subsequent applications. Table 7.2 lists values of Θ_r and Θ_v for a number of diatomic gases. The characteristic temperature of translation has not been tabulated because it varies with the size of the container that holds the gas. For a given gas of molecular mass m, Θ_t can be expressed as

$$\Theta_t = \frac{h^2/8k}{mV^{2/3}} = \frac{3.9 \times 10^{-38}}{mV^{2/3}\text{ g-cm}^2}\ ^\circ\text{K}$$

TABLE 7.2 Some Characteristic Temperatures of Diatomic Gases

Substance	Θ_r, °K	Θ_v, °K
H_2	87.5	6320
HD	65.8	5500
D_2	43.8	4490
HCl	15.2	4330
HBr	12.2	3820
N_2	2.89	3390
CO	2.78	3120
NO	2.45	2745
O_2	2.08	2278
Cl_2	0.351	814
Br_2	0.116	465
I_2	0.0537	309

EXAMPLE 7.2 Evaluate the entropy of nitrogen at 2000°K and 1 atm.

First let us evaluate the corrections as they appear in Eqs. (7.45a) and (7.46). Using data from Tables 7.1 and 7.2 we obtain

$$1 - \frac{\alpha}{2B_e} = 0.996 \qquad 1 - 2x_e = 0.988 \qquad Z_{\text{corr}} = 1.022$$

Clearly the corrections amount to very little in this case. Nevertheless we carry them along. The full partition function is then written in accordance with Eqs. (7.23), (5.28), and (7.45a),

$$Z = Z_e Z_t (Z_r Z_v) = g_e \left(\frac{\pi T}{4\Theta_t}\right)^{3/2} \frac{1}{0.996} \frac{T}{\sigma \Theta_r} \frac{1.022}{1 - \exp(-0.988\ \Theta_v/T)}$$

Now $\Theta_t = 3.98 \times 10^{-38} \times 6.02 \times 10^{23}/28V^{2/3} = 0.856 \times 10^{-15}/V^{2/3}$ and V is given by the ideal-gas law as

$$V = \frac{R^0 T}{p} = 82.05 \frac{\text{atm-cm}^3}{^\circ\text{K-g mole}} 2000^\circ\text{K}/1 \text{ atm}$$
$$= 1.641 \times 10^5 \text{ cm}^3/\text{g mole}$$

Then $g_e = 1$ and $\sigma = 2$ so we obtain for 1 mole of nitrogen,

$$Z = \left[\frac{2000\pi}{4 \times 0.856 \times 10^{-15}}\right]^{3/2} \frac{1.641 \times 10^5}{1.992} \frac{2000(1.022)}{2.89[1 - \exp(-.988\Theta_v/T)]}$$
$$= 1.782 \times 10^{35}$$

The molar entropy is given by Eq. (3.22), corrected in accordance with Sec. 7.1

$$S = kN_A \ln\left(\frac{Ze}{N_A}\right) + \frac{U}{T}$$

where, from Eq. (3.22),

$$\frac{U}{T} = kTN_A \frac{\partial \ln (Ze/N_A)}{\partial T}\bigg|_{V,N} = kN_A\left[\frac{5}{2} + \frac{0.988\Theta_v}{T}\frac{\exp(-.988\Theta_v/T)}{[1-\exp(-.988\Theta_v/T)]}\right]$$

$$= 8.31 \times 10^7\left(2.5 + 1.675\frac{0.187}{0.813}\right) = 2.40 \times 10^8 \text{ ergs/g mole-°K}$$

Thus

$$S = 8.31 \times 10^7 \ln\left[\frac{(1.782)(2.718)}{6.02} \times 10^{35-23}\right] + 2.40 \times 10^8 = 2.525 \times 10^9 \text{ ergs/g mole-°K}$$

or

$$S = 60.2 \frac{\text{cal}}{\text{g mole-°K}}$$

7.4 IDEAL POLYATOMIC GASES

Polyatomic molecules introduce complications that exceed even those we have been discussing. We can make reasonable progress, however, if we initially ignore such problems as anharmonicity, coupling, and electronic excitation, which arise at higher temperatures.

Consider, for example, the problem of describing the vibration of the n atoms in a molecule. A total of $3n$ coordinates is needed to specify the motions of each atom. But our knowledge of the translatory motion of the center of mass, in effect, removes three of these degrees of freedom. By the same token, two more coordinates are given to a description of rotation if the atoms lie on a straight line, and three coordinates are lost if they do not. Consequently, there are $(3n - 5)$ or $(3n - 6)$ modes of vibrational freedom for molecules composed of two or more atoms, depending upon the arrangement. These values correspond with $2(3n - 5)$ and $6(n - 2)$ modes of vibrational-energy storage.

Let us now write partition functions for the polyatomic molecule. The translational partition function is still given by Eq. (5.28) and offers no new problems in more complicated cases. The vibrational and rotational partition functions will, however.[7]

[7] N. Davidson, reference 6, chap. 11, discusses in some detail these and other problems related to polyatomic molecules.

ROTATIONAL PARTITION FUNCTION

The moments of inertia of polyatomic molecules are generally quite large. Accordingly, their characteristic temperatures of rotation, $\Theta_r = \hbar^2/2Ik$, are generally quite low, and we can usually treat them as boltzons. If we leave the nuclear degeneracy out of the rotational partition function, we can write for molecules whose atoms lie on a straight line,

$$Z_r = \frac{T}{\sigma\Theta_r} \tag{5.34b}$$

This can be viewed as the product of two rotational components. Indeed, it would have arisen as such a product had we obtained it using the modified phase integral, Eq. (5.37a),

$$Z_r = \frac{1}{\sigma}\left(\frac{2I_x kT}{\hbar^2}\right)^{1/2}\left(\frac{2I_y kT}{\hbar^2}\right)^{1/2} \tag{7.47}$$

The moments of inertia about the x and y axes, I_x and I_y, would be the same in this case.

The symmetry number, σ, is a device that can be introduced when $T \gg \Theta_r$ to account for the indistinguishability of certain rotational modes. In $CO_2(O{=}C{=}O)$, for example, σ would be two just as it was for a diatomic molecule. For CO it would be only unity. The symmetry number for a three-dimensional tetrahedronal molecule serves to illustrate how a larger value of σ might easily come about.

Figure 7.3 shows a tetrahedronal methane (CH_4) molecule. The four H atoms can be arranged in 4!, or 24, ways. But only 12 of these can be obtained by simple rotation: Every 120 degrees of rotation about *A-A* puts the molecule into a new orientation that is indistinguishable from the last. The same is true about *B-B*, *C-C*, and *D-D*. There are thus 4×3, or 12, "rearrangements" attainable by simple rotations. The remaining $(24 - 12)$, or 12, rearrangements are ones that are not included in the degeneracy $(2l + 1)$. They are not meaningful in this consideration. Twelve is the *symmetry number* for this case and it could have been obtained as a high-temperature limit for very complicated quantum behavior, just as $\sigma = 2$ was obtained in Eq. (7.38) for H_2. Some symmetry numbers are included in Table 7.3 for common molecules.

The partition function for rotation of a "three-dimensional molecule" — one with rotation about three axes — can be obtained using the phase integral. The result is

$$Z_r = \frac{\pi^{1/2}}{\sigma}\left(\frac{2I_x kT}{\hbar^2}\right)^{1/2}\left(\frac{2I_y kT}{\hbar^2}\right)^{1/2}\left(\frac{2I_z kT}{\hbar^2}\right)^{1/2} \tag{7.47a}$$

Fig. 7.3 Axes of symmetry for a methane molecule.

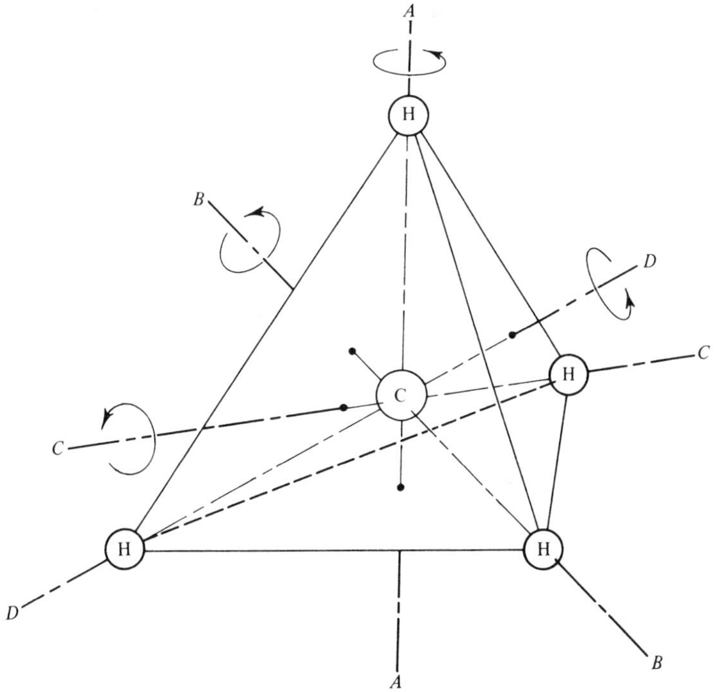

TABLE 7.3 Short List of Some Molecular Properties

Substance	Bond	Spacing, Å	k_e, cm^{-1}	D_0, eV	k_{e_1}, cm^{-1}	k_{e_2}, cm^{-1}	k_{e_3}, cm^{-1}	k_{e_4}, cm^{-1}	Bond angle, 2θ degrees	Symmetry number σ
O_2	O═O	1.207	1580	5.12	—	—	—	—	—	2
N_2	N≡N	1.098	2358	9.76	—	—	—	—	—	2
CO	C═O	1.128	2170	11.09	—	—	—	—	—	1
H_2	H—H	0.742	4405	4.48	—	—	—	—	—	2
CO_2	C═O	1.926	—	—	1343	667	667	2349	O═C═O, 180	2
H_2O	O—H	0.958	—	—	3657	1595	3756	—	H—O—H, 104.5	2
N_2O	O═N	1.184	—	—	1277	589	589	2224	N—NO, 180═	1
	N—N	1.128	—	—						
CH_4	C—H	1.091	—	—	2916 once, 1534 twice, 3019 thrice, 1306 thrice				H—C—H, 109.5	12

where we have incorporated the symmetry number. Clearly the usefulness of this expression to us will depend upon our ability to evaluate the molecular properties I_x, I_y, and I_z. This evaluation is not difficult once we know the masses of the component atoms and their spacings within the molecule.

Figure 7.4 shows how the conventional dynamical formulas for moments of inertia apply to three typical molecules. The derivation of these formulas is elementary and has been left as an exercise

Fig. 7.4 Some moments of inertia for typical molecules.

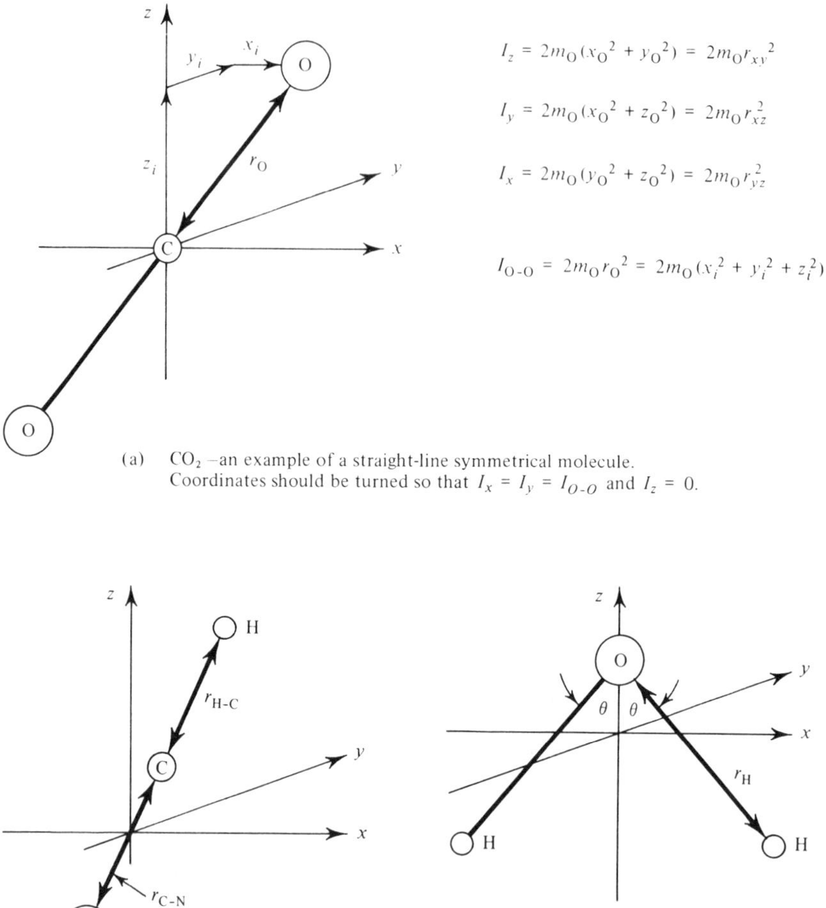

$$I_z = 2m_O(x_O^2 + y_O^2) = 2m_O r_{xy}^2$$

$$I_y = 2m_O(x_O^2 + z_O^2) = 2m_O r_{xz}^2$$

$$I_x = 2m_O(y_O^2 + z_O^2) = 2m_O r_{yz}^2$$

$$I_{O\text{-}O} = 2m_O r_O^2 = 2m_O(x_i^2 + y_i^2 + z_i^2)$$

(a) CO_2—an example of a straight-line symmetrical molecule. Coordinates should be turned so that $I_x = I_y = I_{O\text{-}O}$ and $I_z = 0$.

$$I_{H\text{-}N} = m_C r_{C\text{-}N}^2 + m_H(r_{C\text{-}N} + r_{H\text{-}C})^2 - \frac{[m_C r_{C\text{-}N} + m_H(r_{C\text{-}N} + r_{H\text{-}C})]^2}{m_N + m_C + m_H}$$

(b) HCN—an example of a straight-line asymmetrical molecule.

$$I_x = 2m_H r_H^2 \cos^2\theta \left(1 - \frac{2m_H}{m_O + 2m_H}\right)$$

$$I_z = 2m_H r_H^2 \sin^2\theta$$

$$I_y = I_z + I_x$$

(c) H_2O—an example of a planar molecule.

(Problem 7.14). Table 7.3 lists some typical values for atomic spacings along with certain other molecular constants of interest.

VIBRATIONAL PARTITION FUNCTION

The vibrational partition function for a polyatomic gas is, as we noted earlier in this section, based upon $3n - 5$ or $3n - 6$ modes of vibration, where n is the number of atoms in the molecule. Thus

$$\epsilon_v = \sum_{j=1}^{\substack{3n-5 \\ \text{or } 3n-6}} (v_j + \tfrac{1}{2})h\nu_j$$

and the partition function is given in accordance with Eqs. (5.26) and (5.31) as

$$Z_v = \prod_{j=1}^{\substack{3n-5 \\ \text{or } 3n-6}} \frac{1}{2 \sinh(\Theta_{v_j}/2T)} \tag{7.48}$$

where

$$\Theta_{v_j} \equiv \frac{h\nu_j}{k}$$

In the limit of $\Theta_{v_j}/T \ll 1$,

$$Z_v \rightarrow \prod_{j=1}^{\substack{3n-5 \\ \text{or } 3n-6}} \frac{T}{\Theta_{v_j}} \tag{7.48a}$$

Figure 7.5 illustrates how two molecules of the structure XY_2 might vibrate. If the molecule lies along a straight line, as does CO_2, then there should be $3 \times 3 - 5 = 4$ modes of vibration. If it is planar, as is H_2O, then there should be only $3 \times 3 - 6 = 3$ modes. Figure 7.5(a) shows the four modes for CO_2 and Fig. 7.5(b) shows the three modes appropriate to H_2O. Each of these modes has a wave number of vibration k_{e_j}. Numerical values of the k_{e_j} for typical molecules are also given in Table 7.2.

7.5 IDEAL-GAS MIXTURES

THERMODYNAMIC PROBABILITY AND THE FUNDAMENTAL EQUATION

The defining assumptions for an ideal gas include the stipulation that intermolecular forces act only during collisions — that the molecules are independent at any given instant. Thus we would expect that if gas molecules of several different species are mixed together, the macroscopic behavior of each species will be independent of the other species.

Fig. 7.5 Modes of vibration for two molecules of the type XY_2.

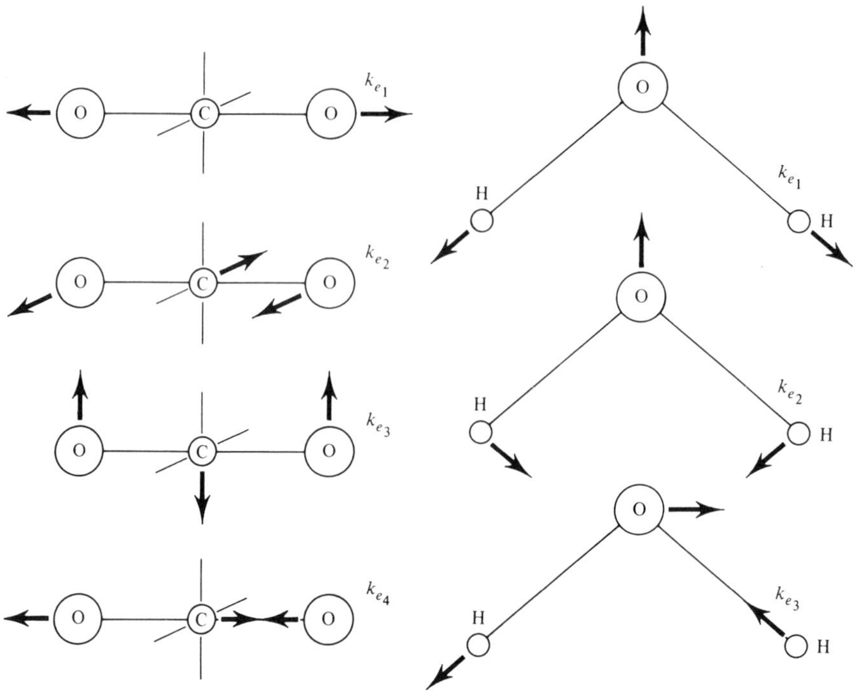

(a) the four modes of vibration of a straight-line molecule, CO_2, with three atoms

(b) the three modes of vibration of a planar molecule, H_2O, with three atoms

If, for example, the independent variables T and V are specified for the entire mixture, the remaining properties should be sums of the properties of the independent species. For r components, $1, 2, \ldots, \alpha, \ldots, r$,

$$p = \sum_{\alpha=1}^{r} p_\alpha \qquad U = \sum_{\alpha=1}^{r} U_\alpha \qquad S = \sum_{\alpha=1}^{r} S_\alpha \qquad \text{etc.}$$

so that such important macroscopic relations as Dalton's law appear as obvious results. In particular, we can write the Helmholtz function as

$$F(T, V, N_1, \ldots, N_r) = \sum_{\alpha=1}^{r} F_\alpha(T, V, N_\alpha) \tag{7.49}$$

This is the fundamental equation for the mixture. It tells us that if we can obtain the fundamental equations for each of the components, then the fundamental equation for the mixture can be written immediately.

Now let us look more carefully at the statistical basis for Eq. (7.49). In developing the statistical-mechanical description of a gas mixture we introduce a formal method that can later be extended to treat more complex systems, such as chemically reacting, and nonideal, gases.

The thermodynamic probability of a mixture of *independent* components is the product of the thermodynamic probabilities of each of these components. We view the gases as being nondegenerate and indistinguishable; thus they should be describable by Eq. (6.8), and W for the mixture is

$$W = \prod_{i=0}^{\infty} \frac{g_{1_i}{}^{N_{1i}}}{N_{1_i}!} \prod_{i=0}^{\infty} \frac{g_{2_i}{}^{N_{2i}}}{N_{2_i}!} \cdots \prod_{i=0}^{\infty} \frac{g_{r_i}{}^{N_{ri}}}{N_{r_i}!} = \prod_{\alpha}^{r} \prod_{i}^{\infty} \frac{g_{\alpha_i}{}^{N_{\alpha i}}}{N_{\alpha_i}!} \tag{7.50}$$

The equations of constraint for this mixture are somewhat more complicated than those for a single component. There are a total of r constraints expressing the conservation of particles,

$$\sum_{i}^{\infty} N_{\alpha_i} = N_\alpha \qquad (\alpha = 1, 2, \ldots, r) \tag{7.51}$$

and a single statement of the conservation of total energy,

$$\sum_{\alpha}^{r} \sum_{i}^{\infty} N_{\alpha_i} \epsilon_{\alpha_i} = U \tag{7.52}$$

The equilibrium distribution of molecules is obtained from Eqs. (7.50), (7.51), and (7.52) using the method of Lagrangian multipliers (Problem 7.15). The result, as we might anticipate, is

$$N_{\alpha_i} = \frac{N_\alpha g_{\alpha i}}{Z_\alpha} \exp\left(-\frac{\epsilon_{\alpha i}}{kT}\right) \tag{7.53}$$

where

$$Z_\alpha = \sum_{i}^{\infty} g_{\alpha i} \exp\left(-\frac{\epsilon_{\alpha i}}{kT}\right) \tag{7.54}$$

The entropy in this case is

$$S \equiv k \ln W = \frac{U}{T} + k \ln\left[\prod_{\alpha}^{r} \left(Z_\alpha \frac{e}{N_\alpha}\right)^{N_\alpha}\right] \tag{7.55}$$

or

$$S = \frac{U}{T} + k \sum_{\alpha}^{r} N_\alpha \ln\left(Z_\alpha \frac{e}{N_\alpha}\right) \tag{7.55a}$$

We can now write the fundamental equation for the mixture in the Helmholtz-function form,

$$F \equiv U - TS = -kT \ln\left[\prod_{\alpha}^{r} \left(Z_\alpha \frac{e}{N_\alpha}\right)^{N_\alpha}\right] \tag{7.56}$$

To verify that this is a proper fundamental equation we should note that, in accordance with Eq. (5.26),

$$Z_\alpha = [Z(T, V)]_{\alpha_t}[Z(T)]_{\alpha_r}[Z(T)]_{\alpha_v}\cdots = [Z(T, V)]_\alpha \tag{7.57}$$

so that the Helmholtz function depends upon the correct variables for it to be the fundamental parameter:

$$F = F(T, V, N_1, \ldots, N_r) \tag{7.58}$$

Equation (7.56) thus should contain all the thermodynamic information for the mixture. It can be written explicitly if the partition function for the component gases can be written.

THERMODYNAMIC FORMULAS

By differentiating the fundamental parameter F with respect to its respective independent variables we can obtain the various equations of state for the mixture. For example, the first derivatives of F as given by Eq. (7.56), with respect to T, V, and N_α give equations for S, p, and μ_α. Of course, $S = -(\partial F/\partial T)$ is simply Eq. (7.55a), which we have already written directly.

The pressure p is

$$p \equiv -\frac{\partial F}{\partial V} = kT\frac{\partial}{\partial V}\left[\ln\left(\prod_\alpha^r Z_\alpha^{N_\alpha}\right)\right] \tag{7.59}$$

Since V is only contained in the translational partition function, and since the translational partition function is always present in Eq. (7.57), we can write, from Eq. (5.28),

$$[Z(T, V)]_{\alpha_t} = V\left(\frac{2\pi m_\alpha kT}{h^2}\right)^{3/2} \tag{5.28a}$$

Using this in Eq. (7.59) we obtain

$$p = \frac{kT\left(\sum_\alpha^r N_\alpha\right)}{V} = \sum_\alpha^r \frac{N_\alpha kT}{V} \tag{7.60}$$

But the well-known "partial pressure" p_α of macroscopic thermodynamics is

$$p_\alpha = \frac{N_\alpha kT}{V} \tag{7.61}$$

so that we have obtained Dalton's law in a formal way,

$$p = \sum_\alpha^r p_\alpha \tag{7.60a}$$

Furthermore, since $N = \sum_{\alpha}^{r} N_\alpha = pV/kT$, we obtain another familiar macroscopic relation,

$$\frac{p_\alpha}{p} = \frac{N_\alpha}{N} = \frac{N_\alpha/N_A}{N/N_A} = x_\alpha \tag{7.62}$$

where N_A is Avogadro's number, N_α/N_A is the number of moles of the α component, and x_α is the mole fraction.

The chemical potential μ_α of the components has particular importance in regard to our study of mixtures, because it is the driving potential for mass transfer. If we write F on a molar basis, then

$$\mu_\alpha = \frac{\partial F}{\partial(N_\alpha/N_A)}$$

and Eq. (7.56) gives

$$\mu_\alpha = -kTN_A \ln (Z_\alpha/N_\alpha) \tag{7.63}$$

Equation (7.63) can easily be put in the form in which it conventionally appears in textbooks on macroscopic thermodynamics[8] (Problem 7.16),

$$\mu_\alpha = R^0T[\phi_\alpha(T) + \ln p_\alpha] \tag{7.64}$$

where

$$\phi_\alpha(T) \equiv \ln \left[\frac{V}{kT[Z(T, V)]_\alpha}\right] \tag{7.65}$$

The function ϕ_α, thus defined, clearly *is* dependent only upon T, because the only V dependence in Z_α is a linear factor [recall Eqs. (7.57) and (5.28a)]. Since ϕ_α is a familiar macroscopic parameter, Eq. (7.65) provides a helpful microscopic basis for it.

ENTROPY OF MIXING AND GIBBS'S PARADOX

The adiabatic mixing of ideal gases is a process that lends itself to a description by statistical mechanics. Suppose that a box of volume, V, contains r gases separated into r subcompartments with $N_1, N_2, \ldots, N_r$ moles in each one as shown in Fig. 7.6(a).

The dividing partitions are lifted and mixing occurs. If the initial temperature is uniformly T, and the partitions are originally placed so the pressures are the same in each one, then N/V must equal N_α/V_α and

$$U_{\text{initial}} = \sum_{\alpha}^{r} U_\alpha = U_{\text{final}}$$

[8]See, for example, M. W. Zemansky, *Heat and Thermodynamics*, 5th ed., McGraw-Hill, Inc., New York, 1968, sec. 16–11.

Fig. 7.6 Mixing of r ideal-gas components.

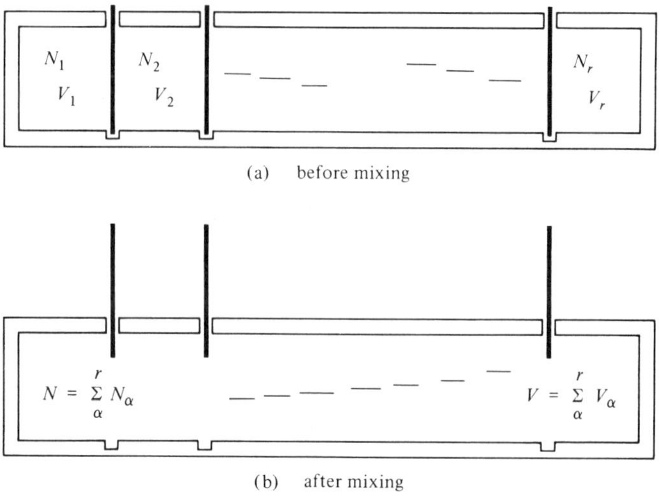

For an ideal gas, $\sum_\alpha^r U_\alpha$ is a function of the temperature T alone. It follows that the final temperature must equal the initial temperature. It likewise follows from

$$p_{\text{final}} = kT\frac{N}{V} = kT\frac{N_\alpha}{V_\alpha}$$

that

$$p_{\text{final}} = p_1 = p_2 = \cdots = p_\alpha = \cdots = p_r$$

This process is a peculiar one in that it progresses spontaneously and irreversibly without degrading the potentials, T and p. What it *does* degrade is the *chemical* potential μ_α. The potentials T, p, and μ_α can be broadly viewed as potentials for heat, work, and mass transfer. No heat or work is transferred in the process, reversibly or otherwise. But the potential for mass transfer μ_α is degraded by the spontaneous flow of mass.

The increase of entropy can be computed for the mixing process with the aid of Eq. (7.55a). First we evaluate the entropy of the unmixed gases

$$S_{\text{initial}} = \frac{U}{T} + k\sum_\alpha^r N_\alpha\left[\ln\left(\frac{Z_\alpha}{N_\alpha}\right) + 1\right] \tag{7.55b}$$

Substituting Eq. (7.63) in this, we obtain

$$S_{\text{initial}} = \frac{U}{T} - \frac{1}{T}\sum_\alpha^r\left(\frac{N_\alpha}{N_A}\mu_{\alpha\text{initial}} - kTN_\alpha\right)$$

and

$$S_{\mathrm{final}} = \frac{U}{T} - \frac{1}{T}\sum_{\alpha}^{r}\left(\frac{N_\alpha}{N_A}\mu_{\alpha\mathrm{final}} - kTN_\alpha\right)$$

Thus the entropy difference is expressible in terms of the changes in the chemical potentials of the components,

$$\Delta S = \frac{1}{T}\sum_{\alpha}^{r}\frac{N_\alpha}{N_A}(\mu_{\alpha\mathrm{initial}} - \mu_{\alpha\mathrm{final}})$$

Finally, we use Eq. (7.64) for μ_α to reduce this to the conventional entropy-of-mixing expression,

$$\Delta S_{\mathrm{mixing}} = -k\sum_{\alpha}^{r} N_\alpha \ln\left(\frac{p_\alpha}{p}\right) \tag{7.66}$$

where p_α now designates a partial pressure and is much less than the pressure in a cell in Fig. 7.6(a). Alternatively,

$$\Delta S_{\mathrm{mixing}} = -\frac{N}{N_{\mathrm{A}}} R^0 \sum_{\alpha}^{r} x_\alpha \ln x_\alpha \tag{7.66a}$$

This is known as the entropy of mixing and it arises from the pressure portion of the chemical potential as given by Eq. (7.64). It is strictly a result of mixing up molecules and in no way relates to $\int \delta Q/T$. We have, in fact, already emphasized this aspect of its character in the context of Eq. (3.18). Let us briefly reconsider that discussion.

Gibbs was perplexed by the following difficulty with regard to the entropy of mixing. If we mix 1 mole of oxygen with 1 mole of helium we obtain Eq. (3.18), $\Delta S = R^0 \ln 4$. If we mix 1 mole of oxygen with 1 mole of nitrogen (which differs relatively little from it) we still find $\Delta S = R^0 \ln 4$. If we mix 1 mole each of two very similar isotopes of oxygen, $\Delta S = R^0 \ln 4$, just as it did before. Finally, let us mix 2 moles of pure oxygen together. Suddenly $\Delta S = 0$! It appears that ΔS somehow varies discontinuously with the degree of similarity of the gases.

This seemingly unreasonable result was called Gibbs's paradox. It was resolved by P. W. Bridgman, but only after the operational viewpoint of quantum mechanics had been established. He pointed out that the mixing of molecules was an operation that was physically meaningful only if means could be devised for determining whether or not an interchange of particles really had occurred. Otherwise, mixing must be viewed as a fictitious operation and cannot be said to have occurred. The uncertainty relation ultimately provides the test that determines whether or not two particles are different enough to be mixed.

EXAMPLE 7.3 Suppose that you have to separate oxygen from nitrogen in atmospheric air. What is the minimum work that would be required to do this in, say, a steady isothermal isobaric process?

We know from the study of macroscopic thermodynamics that the minimum work is given by

$$Wk_{\min} = \Delta g = \Delta h - T\,\Delta s$$

Since the air is approximately ideal, $\Delta h = 0$ in an isothermal process. The change of entropy is given by Eq. (7.66a) with the sign changed and $N = N_A/M_{\text{air}}$ to put the calculation on a unit-weight basis. Then

$$Wk_{\min} \simeq TR_{\text{air}}(x_{N_2} \ln x_{N_2} + x_{O_2} \ln x_{O_2})$$

or

$$Wk_{\min} = 298°\text{K}(2.87 \times 10^6 \text{ ergs/g-°K})[0.79 \ln 0.79 + 0.21 \ln 0.21]$$

where we have used the known fractions of N_2 and O_2 in air. The result is

$$Wk_{\min} = -4.39 \times 10^6 \text{ ergs/g} = -147 \text{ ft-lb}_f/\text{lb}_m$$

where the minus sign indicates that work must be done *on* the system to achieve separation.

This of course is only a limiting value. Real separation processes always involve considerable inefficiency and require much more energy.

7.6 CHEMICAL EQUILIBRIUM OF REACTING MIXTURES

EQUILIBRIUM DISTRIBUTIONS

Our statistical-mechanical description of ideal-gas mixtures must be altered slightly to accommodate the process of chemical reaction. The counting of microstates is not altered by the fact that a reaction is taking place, but the equations of constraint are changed in two important ways.

In the first place, the number of molecules in each component of gas is not conserved. Conservation only applies to the total number of constituent particles participating in the reaction. Furthermore, care must be taken in writing the energy constraint. A difficulty arises because the ground-state energy of a molecule generally differs from the sum of the ground-state energies of the atoms that react to form the molecule.

Consider, as an example that clarifies these changes in the statistical method, the equilibrium of the simple reaction

$$X + Y \rightleftharpoons XY \tag{7.67}$$

The particles X and Y might either be atoms or molecules. The thermodynamic probability for these three components is given by Eq. (7.50),

$$W = \prod_{\alpha} \prod_{i}^{\infty} \frac{g_{\alpha_i}{}^{N_{\alpha i}}}{N_{\alpha_i}!} \qquad \alpha = X, Y, XY \tag{7.68}$$

The conservation-of-mass constraints are

$$\sum_{i=0}^{\infty} N_{X_i} + \sum_{i=0}^{\infty} N_{XY_i} = N_X + N_{XY} = N_{X_t} \tag{7.69}$$

and

$$\sum_{i=0}^{\infty} N_{Y_i} + \sum_{i=0}^{\infty} N_{XY_i} = N_Y + N_{XY} = N_{Y_t} \tag{7.70}$$

where N_{X_t} and N_{Y_t} are the total number of X and Y particles — both combined and free.

The conservation-of-energy constraint is

$$\sum_{i=0}^{\infty} N_{X_i}\epsilon_{X_i} + \sum_{i=0}^{\infty} N_{Y_i}\epsilon_{Y_i} + \sum_{i=0}^{\infty} N_{XY_i}(\epsilon_{XY_i} - D_0) = U \tag{7.71}$$

where D_0 is the "energy of reaction." The energy of reaction, also called the energy of dissociation or formation (recall Sec. 7.3), is the energy required to dissociate the XY molecule. A negative D_0 would be the energy required to form the molecule. The energy of reaction is measured above the ground state of the particle at rest — that is, above the ground state of the particle at the absolute zero of temperature. Figure 7.7 illustrates the relation among the energies of X, Y, and XY for the case of an endothermic (energy consuming) formation of the molecule.

The method of Lagrangian multipliers is again used to maximize the thermodynamic probability subject to the constraints. The result (Problem 7.19) is[9]

$$N_{X_i} = g_{X_i} \exp\left(-\alpha_X - \frac{\epsilon_{X_i}}{kT}\right) \tag{7.72}$$

$$N_{Y_i} = g_{Y_i} \exp\left(-\alpha_Y - \frac{\epsilon_{Y_i}}{kT}\right) \tag{7.73}$$

and

$$N_{XY_i} = g_{XY_i} \exp\left[-\alpha_X - \alpha_Y - \frac{(\epsilon_{XY_i} - D_0)}{kT}\right] \tag{7.74}$$

[9]Care should be taken not to confuse the subscript α with the Lagrangian multiplier α_α.

Fig. 7.7 Relative energies for particles taking part in the endothermic reaction $X + Y \rightleftharpoons XY$. In this case D_0 is positive and is called the energy of dissociation.

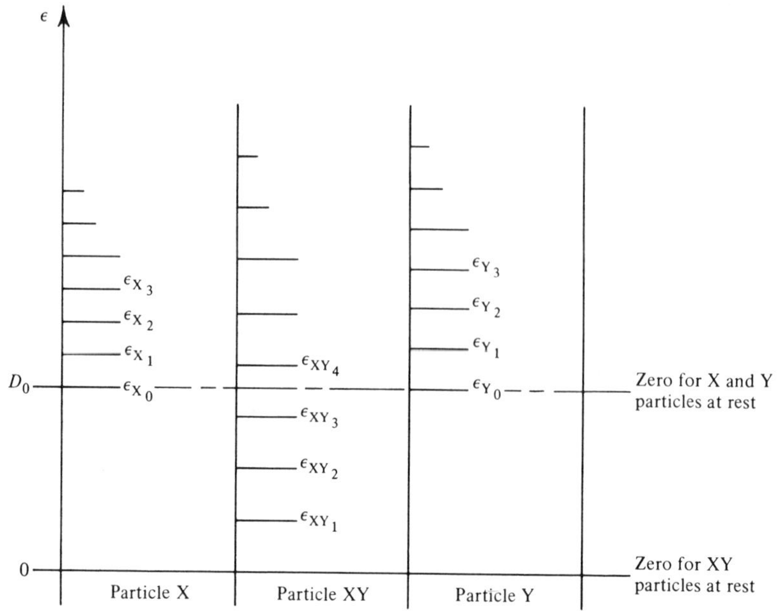

These can be written in the form

$$\frac{N_{X_i}}{N_X} = \frac{g_{X_i}}{Z_X} \exp\left(-\frac{\epsilon_{X_i}}{kT}\right) \qquad \text{etc.}$$

where

$$Z_X = \sum_{i=0}^{\infty} g_{X_i} \exp\left(-\frac{\epsilon_{X_i}}{kT}\right) \qquad \text{etc.} \tag{7.75}$$

We found in chapter 6 that

$$e^{-\alpha} = e^{\mu/kT} \tag{7.76}$$

but in this case there are only two α multipliers, α_X and α_Y. Thus we have

$$N_{X_i} = e^{\mu_X/kT}\left[g_{X_i} \exp\left(-\frac{\epsilon_{X_i}}{kT}\right)\right] \tag{7.77}$$

$$N_{Y_i} = e^{\mu_Y/kT}\left[g_{Y_i} \exp\left(-\frac{\epsilon_{Y_i}}{kT}\right)\right] \tag{7.78}$$

$$N_{XY_i} = e^{(\mu_X+\mu_Y+D_0)/kT}\left[g_{XY_i} \exp\left(-\frac{\epsilon_{XY_i}}{kT}\right)\right] \tag{7.79}$$

Ostensibly, this is the conventional distribution for degenerate boltzons, applicable to each component. However, the equilibrium must be subject to an additional restraint. This is a restraint that restricts the proportion of X, Y, and XY particles, or the extent to which the reaction has progressed. We can obtain this restraint by rearranging Eqs. (7.77), (7.78), and (7.79) and introducing Eq. (7.75),

$$\left(\frac{N_{XY}}{N_X N_Y}\right) \times \left(\frac{Z_X Z_Y}{Z_{XY}}\right) = \exp\left(\frac{\mu_X + \mu_Y + D_0 - \mu_X - \mu_Y}{kT}\right)$$

$$= \exp\left(\frac{D_0}{kT}\right) \quad (7.80)$$

This is the law of mass action in the first form in which we see it. It fixes the equilibrium proportions of particles N_X, N_Y and N_{XY} subject to the constraints $N_X + N_{XY} = N_{X_t}$ and $N_Y + N_{XY} = N_{Y_t}$.

THERMODYNAMIC FORMULAS

Before we move to some of the very important ramifications of the law of mass action it will help to extract the basic thermodynamic relations from the preceding statistical-mechanical development. The entropy is given by the appropriate extension of Eq. (7.4),

$$S = \frac{U + N_{XY} D_0}{T} + k \ln\left[\prod_{\alpha}^{r} \left(\frac{Z_\alpha e}{N_\alpha}\right)^{N_\alpha}\right] \quad (7.81)$$

and the fundamental equation is

$$F(T, V, N_X, N_Y, N_{XY}) \equiv U - TS$$

$$= -kT \ln \prod_{\alpha}^{r} \left(\frac{Z_\alpha e}{N_\alpha}\right)^{N_\alpha} - N_{XY} D_0 \quad (7.82)$$

Some of the results that can be obtained from this include

$$p = -\frac{\partial F}{\partial V} = \frac{kT}{V}(N_X + N_Y + N_{XY}) \quad (7.83)$$

and the chemical potentials

$$\mu_X = \frac{\partial F}{\partial (N_X/N_A)} = -N_A kT\left[\ln\left(\frac{Z_X}{N_X}\right)\right] \quad (7.84)$$

$$\mu_Y = \frac{\partial F}{\partial (N_Y/N_A)} = -N_A kT\left[\ln\left(\frac{Z_Y}{N_Y}\right)\right] \quad (7.85)$$

and

$$\mu_{XY} = \frac{\partial F}{\partial (N_{XY}/N_A)} = -N_A kT\left[\ln\left(\frac{Z_{XY}}{N_{XY}}\right)\right] - N_A D_0 \quad (7.86)$$

With the help of these relations we can recast the law of mass action in the form

$$\mu_{XY} = \mu_X + \mu_Y \tag{7.87}$$

Thus we can anticipate that the equilibrium establishes itself at the point at which the potentials for mass transfer are balanced between the combined and uncombined components.

LAW OF MASS ACTION

The law of mass action provides the means for predicting the extent of reaction as a function of temperature. We wish to put the law into a form that is convenient for this purpose. If we first define a quantity called the *equilibrium constant K*

$$K(T, V) \equiv \frac{Z_{XY}}{Z_X Z_Y} \exp\left(\frac{D_0}{kT}\right) \tag{7.88}$$

then the law of mass action as stated in Eq. (7.80) becomes

$$\frac{N_{XY}}{N_X N_Y} = K(T, V) \tag{7.89}$$

With the inclusion of Eq. (7.61) this can be recast as

$$\frac{p_{XY}}{p_X p_Y} = \frac{V}{kT} K(T, V) = \frac{V}{kT} \frac{Z_{XY}}{Z_X Z_Y} \exp\left(\frac{D_0}{kT}\right) \tag{7.90}$$

But $Z \sim V$, so the terms in Eq. (7.90) are dependent only upon temperature. We therefore define another form of equilibrium constant, $K_p \equiv (V/kT)K$, which, with the help of Eq. (7.62), becomes

$$\frac{p_{XY}}{p_X p_Y} = \frac{x_{XY}}{x_X x_Y} \frac{1}{p} = K_p(T) \tag{7.91}$$

Once K_p is known, and it is plainly calculable, the equilibrium mixture can easily be computed using Eq. (7.91).

GENERAL CHEMICAL REACTIONS

The relations developed thus far in this section have been restricted to the reaction $X + Y \rightleftharpoons XY$ for purposes of simple illustration. All these results can readily be extended to treat a general chemical reaction,

$$0 \rightleftharpoons \sum_{\alpha=1}^{r} \nu_\alpha X_\alpha \tag{7.92}$$

The ν_α's are stoichiometric coefficients and they are positive for the products and negative for the reactants. The X_α's are the participating molecules. In the reaction $2H_2 + O_2 \rightleftharpoons 2H_2O$, for example, $X_1 = H_2$, $\nu_1 = -2$; $X_2 = O_2$, $\nu_2 = -1$; and $X_3 = H_2O$, $\nu_3 = 2$.

We leave as an exercise (Problem 7.21) the development of three forms of the law of mass action for a general reaction,

$$\sum_{\alpha=1}^{r} \nu_\alpha \mu_\alpha = 0 \tag{7.93}$$

$$K(T, V) \equiv \left(\prod_\alpha^r Z_\alpha^{\nu_\alpha}\right) \exp\left(\frac{D_0}{kT}\right) = \prod_\alpha^r N_\alpha^{\nu_\alpha} \tag{7.94}$$

or

$$K_p(T) \equiv \left(\prod_\alpha^r (Z_\alpha kT/V)^{\nu_\alpha}\right) \exp\left(\frac{D_0}{kT}\right) = \prod_\alpha^r p_\alpha^{\nu_\alpha} = \prod_\alpha^r (px_\alpha)^{\nu_\alpha} \tag{7.95}$$

The equilibrium constant, which has thus far been expressed in terms of microscopic parameters, is also expressible in terms of macroscopic variables.[10] Equation (7.93) provides an immediate expression for the law of mass action in terms of the macroscopic variable μ_α, the chemical potential. In pure gases we know that $\mu \equiv g$, the molar Gibbs function, is

$$g = R^0 T(\phi + \ln p) \tag{7.64a}$$

It is thus conventional to define a quantity Δg called the "free energy change of the reaction," which bears some similarity to Eq. (7.93):

$$\Delta g = \sum \nu_\alpha g_\alpha = R^0 T\left(\sum_\alpha^r \nu_\alpha \phi_\alpha + \ln p^{\Sigma \nu_\alpha}\right) \tag{7.96}$$

This is the change in the Gibbs "free energy" of the products (taken as pure components at p and T) as they go to an equilibrium proportion (where the products and reactants are all pure components at p and T).

If we now go to Eq. (7.95) with Eq. (7.64), we can show that

$$\ln K_p(T) = -\sum_\alpha^r \nu_\alpha \phi_\alpha \tag{7.97}$$

This demonstration is left as an exercise (Problem 7.22). It requires recognition of the fact that D_0/kT is counted into the ϕ for one of the reactants [recall Eq. (7.86)]. Then Eq. (7.96) becomes

$$\Delta g = -R^0 T \ln\left[\frac{K_p}{p^{\Sigma \nu_\alpha}}\right] \tag{7.98}$$

But K_p has the dimensions of $p^{\Sigma \nu_\alpha}$. Thus we can get rid of $p^{\Sigma \nu_\alpha}$ by referring Δg to a pressure of unity in the dimensions of pressure in

[10] These relations are conventionally developed in texts on classical thermodynamics. See, e.g., Zemansky, reference 8, chap. 17.

which K_p is expressed. Usually atmospheres are used as the dimensions of pressure and a standard Δg^0 is defined at 1 atm. Thus

$$\Delta g^0 = -R^0 T \ln K_p \tag{7.98a}$$

Equation (7.98a) is convenient because the free energy Δg^0 is often provided directly by experimental chemists, and K_p can be obtained from it.

Another thermodynamic measure of a chemical reaction is the "heat of reaction," or the enthalpy change, Δh, of reaction,

$$\Delta h \equiv \sum_{\alpha}^{r} \nu_\alpha h_\alpha \tag{7.99}$$

The specific enthalpy h_α of the components is easily shown (Problem 7.23) to be

$$h_\alpha = -R^0 T^2 \frac{\partial \phi_\alpha}{\partial T} \tag{7.100}$$

Thus

$$\Delta h = -\sum_{\alpha}^{r} \nu_\alpha R^0 T^2 \frac{\partial \phi_\alpha}{\partial T}$$

so, with the help of Eq. (7.97),

$$\Delta h = R^0 T^2 \frac{d}{dT} \ln K_p(T) \tag{7.101}$$

This important result is called the Van't Hoff equation. It displays a curious kind of physical behavior. It shows that if a reaction is endothermic, Δh is positive and the equilibrium constant rises with temperature. In so doing it tends to shift the equilibrium toward completion. However, completing the reaction *cools* the system and *opposes* the reaction.

By the same token, an exothermic reaction generates heat to raise the temperature, but that reduces K_p in this case and opposes completion of the reaction. Hence there is a tendency in this process for a kind of secondary opposition to arise against completion. This inherently stable behavior is a more general feature of physical systems than we might at first imagine. The stability of chemical reaction indicated by Van't Hoff's equation is predicted by the principle of Le Châtelier and Braun,[11] which addresses the stability of processes in a very general way.

[11] See, e.g., Callen's discussion of the principle of Le Châtelier and Braun, *Thermodynamics*, John Wiley & Sons, Inc., New York, 1960, chaps. 8 and 12.

Table 7.4 presents a few numerical values of Δh and K_p. These data are very often developed experimentally. However, we now have the means for predicting either quantity directly from microscopic parameters.

TABLE 7.4 Some Typical Values of Δh and $K_p(T)$

Reaction	Temperature, °K	Δh, cal/g mole	ln K_p(T), pressure (in atm)
$CO + \frac{1}{2} O_2 \rightleftharpoons CO_2$	298	—	103.768
	500	−67,754	57.622
	1000	−67,601	23.535
	2000	−66,247	6.641
	3000	—	0.972
	5000	—	−3.191
$H_2 + \frac{1}{2} O_2 \rightleftharpoons H_2O$	298	−57,798	92.214
	500	−58,286	52.697
	1000	−59,199	23.169
	2000	−60,296	8.151
	3000	—	3.092
	5000	—	−0.990
$\frac{1}{2} H_2 + OH \rightleftharpoons H_2O$	298	—	106.023
	500	—	60.167
	1000	—	25.973
	2000	—	8.695
	3000	—	2.913
	5000	—	−1.706
$CO + H_2O \rightleftharpoons CO_2 + H_2$	500	−9467.3	4.907
	1000	−8402.6	0.327
	2000	−5951.2	−1.542
$2O \rightleftharpoons O_2$	298	—	186.988
	500	—	105.643
	1000	—	45.163
	2000	—	14.635
	3000	—	4.370
	5000	—	−3.882
$2N \rightleftharpoons N_2$	298	—	367.493
	500	—	213.385
	1000	—	99.140
	2000	—	41.658
	3000	—	22.372
	5000	—	−6.820

DEGREE OF REACTION

The "degree of reaction" is a measure of the extent to which a reaction has progressed. We develop the concept with the aid of the following illustrative example.

EXAMPLE 7.4 Suppose that N_0 moles of H_2S participate in the reaction

$$H_2S + 2H_2O \rightleftharpoons 3H_2 + SO_2$$

Devise a means for expressing K_p in terms of the completeness of reaction.

For this reaction

$$K_p = \frac{x_{H_2}{}^3 x_{SO_2}}{x_{H_2S} x_{H_2O}{}^2} p^{1.0}$$

The "degree of reaction," ϵ, is defined in such a way that

$$\epsilon = \frac{N_0 - N_{H_2S}}{N_0}$$

$$\epsilon = \frac{2N_0 - N_{H_2O}}{2N_0}$$

$$\epsilon = \frac{N_{H_2}}{3N_0}$$

$$\epsilon = \frac{N_{SO_2}}{N_0}$$

It is, in other words, the percentage departure of the reaction from the state of pure reactants, and it can be expressed in terms of N_0 moles of *any* of the participating components. Our choice of H_2S was arbitrary in this example. The reaction is complete when $\epsilon = 1$.

We can then express the mole fractions in terms of ϵ:

$$x_{H_2S} = \frac{1 - \epsilon}{3 + \epsilon}$$

$$x_{H_2O} = \frac{2 - 2\epsilon}{3 + \epsilon}$$

$$x_{H_2} = \frac{3\epsilon}{3 + \epsilon}$$

$$x_{SO_2} = \frac{\epsilon}{3 + \epsilon}$$

so that

$$K_p = \frac{27\epsilon^4}{(3 + \epsilon)(1 - \epsilon)(2 - 2\epsilon)^2} p$$

This result illustrates how the equilibrium mixture can easily be calculated once the equilibrium constant is known.

7.7 IDEAL DISSOCIATING GASES

"Dissociation" is merely a name that is given to a particular class of chemical reactions: the dismantling of molecules that occurs when they are exposed to very high temperatures. Dissociation does not differ in principle from any other kind of chemical reaction. However, there is great practical interest in dissociation, as it occurs in conjunction with heat-transfer and fluid-flow problems for a variety of aerospace situations. The phenomenon of dissociation requires special handling if it is to remain tractable within the context of a complicated flow problem. In particular, we need to introduce approximate means for handling dissociation if such problems are to find usable solutions.

DISSOCIATING DIATOMIC GAS

In the simple dissociation process $X_2 \rightleftharpoons X + X$, ϵ is called the "degree of dissociation" and is defined in the same way as the degree of reaction:

$$\epsilon \equiv \frac{N_X}{N_{X_t}} = \frac{\text{number of dissociated X atoms}}{\text{total number of X atoms in the mixture}} \tag{7.102}$$

The equilibrium state of a dissociating diatomic gas is dictated by the constraints

$$\frac{N_{X_2}}{N_X{}^2} = \frac{Z_{X_2}}{(Z_X)^2} \exp\left(\frac{D_0}{kT}\right) \tag{7.103}$$

and

$$N_X + 2N_{X_2} = N_{X_t} \tag{7.104}$$

We should be aware that Z_{X_2} as we use it here is defined somewhat differently than it was in the context of Eqs. (7.19) and (7.21). At that point we absorbed exp (D_0/kT) into the definition of the partition function. Here (as in Sec. 7.6) it proves useful, for purposes of computation, to separate it. Equations (7.102), (7.103), and (7.104) then combine to form

$$\frac{1-\epsilon}{\epsilon^2} = 2N_{X_t} \frac{Z_{X_2}}{Z_X^2} \exp\left(\frac{D_0}{kT}\right) \tag{7.105}$$

The characteristic temperature of dissociation Θ_D can be defined just as any other characteristic temperature,

$$\Theta_D \equiv \frac{D_0}{k} \tag{7.106}$$

and the mass density of the mixture is

$$\rho = \frac{m_X N_{X_t}}{V} \qquad (7.107)$$

When these expressions are included in Eq. (7.105) it becomes

$$\frac{\epsilon^2}{1-\epsilon} = \frac{m_X}{2\rho V} \frac{Z_X^2}{Z_{X_2}} \exp\left(-\frac{\Theta_D}{T}\right) \qquad (7.108)$$

And since $Z_X{}^2/VZ_{X_2}$ is independent of V, we find that $\epsilon = \epsilon(\rho, T)$.

Equation (7.108) is the governing equation for the dissociation equilibrium of a symmetric diatomic gas. It is, of course, the law of mass action for dissociation. The characteristic temperature, which appears in an important way in this expression, proves to be a temperature at which dissociation is virtually complete. It is often quite high — on the order of 10^5 °K for N_2 and O_2 — but partial dissociation still plays an important role at temperatures of practical interest.

The fundamental equation for dissociation equilibrium is merely a special case of Eq. (7.82),

$$F(T, V, N_X, N_{X_2}) = -kT \ln\left[\left(\frac{Z_X e}{N_X}\right)^{N_X}\left(\frac{Z_{X_2} e}{N_{X_2}}\right)^{N_{X_2}}\right] - N_{X_2} D_0 \qquad (7.109)$$

We are particularly interested in two of the thermodynamic properties obtainable from the fundamental equation p and U or c_v. The pressure is

$$p = -\frac{\partial F}{\partial V} = \frac{kT}{V}(N_X + N_{X_2}) \qquad (7.110)$$

With the inclusion of Eqs. (7.102), (7.104), and (7.107) this becomes

$$p = (1+\epsilon)\rho R_{X_2} T \qquad (7.111)$$

where $R_{X_2} = R^0/M_{X_2}$ or $k/2m_X$. This is very similar to the ideal-gas law for pure X_2 with a "compressibility factor," Z. Unlike the compressibility factor, $1 + \epsilon(\rho, T)$ is bounded between 1.0 (no dissociation) and 2.0 (complete dissociation).

The internal energy U of a dissociating gas is

$$U \equiv F + TS = F - T\frac{\partial F}{\partial T} = -T^2\left[\frac{\partial}{\partial T}\left(\frac{F}{T}\right)\right]$$

from which we obtain

$$U = kT^2\left(N_X \frac{\partial}{\partial T}\ln Z_X + N_{X_2}\frac{\partial}{\partial T}\ln Z_{X_2}\right) - N_{X_2} D_0 \qquad (7.112)$$

Introducing ϵ, Θ_D, and ρ into this expression we can obtain the following expression for the specific internal energy:

$$u \equiv \frac{U}{\rho V} = R_{X_2}T^2\left[2\epsilon\frac{\partial}{\partial T}\ln Z_X + (1-\epsilon)\frac{\partial}{\partial T}\ln Z_{X_2}\right] - (1-\epsilon)R_{X_2}\Theta_D \quad (7.113)$$

But we have, in accordance with Eq. (3.23),

$$u = u(T) = RT^2\frac{\partial}{\partial T}\ln Z$$

so, because $R_X = 2R_{X_2}$, Eq. (7.113) takes the form

$$u = \epsilon u_X + (1-\epsilon)u_{X_2} - (1-\epsilon)R_{X_2}\Theta_D \quad (7.114)$$

Equation (7.114) is a curious result in that $\epsilon = \epsilon(\rho, T)$ introduces ρ dependence into u. Thus u depends upon both ρ and T, instead of just T alone as it would for a nonreacting gas.

With Eq. (7.114) the calculation of u would seem to become quite a simple matter because the internal energies of the components X and X_2 can usually be obtained easily. Consider, for example, the dissociation of an ideal gas that obeys the rigid-rotor harmonic-oscillator approximation and for which the electronic contribution is in the ground state. We are able to improve upon this representation shortly, but for this case

$$u_{X_2} = \tfrac{7}{2}R_{X_2}T \quad (7.115)$$

and

$$u_X = \tfrac{3}{2}R_XT \quad \text{or} \quad 3R_{X_2}T \quad (7.116)$$

so, from Eq. (7.114), we finally get

$$u(\rho, T) = R_{X_2}\left[\frac{7-\epsilon}{2}T - (1-\epsilon)\Theta_D\right] \quad (7.117)$$

Despite its appearance, Eq. (7.117) is actually fairly cumbersome for use in connection with fluid-flow problems because $\epsilon(\rho, T)$ has to be obtained from Eq. (7.108) or some comparably intractable relationship.

LIGHTHILL IDEAL DISSOCIATING GAS

We clearly need some means for dealing with $\epsilon = \epsilon(\rho, T)$. In 1957 Lighthill[12] proposed an approximate form of Eq. (7.108) which places

[12] M. J. Lighthill, "Dynamics of a Dissociating Gas," *J. Fluid Mech.*, **2**, No. 1 (1957), part I. The interested reader will find that Lighthill's textbook style and clear insight makes his paper well worth studying.

the analysis of dissociation equilibrium in fluid-flow problems within grasp. He began by defining a "characteristic density for dissociation" ρ_D. With reference to Eq. (7.108) this is

$$\rho_D \equiv \frac{m_X}{2V} \frac{Z_X^2}{Z_{X_2}} \tag{7.118}$$

so that

$$\frac{\epsilon^2}{1-\epsilon} = \frac{\rho_D}{\rho} \exp\left(-\frac{\Theta_D}{T}\right) \tag{7.119}$$

For a rigid-rotor harmonic-oscillator molecule, in which translation and rotation are fully excited but vibration is not, the ratio Z_X^2/Z_{X_2} can be written as follows:

$$\frac{Z_X^2}{Z_{X_2}} = \frac{(Z_X)_t^2}{(Z_{X_2})_t} \frac{1}{(Z_{X_2})_r} \frac{1}{(Z_{X_2})_v} \frac{(Z_X)_e^2}{(Z_{X_2})_e}$$

Equation (5.28) is then used to evaluate the translational partition functions. Equation (5.34b) in Sec. 7.4 expresses $(Z_{X_2})_r$ using a symmetry number of 2, because we are primarily interested in air, and the oxygen and nitrogen atoms in their molecules are indistinguishable. We noted in Sec. 7.3 that the vibrational partition function must be modified to account for the absorption of the ground-level vibrational energy into the dissociation energy. Accordingly, $(Z_{X_2})_v$ is expressed by Eq. (7.20a). Thus

$$\frac{Z_X^2}{Z_{X_2}} = \underbrace{\left\{V\left(\frac{\pi m_X kT}{h^2}\right)^{3/2}\right\}}_{\text{translation}} \underbrace{\left\{\frac{2\Theta_r}{T}\right\}}_{\text{rotation}} \underbrace{\left\{1 - \exp\left(-\frac{\Theta_v}{T}\right)\right\}}_{\text{vibration}} \underbrace{\left\{\frac{Z^2{}_{X_e}}{(Z_{X_2})_e}\right\}}_{\text{electronic}}$$

Since electronic excitation usually occurs at temperatures $> 10^5$ °K, the electronic partition functions can frequently be replaced with the ground-state degeneracies [recall Eq. 7.23)]. When this is the case, Eq. (7.118) becomes

$$\rho_D = m_X\left(\frac{\pi m_X k}{h^2}\right)^{3/2} \Theta_r \sqrt{T}\,(1 - e^{-\Theta_v/T}) \frac{(g_X)_e^2}{(g_{X_2})_e} \tag{7.120}$$

Numerical values of ρ_D are, fortuitously, very insensitive to temperature. The values computed by Lighthill are given in Table 7.5 to illustrate this for oxygen and nitrogen. In each case an approximately constant ρ_D is suggested for computation.

The basis of Lighthill's approximate model is to replace ρ_D with this constant value in Eq. (7.119). This is, of course, a vast simplification over the law of mass action for dissociation given by Eq. (7.108). It can, in fact, be plotted once and for all for any dissociation process on the dimensionless coordinates, ϵ versus T/Θ_D versus

TABLE 7.5 Characteristic Densities of Oxygen and Nitrogen in the Temperature Range 1000°K $\leq$ T $\leq$ 7000°K

	Θ_D, °K	ρ_D (g/cm³) when T (°K) is:							Suggested constant ρ_D, g/cm³
		1000	2000	3000	4000	5000	6000	7000	
Oxygen	59,000	145	170	166	156	144	133	123	~150
Nitrogen	113,000	113	135	136	133	128	123	118	~130

ρ_D/ρ. This has been done in Fig. 7.8, and the result shows that dissociation tends to be complete at temperatures well below Θ_D for ρ_D's of practical interest.

Lighthill's model leads to an improved expression for the specific internal energy because it carries with it the implication that the term $(7 - \epsilon)/2$ in Eq. (7.117) should be approximated as 3. To show how this comes about let us write Eq. (7.118) as

$$\frac{\partial}{\partial T} \ln Z_{X_2} \simeq 2 \frac{\partial}{\partial T} \ln Z_X \tag{7.121}$$

where we take ρ_D to be approximately independent of T. Then, in accordance with Eq. (3.23),

$$u_{X_2}(T) \simeq 2u_X(T) \tag{7.121a}$$

Equation (7.121a) disagrees with Eqs. (7.115) and (7.116), but it would agree perfectly if the vibration of X_2 were only half-excited. This, in

Fig. 7.8 Graphical representation of Lighthill's approximate equilibrium dissociation equation, $\epsilon^2/(1-\epsilon) = (\rho_D/\rho) \exp(-\Theta_D/T)$.

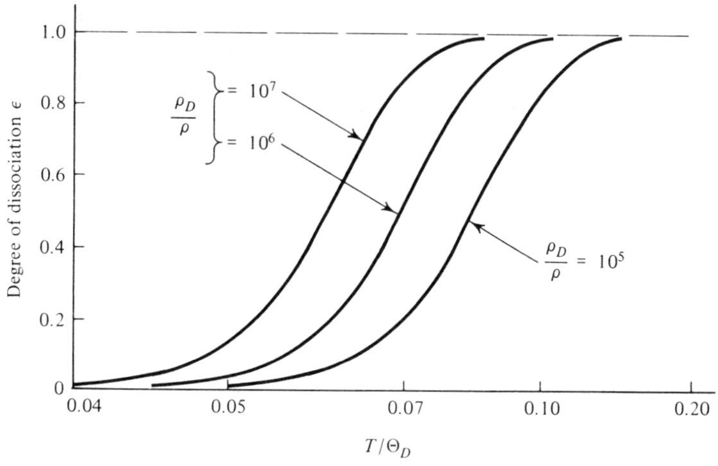

all probability, is more reasonable than taking it to be fully excited. Thus

$$u_{X_2}(T) \simeq 2u_X(T) = 3R_{X_2}T \tag{7.121b}$$

and Eq. (7.114) becomes

$$u(\rho, T) = R_{X_2}[3T - (1 - \epsilon)\Theta_D] \tag{7.122}$$

This is both more accurate, and far simpler, than Eq. (7.117).

The pressure, Eq. (7.111), is unchanged under Lighthill's approximation. Other properties have been worked out in detail by Lighthill. Vincenti and Kruger[13] also present a detailed discussion of ideal dissociating gas properties.

7.8 IDEAL IONIZING GASES

IONIZATION

Dissociation is not the only way in which a particle can go to pieces at elevated temperature. At generally higher (but often overlapping) temperatures, orbiting electrons can be excited to the point that they, too, begin to separate from their parent atoms. This process is called ionization. In the monatomic gas helium, for example, the reaction

$$\mathrm{He} \rightleftharpoons \mathrm{He}^+ + e$$

in which one electron is stripped off, is 2 percent complete at about 15,000° K.

Whereas a dissociating gas consists of neutral particles, the stripped helium atoms are positively charged and the free electrons are negative. When the total negative electronic charge of such a mixture (or *plasma*, as it is often called) equals the total positive charge of the ionized particles (atoms or molecules), the plasma is said to be *neutral*. The thermodynamic analysis of ionized gases may be quite different from that of dissociated gases, as a consequence of the difference in interaction among charged particles as compared with that among neutral particles. Of course this is not the case when gases can be treated as ideal and interactions can be neglected. Still, ionization reactions have other special features which are important even for ideal gases and which we shall want to treat.

[13]W. G. Vincenti and C. H. Kruger, Jr., *Introduction to Physical Gas Dynamics*, John Wiley & Sons, Inc., New York, 1965, chap. V.

IONIZATION EQUILIBRIUM OF SINGLY IONIZED GASES

As the temperature of a plasma increases first one electron will be lost to a particle, then two, and so forth:

$$X \rightleftharpoons X^{+} + e \rightleftharpoons X^{++} + e + e \rightleftharpoons \cdots \tag{7.123}$$

The particle X^{+} with one electron removed is said to be *singly ionized*, X^{++} is *doubly ionized*, and so on. We restrict this discussion to a singly ionized neutral plasma.

We also restrict our discussion to ideal gases; so the single ionization reaction, $X \rightleftharpoons X^{+} + e$, becomes formally identical to the dissociation reaction and the results in the preceding section are directly applicable. The law of mass action, for example, can be written down directly in form similar to Eq. (7.108),

$$\frac{\epsilon^2}{1-\epsilon} = \frac{m_X}{\rho V} \frac{(Z_{X^+})Z_e}{Z_X} \exp\left(-\frac{\Theta_I}{T}\right) \tag{7.124}$$

where ϵ is the "degree of ionization,"

$$\epsilon \equiv \frac{N_{X^+}}{N_{X_t}} = \frac{\text{number of ionized particles}}{\text{number or originally neutral particles}} \tag{7.125}$$

the characteristic temperature for ionization Θ_I is I/k; and I is the ionization energy of the particle. Table 7.6 presents numerical values of Θ_I for typical atoms. These temperatures all exceed 10^5 °K, but ionization, like dissociation, becomes important while T is still well below Θ_I.

TABLE 7.6 Electronic Energy States of Gases

Gas	Atomic weight	Θ_I, °K	$(Z_X)_e$ (T in °K)	$(Z_{X^+})_e$ (T in °K)
Hydrogen	1.008	157,800	2	1
Helium	4.003	285,300	1	2
Nitrogen	14.01	168,800	4 + 10 exp (−27,700/T)	9 + 5 exp (−22,000/T)
Oxygen	16.000	158,100	9 + 5 exp (−22,900/T)	4 + 10 exp (−38,600/T)
Neon	20.183	250,200	1	4 + 2 exp (−1,126/T)
Argon	39.944	182,840	1	4 + 2 exp (−2,063/T)
Xenon	131.3	140,200	1	4 + 2 exp (−15,200/T)

Once again, the degree of ionization depends upon both ρ and T in accordance with Eq. (7.124). The dependence again is expressed by writing out the partition functions. Since the only contribution to $(Z_e)_{\text{int}}$ is the ground-level degeneracy of an electron $g_{e_0} = 2$, we have

$$Z_e = (Z_e)_{\text{int}}(Z_e)_t = 2V\left(\frac{2\pi m_e kT}{h^2}\right)^{3/2} \tag{7.126}$$

Furthermore, $(Z_{X^+})_t \simeq (Z_X)_t$ since the masses of X and X^+ are nearly the same. Thus Eq. (7.124) becomes

$$\frac{\epsilon^2}{1-\epsilon} = 2\frac{m_X}{\rho}\left(\frac{2\pi m_e kT}{h^2}\right)^{3/2}\frac{(Z_{X^+})_{\text{int}}}{(Z_X)_{\text{int}}}\exp\left(-\frac{\Theta_I}{T}\right) \tag{7.127}$$

The pressure expression that we developed earlier,

$$p = \frac{kT}{V}(N_X + N_{X^+} + N_e) \tag{7.83a}$$

is true for any ideal gas mixture. In this case it becomes

$$p = (1+\epsilon)\frac{\rho k}{m_X}T \tag{7.128}$$

Substitution of this expression in Eq. (7.127) gives the conventional form of the law of mass action for an equilibrium ionized gas,

$$\frac{\epsilon^2}{1-\epsilon^2} = 2\left(\frac{2\pi m_e}{h^2}\right)^{3/2}\frac{(kT)^{5/2}}{p}\frac{(Z_{X^+})_{\text{int}}}{(Z_X)_{\text{int}}}\exp\left(-\frac{\Theta_I}{T}\right) \tag{7.129}$$

Since the temperature dependence of Z_{X^+} and Z_X is roughly the same, we can see that the degree of ionization should rise sharply with temperature but will drop off with pressure. Figure 7.9 illustrates this behavior for singly ionized helium.

SAHA EQUATION FOR SINGLY IONIZED MONATOMIC GASES

For monatomic gases the internal partition functions reduce to Z_e, and this in turn is usually a constant g_{e_0}. Thus Eq. (7.129) takes the form

$$\frac{\epsilon^2}{1-\epsilon^2} = C\frac{T^{5/2}}{p}\exp\left(-\frac{\Theta_I}{T}\right) \tag{7.130}$$

where C is a constant, expressible as

$$C \equiv 2\left(\frac{2\pi m_e}{h^2}\right)^{3/2}k^{5/2}\frac{(Z_{X^+})_{\text{int}}}{(Z_X)_{\text{int}}} \tag{7.131}$$

Equation (7.130) is named the Saha equation after the man who derived it in 1920 on the basis of purely classical thermodynamic arguments.

As a numerical example, let us apply the Saha equation to helium. Using data from Table 7.4 we obtain $C = 0.760 \times 10^6 (°\text{K})^{-5/2}$ (atm). Equation (7.130) then becomes

$$\epsilon = \left[1 + \frac{1}{0.760 \times 10^6}\frac{p}{T^{5/2}}\exp\left(\frac{285{,}300}{T}\right)\right]^{-1/2} \tag{7.132}$$

Fig. 7.9 Degree of ionization for singly ionized helium.

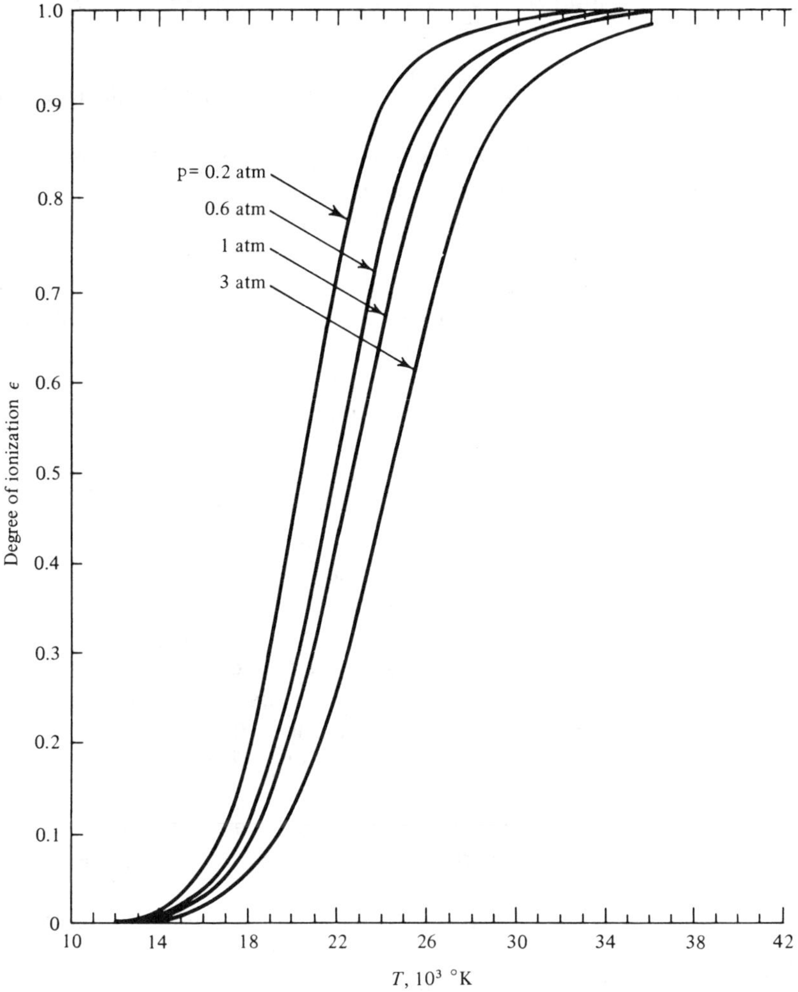

Equation (7.132) is the expression that was used to plot the curves shown in Fig. 7.9.

In employing the Saha equation care must be taken as to whether the electronic energy levels are indeed unexcited. If they *are* excited, a modified form of the Saha equation accounting for these excitations should be used.[14] Furthermore, since the Saha equation is applicable only to ionized gases at thermodynamic equilibrium, it must be modified again for systems that are not at equilibrium.

The analysis presented here may be readily extended to

multiple ionization processes.[14] It is also possible to develop an approximate analysis for ionizing gases similar to that of the Lighthill dissociating gas.[15]

Problems 7.1 Complete the missing steps in the derivation of Eqs.(7.3a) and (7.5).

7.2 Prepare a table expressing the thermodynamic properties of an ideal monatomic gas, based upon both distinguishable and indistinguishable classical particles. Express extensive properties on a molar basis. Show any derivation not worked out in the text. Discuss the practical ramifications of the indistinguishable character of particles in the evaluation of thermodynamic properties. Consider T and V to be independent variables.

7.3 For any ideal monatomic gas of your choosing, evaluate the absolute entropy at some point (a) using classical thermodynamics and experimental data, and (b) using the Sackur–Tetrode equation. Do not choose mercury vapor.

7.4 The fundamental equation for an ideal monatomic gas was given by Eq. (1.23), but certain constants were not evaluated. Lump the constants together and show how to obtain a numerical value for them for the case in which $S_0 = 0$.

7.5 Verify Eqs. (7.23) and (7.24).

7.6 Obtain the *classical* specific heat for diatomic molecules that rotate, translate, and vibrate, using equation (7.26).

7.7 A system of N linear harmonic oscillators at 300°K has a fundamental frequency of 10^{13} rad/sec. Use the Boltzmann distribution to determine what fraction of the oscillators occupy each of (at least) the first four energy levels. Does your result appear to be reasonable (a) in view of the value of Θ_v; (b) in view of the constraint, $\sum N_i = N$?

7.8 Compute the characteristic temperatures of translation, rotation, and vibration for nitrogen at room temperature. (This requires finding data for certain molecular properties.)

7.9 Verify Eq. (7.31).

7.10 Derive the correct form of Eq. (7.38a) for a general value of spin degeneracy.

7.11 Evaluate the molar Gibbs function for oxygen at 4000°K and 1 atm. Report the portions attributable to translation and to the internal modes of energy storage. What percentage is contributed by Z_{corr}? How does the portion attributable to translation compare with a value of g predicted by the Sackur–Tetrode equation?

[14] A. B. Cambel, D. P. Duclos, and T. P. Anderson, *Real Gases*, Academic Press, Inc., New York, 1963.

[15] D. P. Duclos, D. P. Aeschliman, and A. B. Cambel, *Am. Rocket Soc. J.* **32,** 641 (1962).

7.12 Compute c_v for the gas in Problem 7.11. Consider the correction terms.

7.13 Redo Example 7.2 at 50°K. What terms are negligible in this case and why?

7.14 Derive the moments of inertia shown in Fig. 7.4. Evaluate I_{A-A}, I_{B-B}, I_{D-D}, and I_{C-C} numerically for methane (Fig. 7.3).

7.15 Derive Eqs. (7.53) and (7.54) taking care to evaluate the multiplier β.

7.16 Verify Eqs. (7.64) and (7.65). Compare Eq. (7.65) with the equivalent expression obtainable from macroscopic thermodynamics.

7.17 A new gaseous organic compound, called krypto-olefacto, can be made by either of two processes. One day it is discovered that a certain bootlegger from the hills of Eastern Kentucky—Nimrod Probiscus, by name—can smell the difference between the products of the two processes. However, no other test will make the distinction. Thereafter the gas is named k-I or k-II, according to the process that makes it. Suppose that 2 moles of k-I and 3 moles of k-II at the same temperature and pressure are mixed adiabatically. What would ΔS_{mixing} be in cal /°K? Now assume that Nimrod dies a moment before mixing occurs and compute ΔS_{mixing}. Discuss your answer fully.

7.18 n_1 moles of an ideal gas at pressure p_1 and temperature T are separated from n_2 moles of another ideal gas at pressure p_2 and temperature T, in an adjoining container, separated by a membrane. The membrane is then broken.

(a) What is the final pressure of the mixture?
(b) What is ΔS_{mixing}?
(c) What would ΔS_{mixing} have been if the gases had been identical?
(d) Prove that ΔS_{mixing} for two different gases is the same as the sum of the entropy increases associated with two free expansions.

7.19 Verify Eqs. (7.72) through (7.75).

7.20 Verify Eqs. (7.83) through (7.87).

7.21 Verify Eqs. (7.93), (7.94), and (7.95).

7.22 Verify Eq. (7.97).

7.23 Verify Eqs. (7.100) and (7.101).

7.24 Use the data in Table 7.4 to verify Van't Hoff's equation, graphically.

7.25 Determine the equilibrium composition of H_2O after it has been heated to 2800°K. Assume that the H_2O decomposes into H_2, O_2, and OH but that there is no free hydrogen. Note that there are two equilibrium constants, each expressible in terms of two degrees of reaction — one for decomposition into H_2 and O_2 and the other for decomposition into H_2 and OH.

7.26 Check Lighthill's computation of ρ_D for oxygen. $(Z_{O_2})_e \simeq 3$ and $(Z_O)_e \simeq 9$. $\Theta_r = 2.08°K$ and $\Theta_v = 2230°K$.

7.27 Consider an isentropic compression of argon from an original temperature of 300°K and $p = 1.536 \times 10^{-4}$ atm. Plot p versus T in the range 5000 to 15,000°K (a) for the nonionized case, and (b) for the ionizing process. Present the equation relating p, T, and ϵ, and indicate the degree of ionization at several points on the curve. Discuss the comparison.

7.28 Derive an expression for the speed of sound in a singly ionized monatomic gas for which the electronic states are unexcited. The expression should give the sound speed as a function of ϵ and T. Compare the result with the sound speed of the same gas without ionization.

8 statistical-mechanical ensembles

8.1 ENSEMBLE CONCEPT

LIMITATIONS OF PREVIOUS METHODS

The "actuarial" methods of statistical mechanics are farther reaching than the preceding chapters might indicate. By concentrating upon the behavior of systems of independent particles, we have been limited to the treatment of a class of fairly simple systems — ideal gases, for example. By taking particles to be independent we could focus attention on the single particle. The important concepts have been the energy levels of *isolated* particles, the distribution functions for *independent* particles, and so on.

And, up to now, we have always looked at a group of particles and asked: What is its most probable macrostate? We then accepted this as the macrostate that would occur in nature. It is not really the one that occurs, however. There are myriads of other macrostates that are almost as probable and very nearly identical in their properties. Only those macrostates that differ perceptably from the most probable one become extremely improbable.

Furthermore, we have imposed constraints of constant energy and (for all instances except the photon gas) of a constant number of particles. These constraints limited the scope of the problems that we could deal with.

The statistical mechanics that has been the subject of our interest up to now is, in fact, a special case of a much more general theory which, for example, treats interdependent particles. We have chosen to discuss the special case first because it offers a direct, convenient analysis for a wide range of important practical problems, as we found in chapter 7. Beyond this, it also provided a simple structure upon which we can now build the more general theory in a reasonable pedagogical sequence.

In contrast with our previous considerations, in which we looked at a single system or group of particles, we now want to make a statistical analysis of a large number or "ensemble," of identical systems. It is therefore called the *ensemble theory*.

SYSTEMS AND ENSEMBLES

Conjecture now that our world is duplicated in many parallel universes. Any system — say a container of gas — is reproduced identically in each universe with the exception of random fluctuations on the microscopic level. We then ask how the ensemble of parallel systems behaves.

The *ergodic hypothesis* (or the ergodic "surmise") proposes that this ensemble exhibits the same average properties in space as a single system exhibits in time. However, it is possible, as we soon discover, to relax the constraints that dictate the behavior of a single system when we treat the ensemble of parallel systems. Specifically, the ergodic hypothesis says that the ensemble average of any property is the same as the time average of that property for the single system.

One way to calculate the actual behavior of a system is to calculate its time-average behavior over a period that is much longer than, say, the average period between molecular collisions. Such a computation is, of course, impractical because of the massive difficulties involved in treating so many particles. The use of classical mechanics does provide some important criteria for the general characteristics of large systems of particles. These methods were especially evident in the early development of statistical mechanics,[1] although we do not elaborate these ideas here. There are more complications, however. For example, there are questions as to the uniqueness of the resulting average with respect to the initial

[1]See, especially, J. W. Gibbs, *Elementary Principles in Statistical Mechanics*, Yale University Press, New Haven, 1902; Dover Publications, Inc., New York, 1960.

conditions and with respect to the length of time over which the average is made.

Faced with these difficulties, Gibbs[2] apparently concluded that it is not fruitful to try to form time averages of a single ensemble. Instead he chose to limit consideration to an "ergodic ensemble" of systems — a collection of systems that obeys the ergodic hypothesis — at the present moment. The simplification lies in the fact that the average behavior of a suitably chosen collection of systems is mathematically easier to treat than the complex time-dependent behavior of a single system.

By way of illustration let us consider the problem of two efficiency experts who want to learn whether they can speed the noontime movement of customers through a crowded downtown lunchroom by charging money for the second cup of coffee. They must learn how much expensive seating capacity is lost to customers who leisurely sip the free cup that is now made available. One expert identifies 10 customers as they enter and he records the time they spend eating and the time they spend sipping. He is observing the behavior in time of a "system" of 10 people. The second expert appears at the time of peak loading and photographs the entire lunchroom from a balcony. He then has information as to the state of the "ensemble" of all systems of 10 people, and can obtain the same information by establishing the percentages of eating and sipping customers at the instant. The observation made by the second expert is simpler and it provides more information.

This example can probably be as helpful to us by virtue of its weakness as by its strength. The weakness lies in that the ensemble might not be ergodic. It might, for example, turn out that the "system" that enters at 11:50 A.M. is composed of representatives from middle management who have come to conduct business over a slow lunch. The 12:10 P.M. arrivals, on the other hand, might be wage earners with a 40-minute lunch hour. The lack of ergodicity in this case could lead to disagreement between the experts and result in a money-saving strategy that would only hurt business by driving away the hurried lower-income group.

BASIC POSTULATES

The first postulate in Gibbs's approach is, then, that whatever ensemble we set up must be ergodic. And this in turn entails the

[2]J. W. Gibbs was this country's first Ph.D. in mechanical engineering (Yale, 1863). Although his thesis dealt with the forming of gear teeth, he is known to us as the greatest mathematical physicist this country has produced. He formulated vector analysis, chemical thermodynamics, and the foundations of statistical mechanics, among other accomplishments. His organizational work on statistical mechanics followed Boltzmann's introductory work by about a quarter of a century.

assumption that ergodic systems exist. Boltzmann introduced the idea of an ergodic system and invented the name in 1887. He combined the Greek *ergon* (meaning "work") with *hodos* (meaning "path") to evoke the idea of a constant-energy path. Strictly speaking, an ergodic *system* is one that "moves" on a constant-energy surface in phase space and whose path sooner or later touches all points in that surface.

The surmise that an ergodic system of particles exists is intuitively compelling but, interestingly enough, is not strictly true. There exist states of motion of the molecules of a gas, for example, that cannot be reached from other states of motion. Consider, for example, a group of particles in a box, all bouncing back and forth in the same vector direction but in different paths and at different speeds. No collisions would occur to disrupt the configuration, and other configurations would be unreachable. Fortunately, such exceptions are sufficiently remote that they do not seriously alter the usability of the ergodic surmise in problems of gas behavior.

Actually some mathematical justification of the ergodic surmise can be made under certain abstract assumptions that cannot be fully justified physically. Khinchin,[3] for example, does this for an infinitely long time average. Nevertheless, the final justification for assuming the systems in an ensemble to be ergodic is really to be found in the success of the theory that can be erected on the assumption.

The principle of equal a priori probabilities, which we have used extensively in the past, must be used again in the following way: Each system in the ensemble must be in one of the microstates consistent with the imposed constraints and each of these system microstates is equally likely. Sometimes this is called the fundamental postulate of statistical mechanics because it is so broadly used. It took a slightly different form in Sec. 3.1.

In the formulation of classical statistical mechanics, which employs the classical laws of motion, this completes the assumptions needed. In the formulation of quantum statistical mechanics, however, an additional postulate is needed. This is the so-called *principle of random phases*. It says that "equilibrium may be regarded as an *incoherent* superposition of (quantum) eigenstates."[4] Reference should be made to more advanced textbooks for detailed discussion of this idea.

[3]A. I. Khinchin, *Mathematical Foundations of Statistical Mechanics* (English translation from Russian), Dover Publications, Inc., New York, 1949.

[4]K. Huang, *Statistical Mechanics*, John Wiley & Sons, Inc., New York, 1966, p. 185.

TYPES OF ENSEMBLES

Since an ensemble is a collection of identical systems subject to a common set of constraints, our choice of constraints will result in different kinds of ensembles. There are three conventional kinds of constraints commonly used to define three conventional kinds of ensembles. The latter are called the *microcanonical*, the *canonical*, and the *grand canonical ensembles*. The recurring word "canonical" assumes different meanings in a dictionary. It is used here in the sense of something that is authorized, recognized, or accepted. Gibbs probably intended to designate these ensembles as basic types. Although the physical model for each ensemble differs, we shall find that the utility of each is identical and universal insofar as macroscopic properties are concerned. The choice of a particular ensemble is therefore made on the basis that it will yield the most convenient results, mathematically.

The macroscopic constraints that specify the three kinds of ensemble are shown schematically in Fig. 8.1. The three ensembles are

microcanonical ensemble:	Each system is isolated from the others by rigid, adiabatic, impermeable walls; U, V, and N for each system are constants.
canonical ensemble:	Each system is closed to the others by rigid, diathermal, impermeable walls; T, V, and N for each system are constants, but energy can be exchanged—it is as though the systems were immersed in an isothermal bath.
grand canonical ensemble:	Each system is separated by rigid, permeable, diathermal walls. T, V, and μ for each system are constants, but both energy and mass can be exchanged; this is like a canonical ensemble made up of leaky boxes.

Let us now explore the properties and the consequences of each of these ensembles in greater detail.

Fig. 8.1 Schematic representation of three kinds of ensembles.

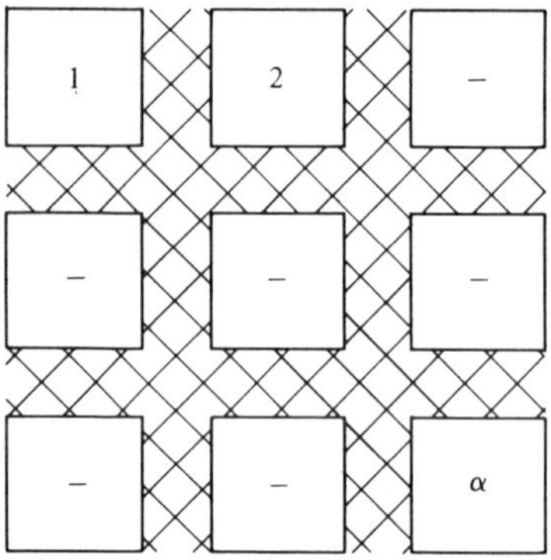

$$U_1 = U_2 = \cdots = U_\alpha$$
$$V_1 = V_2 = \cdots = V_\alpha$$
$$N_1 = N_2 = \cdots = N_\alpha$$

(a) A microcanonical ensemble. Systems 1 through α are totally isolated from one another.

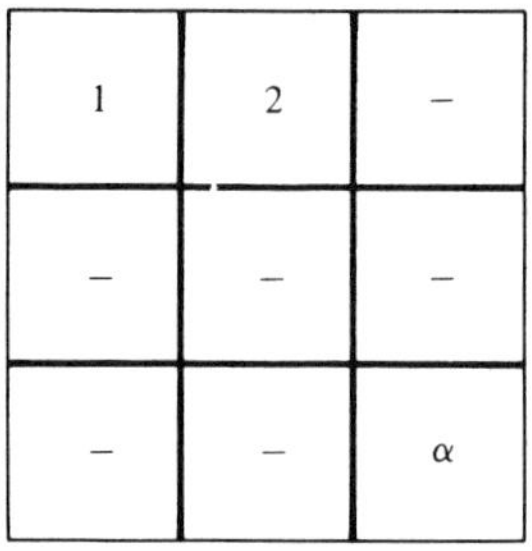

$$T_1 = T_2 = \cdots = T_\alpha$$
$$V_1 = V_2 = \cdots = V_\alpha$$
$$N_1 = N_2 = \cdots = N_\alpha$$
$$U_{\text{total}} = \text{constant}$$

(b) A canonical ensemble. Systems 1 through α are closed to one another, but energy can be exchanged among systems.

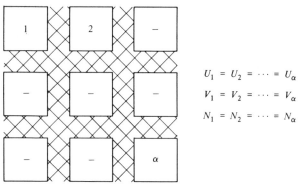

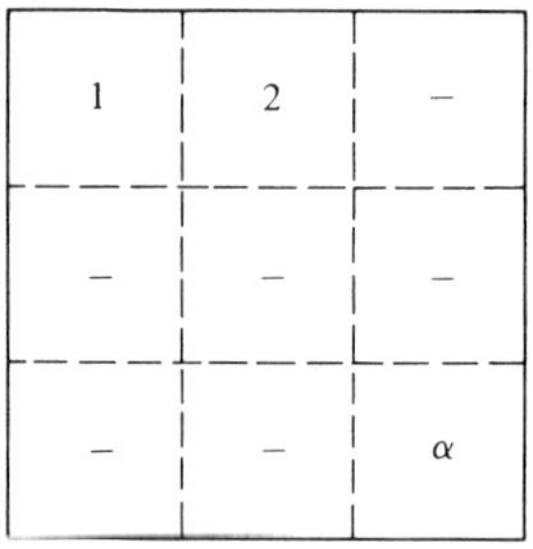

$$T_1 = T_2 = \cdots = T_\alpha$$
$$V_1 = V_2 = \cdots = V_\alpha$$
$$\mu_1 = \mu_2 = \cdots = \mu_\alpha$$
$$U_{\text{total}} \text{ and } N_{\text{total}} = \text{constants}$$

(c) A grand canonical ensemble. Energy and mass can be exchanged among systems.

8.2 MICROCANONICAL ENSEMBLE

ANALYTICAL DESCRIPTION OF THE ENSEMBLE

We wish now to build a statistical description of the microcanonical ensemble that we have just described in physical terms. The starting point for the statistical description of a system of particles has been the thermodynamic probability, W, in the past. This was defined in such a way that

$$\mathcal{P}(\text{macrostate}) = \frac{W}{\Omega} \tag{3.4a}$$

where

$$\Omega \equiv \text{total number of possible microstates}$$

As we look at the microcanonical ensemble that we have constructed, our attention moves naturally to this quantity Ω which we have heretofore had no need to identify with a special symbol. When we wrote the total number of microstates consistent with the macrostate of a single system of independent particles (of boltzons, in this case), this result was

$$W = \prod_{i=0}^{\infty} \frac{g_i^{N_i}}{N_i!} \tag{6.8a}$$

But the number of microstates consistent with the macrostate of the *ensemble* is Ω, which would be written as

$$\Omega = \sum_{\{N\}} \prod_{i=0}^{\infty} \frac{g_i^{N_i}}{N_i!} \tag{8.1}$$

where the $\sum_{\{N\}}$ indicates a sum over all possible sets of the N_i's consistent with the usual constraints on the number of particles and the energy of the system,

$$\sum_{i=0}^{\infty} N_i = N \tag{3.1}$$

and

$$\sum_{i=0}^{\infty} \epsilon_i N_i = U \tag{3.2}$$

The last constraint is actually imprecise because the Heisenberg uncertainty principle allows some variability in the energy.

There is a distribution function f of system energies in the ensemble, in the variable U. It is formed in such a way that

$$\Omega = \int_{-\infty}^{\infty} f(U)\, dU \simeq \int_{U-\delta U/2}^{U+\delta U/2} f(U)\, dU \tag{8.2}$$

where δU is the variability required by the uncertainty relation. The spread of energy in an ensemble composed of large systems is very, very narrow, as implied in Fig. 8.2. It actually approaches the Dirac delta function as the size of the system becomes large,

$$f(U) \rightarrow \Omega[\delta(U - U_0)] \equiv \begin{cases} \infty & U = U_0 \\ 0 & U \neq U_0 \end{cases} \tag{8.3}$$

where U_0 is used here to designate the value specified by Eq. (3.2) and $\int_{-\infty}^{\infty} \delta\, dU = 1$, by definition of δ.

BASIC THERMODYNAMIC RELATION

We call the sum Ω the *microcanonical partition function* and seek to form a basic relation between it and macroscopic system variables. The explicit evaluation of Ω is formidable, but we can see immediately that

$$\Omega = \Omega(U, V, N) \tag{8.4}$$

Fig. 8.2 Distribution of energies of systems in a microcanonical ensemble.

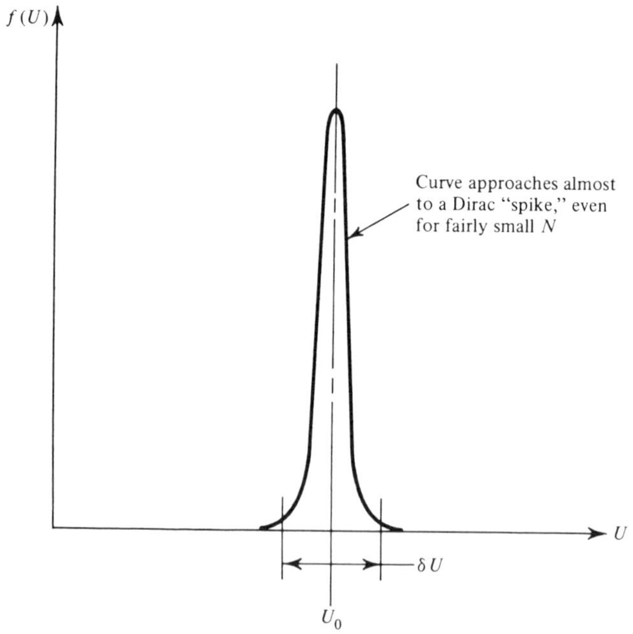

The independent variables U and N necessarily enter through the constraints, and V is introduced in the specification of the energy levels. If we note that N can be viewed either as the number of moles or the number of molecules (since the two differ only by the conversion factor N_A), then we see that Eq. (8.4) has the same independent variables as the entropy form of the macroscopic fundamental equation,

$$S = S(U, V, N) \tag{1.4a}$$

The entropy must therefore be related through U, V, and N to Ω. By constructing arguments similar to those set forth in Sec. 3.5, we arrive at the following relation, similar in form to Eq. (3.17), for S in terms of Ω,

$$S(U, V, N) = k \ln \Omega(U, V, N) \tag{8.5}$$

This is the basic relationship relating the macroscopic properties of a system from a microcanonical ensemble with the statistical properties of the ensemble.

The difficulties of evaluating Ω can be largely overcome once we can accept a startling mathematical result, that

$$\ln \Omega = \ln (W_1 + W_2 + \cdots) \simeq \ln W_{\max} \tag{8.6}$$

where $W_{\max}$ is the largest W in the series. The fact that the largest W can be used to characterize the whole of Ω is central to our method and requires some justification. Instead of dwelling on a rigorous mathematical deduction[5] that would be beyond our scope here, we shall use a fairly evocative physical argument presented by Rushbrooke.[6]

Suppose that there were as many as N systems with W's comparable to $W_{\max}$ (i.e., as many such systems as there are particles in any one system). This is a very large number but it pales against the magnitude of W, which is perfectly enormous by any stretch of the imagination. Then

$$\ln \Omega \simeq \ln NW_{\max} + \text{(much smaller terms)}$$

or

$$\ln \Omega \simeq \ln W_{\max} + \ln N$$

[5]The "method of steepest descents" developed by Darwin and Fowler is described in detail by R. H. Fowler and E. A. Guggenheim, *Statistical Thermodynamics*, Cambridge University Press, New York, 1939.

[6]G. S. Rushbrooke, *Introduction to Statistical Mechanics*, Oxford University Press, New York, 1949, chap. 2, sec. 6.

But ln N is very small in comparison with ln W_{max}. Thus, to a degree of approximation that becomes increasingly accurate with increasing N, we obtain

$$\ln \Omega = \ln W_{max} \tag{8.6a}$$

It also follows, then, that Eq. (8.5) reduces, in effect, to Eq. (3.17),

$$S = k \ln \Omega = k \ln W_{max}$$

in which the constant, k, must again be Boltzmann's constant.

EXAMPLE 8.1 Show that in the microcanonical ensemble formulation, the Helmholtz function of a system can be expressed as

$$F = U^2 \frac{\partial}{\partial U}\left(\frac{\ln \Omega}{U}\right)_{V,N} \Big/ \left(\frac{\partial \ln \Omega}{\partial U}\right)_{V,N}$$

In the microcanonical formulation, the Helmholtz function takes the following functional form:

$$F(U, V, N) = U - T(U, V, N)S(U, V, N)$$

Since $S = k \ln \Omega(U, V, N)$ and $1/T = (\partial S/\partial U)_{V,N}$, there follows:

$$F = T\left(\frac{U}{T} - S\right) = \left(U\frac{\partial S}{\partial U} - S\right) \Big/ \left(\frac{\partial S}{\partial U}\right)$$

$$= U^2 \frac{\partial}{\partial U}\left(\frac{S}{U}\right) \Big/ \frac{\partial S}{\partial U}$$

$$= U^2 \frac{\partial}{\partial U}\left(\frac{\ln \Omega}{U}\right)_{V,N} \Big/ \left(\frac{\partial \ln \Omega}{\partial U}\right)_{V,N}$$

In Sec. 3.5, we argued in purely physical terms that maximizing W would result in an *equilibrium* situation. Now, with the help of ensemble arguments, we see that this conclusion is deducible in a purely mathematical way. The elementary statistical mechanics that we developed in chapter 3 is, therefore, based upon the microcanonical ensemble. We have already shown the universal character of this method, although we restricted its application to independent particles. This universality might seem odd in view of the fact that the systems of the ensemble are isolated from one another. This paradoxical situation is discussed further in Sec. 8.5.

The microcanonical ensemble is universal in its applications, but there are other equally general approaches. The canonical and grand canonical ensemble approaches are, in particular, based upon broader models, and they result in simpler descriptions of many practical situations. Let us consider the canonical ensemble next.

8.3 CANONICAL ENSEMBLE

ANALYTICAL DESCRIPTION OF THE ENSEMBLE

The canonical ensemble provides a more utilitarian fundamental equation for a variety of problems than the microcanonical equation did. In this case, let us first recall one of the Massieu-function forms of the fundamental equation,

$$F = F\left(\frac{1}{T}, V, N\right) \tag{1.48a}$$

In certain cases of practical interest, this provides some advantage over Eq. (1.4a), which we used in Sec. 8.2. By replacing U (which often is not measured or "given" information) with $1/T$, we obtain a more usable fundamental equation. This kind of improvement was, of course, the whole purpose of the Legendre transforms that we discussed in chapter 1.

In the canonical ensemble, as we already have noted in the context of Fig. 8.1(b), the constant-energy constraint is replaced with a constant-temperature constraint. We therefore look for a fundamental microscopic parameter that will have the same property dependence as Eq. (1.48a). This will require more development than it did in Section 8.2, however.

There will now be a broader distribution of the number of individual systems $\tilde{N}_j$ at each energy level U_j. This is illustrated roughly in Fig. 8.3, which shows the distribution function $f(U_j)$ for U_j in an ensemble composed of $\tilde{N}$ systems. As $\tilde{N}$ increases, $f(U_j)$ tends (as we shall show shortly) to become increasingly spiked. We now will establish this distribution in much the same way as we established the distribution of the energies of the individual particles in chapters 3 and 6. It should be emphasized, however, that the systems in an ensemble are distinguishable owing to the macroscopic nature of each system.

The thermodynamic probability, $\tilde{W}$, of a macrostate of the canonical ensemble is

$$\tilde{W} = \tilde{N}!\left[\prod_{j=0}^{\infty} \frac{G_j^{\tilde{N}_j}}{\tilde{N}_j!}\right] \tag{8.7}$$

where the "degeneracy," G_j, is the number of ways in which any one of the $\tilde{N}_j$ systems in the ensemble can occupy its particular energy level U_j. But this G_j is exactly what we formerly called the thermodynamic probability W_j of a system with energy U_j. Thus, if we once more restrict ourselves to a consideration of independent

Fig. 8.3 Energy distribution of the $\tilde{N}$ systems comprising a canonical ensemble. (The sketch is scaled so that it can only apply to relatively small $\tilde{N}$'s.)

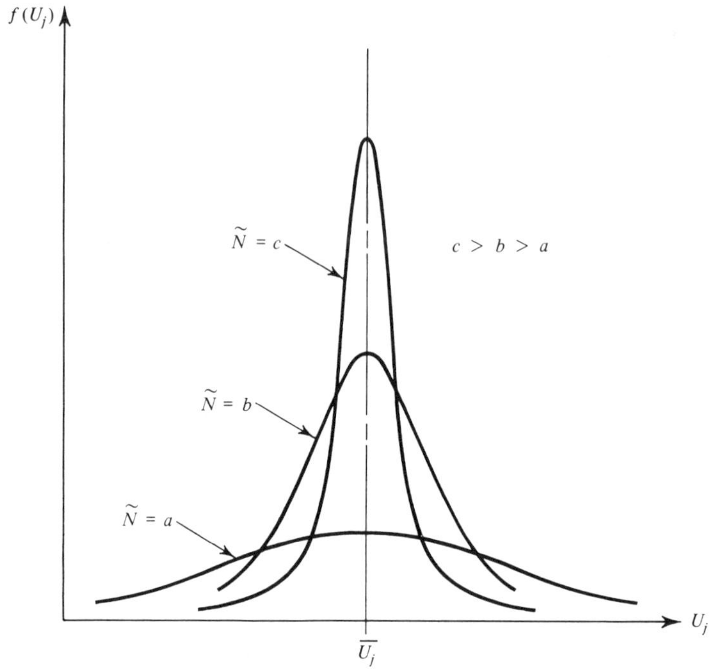

and indistinguishable boltzons[7] for illustrative purposes, we can write, in accordance with Eqs. (6.1) and (6.4),

$$G_j \equiv W_j = \prod_{i=0}^{\infty} \frac{g_i^{N_{ij}}}{N_{ij}!} \tag{8.8}$$

where N_{ij} is the number of particles with the particle energy ϵ_i in a system with the system energy U_j. The N_{ij}'s obey the constraints

$$\sum_{i=0}^{\infty} N_{ij} = N_j = N \tag{8.9}$$

where N is the number of particles in each of the $\tilde{N}$ systems, and

$$\sum_{i=0}^{\infty} N_{ij}\epsilon_i = U_j \tag{8.10}$$

Thus we are erecting a kind of second-order thermodynamic probability $\tilde{W}$ and we need to write another set of constraints on the

[7]Note that this assumption, and Eq. (8.8), is restrictive. Equation (8.7), on the other hand, is true for any canonical ensemble because the systems of which it is composed are always independent of one another.

whole ensemble. The total number of systems and the total energy of all systems are fixed, so these constraints are

$$\sum_{j=0}^{\infty} \tilde{N}_j = \tilde{N} \tag{8.11}$$

and

$$\sum_{j=0}^{\infty} \tilde{N}_j U_j = \tilde{U} \tag{8.12}$$

The problem of maximizing $\tilde{W}$, as given by Eq. (8.7) and subject to the constraining Eqs. (8.11) and (8.12), differs in no way from the problem of obtaining W_B that we solved in Sec. 6.1. The answer [recall Eq. (6.5a)] is

$$\frac{\tilde{N}_j}{\tilde{N}} = \frac{G_j \exp(-\beta U_j)}{\sum_j G_j \exp(-\beta U_j)} \tag{8.13}$$

where the identification of β with $1/kT$ has yet to be made. The argument that this distribution of systems corresponds with the equilibrium ensemble is no different from the justification of Eq. (8.6).

The denominator of Eq. (8.13) very closely resembles the partition function. We call it the canonical partition function Q

$$Q \equiv \sum_{j=0}^{\infty} G_j \exp(-\beta U_j) \tag{8.14}$$

or, for the special case of systems that are composed of independent and indistinguishable boltzons,

$$Q = \sum_{j=0}^{\infty} \prod_{i=0}^{\infty} \frac{g_i^{N_{ij}}}{N_{ij}!} \exp\left(-\beta \sum_{i=0}^{\infty} N_{ij}\epsilon_i\right) \tag{8.15}$$

The canonical partition function is the fundamental parameter for the canonical ensemble. After we first see how to evaluate β, we then show how the remaining thermodynamic information can be obtained from it.

IDENTIFICATION OF β

The average of the system energies $\overline{U_j} = \tilde{U}/\tilde{N}$. Thus

$$\overline{U_j} = \sum_{j=0}^{\infty} \frac{\tilde{N}_j}{\tilde{N}} U_j \tag{8.12a}$$

or

$$\overline{U_j} = \frac{\sum_{j=0}^{\infty} U_j G_j \exp(-\beta U_j)}{Q} \tag{8.16}$$

or

$$\overline{U_j} = -\left.\frac{\partial \ln Q}{\partial \beta}\right|_{V,N} \tag{8.17}$$

The similarity of Eq. (8.17) to (6.26) implies a relationship between Q and the q potential discussed in chapter 6. The resemblance between these two fundamental parameters is more than skin deep and will merit discussion later.

Next consider a differential change in ln Q, where $Q = Q(\beta, V, N)$ but N = constant:

$$d \ln Q = \left.\frac{\partial \ln Q}{\partial \beta}\right|_{N,V} d\beta + \left.\frac{\partial \ln Q}{\partial V}\right|_{\beta,N} dV \tag{8.18}$$

But the coefficient of $d\beta$ is $-U$, in accordance with Eq. (8.17), and

$$\left.\frac{\partial \ln Q}{\partial V}\right|_{\beta,N} dV = \frac{1}{Q}\frac{\partial}{\partial V}\left.\left\{\sum_{j=0}^{\infty} G_j \exp(-\beta U_j)\right\}\right|_{\beta,N} dV$$

or, since U_j depends upon V, in general

$$\left.\frac{\partial \ln Q}{\partial V}\right|_{\beta,N} dV = -\beta \sum_{j=0}^{\infty} dU_j \frac{G_j}{Q} \exp(-\beta U_j) = -\beta \sum_{j=0}^{\infty} \frac{\tilde{N}_j}{\tilde{N}} dU_j$$

Therefore, in accordance with the definition of an average,

$$\left.\frac{\partial \ln Q}{\partial V}\right|_{\beta,N} dV = -\beta(\overline{dU})$$

and Eq. (8.18) becomes

$$d \ln Q = -[U\, d\beta + \beta(\overline{dU})] \tag{8.19}$$

or

$$dU = \frac{1}{\beta} d(\ln Q + \beta U) + (\overline{dU}) \tag{8.19a}$$

The physical meaning of $(\overline{dU})$ can be clarified in the following way. First we write Eq. (8.12a) in the form

$$dU = d\sum_j \frac{\tilde{N}_j}{\tilde{N}} U_j = \sum_j U_j d\left(\frac{\tilde{N}_j}{\tilde{N}}\right) + \sum_j \frac{\tilde{N}_j}{\tilde{N}} dU_j$$

or

$$dU = \sum_j U_j d\left(\frac{\tilde{N}_j}{\tilde{N}}\right) + \overline{dU} \tag{8.20}$$

Thus two kinds of effects give rise to variations in U. Either the proportions of systems with different U_j's vary, or the system energies vary without changing the set of numbers, $\tilde{N}_j$. We considered this situation for the N_i's and ϵ_i's *within* a system when we discussed Eq. (6.14) in Sec. 6.1. The same kind of inferences can be drawn from Eq. (8.20) as were drawn then.

Changes in U without associated changes in the distribution numbers will leave $\tilde{W}$ unchanged. The entropy of the ensemble will also be unchanged, and $\overline{dU}$ will necessarily correspond with *isentropic* changes of state. We know, of course, that the U_j's depend upon volume, so $\overline{dU}$ is then the isentropic work $-p\,dV$. In the light of the well-established thermodynamic relation

$$dU = T\,dS - p\,dV$$

we can then make the identification

$$T\,dS = \sum_j U_j d\left(\frac{\tilde{N}_j}{\tilde{N}}\right) = \frac{1}{\beta} d(\ln Q + \beta U) \tag{8.21}$$

Equation (8.21) now makes possible the identification of T and S,

$$T \sim \beta^{-1} \tag{8.22}$$

and

$$S \sim \ln Q + \beta U \tag{8.23}$$

The constant of proportionality in Eq. (8.22) will represent no more than a *scaling* of temperature. We choose the constant as the inverse Boltzmann constant because in later computations this results in a proper statement of the ideal-gas law. Thus

$$\beta = \frac{1}{kT} \tag{2.34}$$

as before.

BASIC THERMODYNAMIC RELATION

Since k is the constant of proportionality in Eq. (8.23), it can be written as

$$S = k \ln Q + \frac{U}{T} \tag{8.24}$$

and because $F \equiv U - TS$, we can write at once

$$-\frac{F}{T} = k \ln Q \tag{8.25}$$

This result strongly resembles Eq. (8.5), the fundamental equation for the microcanonical ensemble. The canonical partition function Q, depends upon β or $1/T$, upon V through the U_j's, and upon N. Thus Eq. (8.25) is in the form of the fundamental equation that results from subjecting $S = S(U, V, N)$ to a Legendre transform that replaces U. That is,

$$-\frac{F}{T}\left(\frac{1}{T}, V, N\right) = k \ln Q\left(\frac{1}{T}, V, N\right) \tag{8.25a}$$

The independent variables can be transformed (as was required in Problem 1.3) to give a fundamental equation of the form

$$F(T, V, N) = -kT \ln Q(T, V, N) \tag{8.25b}$$

EQUIVALENCE OF THE MICROCANONICAL AND CANONICAL ENSEMBLE METHODS

The two methods are identical if system energies and other properties do not fluctuate appreciably from the ensemble average values. This is indeed the case, but we shall defer consideration of this and other aspects of fluctuation theory to Sec. 8.5. At this point we only want to show that the fundamental equation for the canonical ensemble [Eq. (8.25)] is the same as that for the microcanonical ensemble for a system of independent and indistinguishable boltzons:

$$F(T, V, N) = -NkT \ln [(e/N)\, Z(T, V, N)] \tag{7.8}$$

If we introduce the definition

$$q_j \equiv G_j \exp\left(-\frac{U_j}{kT}\right) = \prod_i \frac{g_i^{N_{ij}}}{N_{ij}!} \exp\left(-\frac{\sum_i N_{ij}\epsilon_i}{kT}\right) \tag{8.26}$$

into Eq. (8.15) we obtain

$$Q = \sum_j q_j \tag{8.27}$$

But a comparison of Eqs. (8.26) and (8.13) reveals that

$$q_j = \tilde{N}_j Q/\tilde{N}$$

and we have already noted that the distribution function of, say, U_j is extremely steep for large $\tilde{N}$ (recall Fig. 8.3). Thus $\tilde{N}_j/\tilde{N}$ is large only for U_j very close to U and small elsewhere. This makes it possible to introduce the approximation

$$\ln Q \simeq \ln (q_j)_{\max} \tag{8.28}$$

which is not unlike our earlier approximation, $\ln \Omega \simeq \ln W_{\max}$.

To obtain a particular set of N_{ij}'s that would maximize q_j, we again employ the method of Lagrangian multipliers. The function to be maximized is given in the right-hand side of Eq. (8.26) and the single constraint is

$$\sum_i N_{ij} = N_j = N \tag{8.9}$$

The resulting $q = (q_j)_{\max}$ is such that

$$\ln Q = \ln q = N_j \ln \left[(e/N) \sum_i g_i \exp\left(-\frac{\epsilon_i}{kT}\right)\right] \tag{8.29}$$

but the summation is merely Z. Therefore,

$$\ln Q = N \ln (Ze/N) = \ln (Ze/N)^N \tag{8.30}$$

or

$$Q = (Ze/N)^N \tag{8.31}$$

and the fundamental equation, Eq. (8.25b), becomes

$$F = -kT \ln Q = -NkT \ln (Ze/N) \tag{7.8a}$$

We see then that in the special case of independent particles the same fundamental equation results from either the microcanonical ensemble or the canonical ensemble. The advantage we have secured through the use of the canonical ensemble is the formulation of a fundamental equation that will prove to be much more adaptable to systems of dependent particles (in which Z cannot be used). The canonical ensemble is, perhaps, more elaborate in its logical understructure but it is more general and it is convenient to use. Chapters 9 and 10 provide illustrations of the way in which dependent particles can be treated by using canonical ensemble methods.

8.4 GRAND CANONICAL ENSEMBLE

The grand canonical ensemble has the broadest logical structure of the three that we discuss in this chapter. Not only does it abandon the constraint of constant energy, but it releases the constraint of a constant number of particles as well. This change also leads to certain mathematical simplifications in the evaluation of thermodynamic properties. However, the concepts and formulations involved in developing the fundamental equation are still more elaborate in this ensemble than in the preceding two ensembles. Consequently, the grand-canonical-ensemble formulation of statistical mechanics has seen less practical use than it rightly deserves.

STATISTICAL DESCRIPTION OF THE GRAND CANONICAL ENSEMBLE

As we have already seen within the context of Fig. 8.1(c), the grand canonical ensemble has a constant total energy $\tilde{U}$ and number of particles N_t, within it. The individual systems are possessed of variable energies, U_k, and particles N_j, although their temperatures and chemical potentials are equal to one another.

Figure 8.4 shows how the system energies and the number of particles within the systems might be distributed within a grand canonical ensemble. As was the case in Fig. 8.3, we should anticipate that the distribution would normally be quite steep. The thermodynamic probability of the ensemble $\tilde{W}$ is more complicated in this case,

$$\tilde{W} = \prod_j \prod_k \frac{G_{jk}^{\tilde{N}_{jk}}}{\tilde{N}_{jk}!} \tag{8.32}$$

Fig. 8.4 Distribution of system energies, and particle populations of systems, in a grand canonical ensemble.

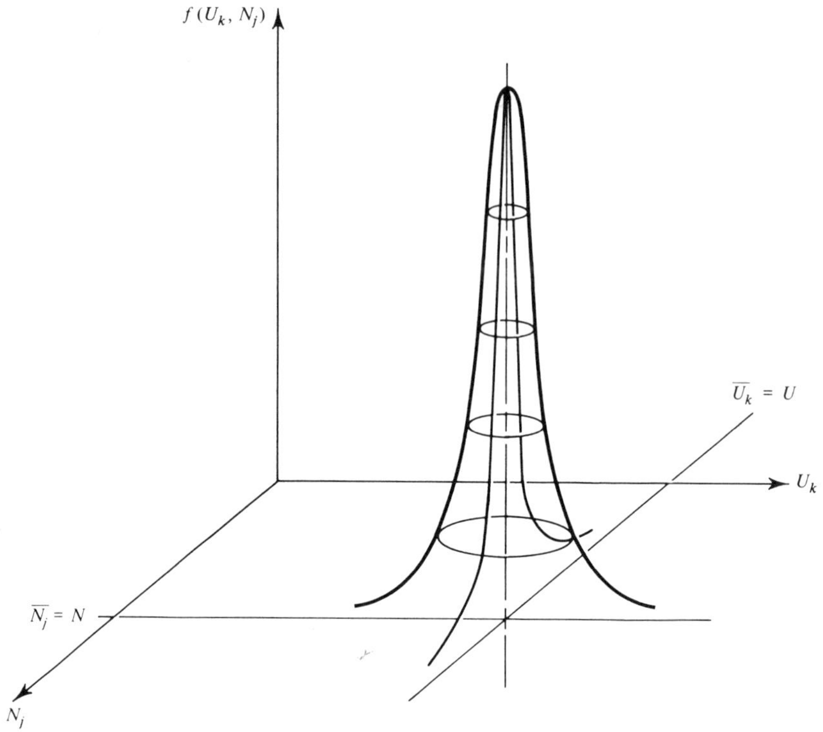

where $\tilde{N}_{jk}$ is the number of systems having *both* an energy U_k and a number of particles N_j. There are three constraints on $\tilde{W}$. The first is

$$\sum_j \sum_k \tilde{N}_{jk} N_j = N_t \qquad \text{total number of particles in the ensemble} \tag{8.33}$$

and, because

$$\sum_j \tilde{N}_{jk} = \tilde{N}_k \qquad \text{number of systems with energy } U_k$$

the second can be written as

$$\sum_j \sum_k \tilde{N}_{jk} U_k = \tilde{U} \qquad \text{energy of the ensemble} \tag{8.34}$$

Again, the total number of systems is fixed and the third constraint is thus

$$\sum_j \sum_k \tilde{N}_{jk} = \tilde{N} \tag{8.35}$$

The degeneracies G_{jk} are again the thermodynamic probabilities of the individual systems. For a system of independent and indistinguishable boltzons, they would be

$$G_{jk} = \prod_i \frac{g_i^{N_{i_{jk}}}}{N_{i_{jk}}!} \tag{8.36}$$

where the subscript jk on $N_{i_{jk}}$ serves to identify that set of N_i's as belonging to the particular subgroup of systems with N_j particles and energy U_k. It differs in this respect from a summation index. Accordingly, the constraints on the individual systems are

$$\sum_i N_{i_{jk}} = N_j \tag{8.37}$$

and

$$\sum_i N_{i_{jk}}\epsilon_i = U_k \tag{8.38}$$

The method of Lagrangian multipliers yields, in this case,

$$\frac{\tilde{N}_{jk}}{\tilde{N}} = \frac{G_{jk} \exp(-\alpha N_j - \beta U_k)}{\sum_j \sum_k G_{jk} \exp(-\alpha N_j - \beta U_k)} \tag{8.39}$$

The multiplier β can be identified as $1/kT$ by comparing Eq. (8.39) with the canonical distribution, Eq. (8.13). If all the N_j's are identical — if no "leakage" occurs within the grand canonical ensemble — the two results are the same. The identification of α is made later. We can write

$$\alpha \equiv -\frac{\mu}{kT}$$

at the present, in anticipation of the fact that we will prove μ equal to the chemical potential. The motivation for supposing that α takes this form is Eq. (6.21).

The grand canonical distribution or distribution of U_k's and N_j's can then be rewritten as

$$\frac{\tilde{N}_{jk}}{\tilde{N}} = \frac{G_{jk} \exp[(\mu N_j - U_k)/kT]}{\sum_j \sum_k G_{jk} \exp[(\mu N_j - U_k)/kT]} \tag{8.40}$$

This result can be readily redeveloped (Problem 8.3) for the case in which two species a and b of particles occupy the systems. The resulting distribution accrues a third subscript l, since N_{a_j}, N_{b_l}, and U_k are all randomly distributed. Thus

$$\frac{\tilde{N}_{jkl}}{\tilde{N}} = \frac{G_{jkl} \exp[(\mu_a N_{a_j} + \mu_b N_{b_l} - U_k)/kT]}{\sum_j \sum_k \sum_l G_{jkl} \exp[(\mu_a N_{a_j} + \mu_b N_{b_l} - U_k)/kT]} \tag{8.41}$$

Other applications of this kind can readily be generated. We should also note that Eq. (8.40) is not restricted to the treatment of boltzons. It should become clear in a moment that our treatment of Fermi–Dirac and Bose–Einstein gases in Sec. 6.2 was but a special case of the use of the grand canonical ensemble.[8]

The denominator in Eq. (8.40) is analogous to the partition function and the canonical partition function in its role in the grand canonical ensemble. It is given the symbol Q_G and called the grand canonical partition function,

$$Q_G \equiv \sum_j \sum_k G_{jk} \exp\left(\frac{\mu N_j - U_k}{kT}\right) \tag{8.42}$$

and for systems of independent and indistinguishable particles it takes the form

$$Q_G = \sum_j \sum_k \prod_i \frac{g_i^{N_{ijk}}}{N_{ijk}!} \exp\left[\frac{\sum_i N_{ijk}(\mu - \epsilon_i)}{kT}\right] \tag{8.42a}$$

BASIC THERMODYNAMIC RELATION

It would be possible to put Q_G in its proper form as a fundamental parameter for the grand canonical ensemble without recourse to other ensemble formulations. For convenience we instead employ an approach that makes reference to the preceding developments, and in so doing we can clarify the interrelation between the canonical and grand canonical ensembles.

We can rewrite Eq. (8.42) as

$$Q_G = \sum_j \exp\left(\frac{\mu N_j}{kT}\right)\left[\sum_k G_{jk} \exp\left(-\frac{U_k}{kT}\right)\right]$$

The quantity in brackets is, in accordance with Eq. (8.14), the canonical partition function Q_j for systems composed of N_j particles. Thus

$$Q_G = \sum_j Q_j \exp\left(\frac{\mu N_j}{kT}\right) \tag{8.43}$$

It can again be proved that this summation can be replaced with its largest element, $Q \exp(\mu N/kT)$,

$$Q_G = Q \exp\left(\frac{\mu N}{kT}\right) \tag{8.44}$$

[8] L. D. Landau and E. M. Lifshitz, for example, use the grand canonical ensemble *directly* to develop quantum statistics. See *Statistical Physics* (English translation from Russian), Addison-Wesley Publishing Company, Inc., Reading, Mass., 1958, p. 152.

The problem of determining the value of N for which the summand of Eq. (8.43) is truly maximum must still be solved. This can be done very easily: we differentiate the logarithms of Q exp $(\mu N/kT)$ with respect to N, and require that this differential must vanish. This resulting condition is

$$\frac{\partial \ln Q}{\partial N} = -\frac{\mu}{kT}$$

But we know that N should be the *average* number of particles in the systems, so from Eq. (8.25) we have

$$\frac{\partial \ln Q}{\partial N} = -\frac{1}{kT}\frac{\partial F}{\partial N}$$

We showed by perfectly general arguments at the end of Sec. 6.1 that

$$\frac{\partial F}{\partial N} = \text{chemical potential}$$

It follows that the μ that we should be using in Eq. (8.44), and in the development preceding it, should indeed be the chemical potential.

Finally, then, Eq. (8.44) can be recast in the form

$$\ln Q_G = \ln Q_{\max} + \frac{\mu N}{kT} = -\frac{F}{kT} + \frac{G}{kT} = \frac{pV}{kT}$$

or

$$\ln Q_G = \frac{pV}{kT} \qquad \text{compressibility factor} \tag{8.45}$$

At this point we should recall that Eq. (6.28) is of exactly this form,

$$q = \frac{pV}{kT} \tag{6.28}$$

The q potential and the logarithm of the grand canonical partition function are therefore exactly the same thing.[9] Like the q potential, $\ln Q_G$ may be viewed as the fundamental parameter. The Legendre transform of $S(U, V, N)$ into terms of $(1/T, V, \mu/T)$ is, as we found in Example 1.3,

$$S(U, V, N) \Rightarrow \frac{pV}{T}\left(\frac{1}{T}, V, \frac{\mu}{T}\right)$$

so that pV/T is the fundamental *macroscopic* parameter corresponding with the *microscopic* parameter, $\ln Q_G$. For convenience of computation we can recast this (Problem 8.4) as

$$pV(T, V, \mu) = kT \ln Q_G(T, V, \mu) \tag{8.46}$$

[9]And they, in turn, can be thought of as the compressibility factor per particle in a mole of gas.

Then in precise analogy with Eqs. (6.25), (6.26), and (6.27) we obtain

$$N = \left.\frac{\partial(pV)}{\partial\mu}\right|_{T,V} = kT\left.\frac{\partial \ln Q_G}{\partial\mu}\right|_{T,V} \tag{8.47}$$

$$U = \mu N - pV + TS = kT^2\left.\frac{\partial \ln Q_G}{\partial T}\right|_{V,\mu} \tag{8.48}$$

and

$$p = \left.\frac{\partial(pV)}{\partial V}\right|_{T,\mu} = kT\left.\frac{\partial \ln Q_G}{\partial V}\right|_{T,\mu} \tag{8.49}$$

The entropy is obtainable directly from Eq. (8.46),

$$S - \left.\frac{\partial(pV)}{\partial T}\right|_{V,\mu} = \frac{\partial}{\partial T}(kT \ln Q_G)_{V,\mu} \tag{8.50}$$

EXAMPLE 8.2 Verify the following unfamiliar Maxwell relation using the grand canonical partition function,

$$\left.\frac{\partial p}{\partial\mu}\right|_{T,V} = \left.\frac{\partial N}{\partial V}\right|_{T,\mu}$$

From Eqs. (8.49) and (8.47) we have

$$p = kT\left.\frac{\partial \ln Q_G}{\partial V}\right|_{T,\mu} \qquad \text{and} \qquad N = kT\left.\frac{\partial \ln Q_G}{\partial\mu}\right|_{T,V}$$

Therefore,

$$\left.\frac{\partial p}{\partial\mu}\right|_{T,V} = kT\frac{\partial^2 \ln Q_G}{\partial\mu\,\partial V} \qquad \text{and} \qquad \left.\frac{\partial N}{\partial V}\right|_{T,\mu} = kT\frac{\partial^2 \ln Q_G}{\partial V\,\partial\mu}$$

and since the order of partial differentiation does not alter the result, the Maxwell relation is proved.

EXAMPLE 8.3 Obtain the expression for the grand partition function for a single-phase, multicomponent system composed of independent and indistinguishable particles. Use this expression to evaluate pV.

The desired partition function is written for two components in the denominator of the right-hand side of Eq. (8.41). For α components it would be

$$Q_G = \sum_j \cdots \sum_l \sum_k G_{j,\ldots,l,k} \exp\left(\frac{\mu_1 N_{1_j} + \cdots + \mu_\alpha N_{\alpha l} - U_k}{kT}\right)$$

where, for independent and indistinguishable particles,

$$G_{j,\ldots,l,k} = \prod_{m=1}^{\alpha}\left(\prod_i \frac{g_i^{N_{imk}}}{N_{imk}!}\right)$$

Thus

$$\ln Q_G = \ln \sum_k \left[\prod_{m=1}^{\alpha} Q^m \exp\left(\frac{\mu_m N_m}{kT}\right) \right]$$

$$\simeq \ln \left[\prod_{m=1}^{\alpha} Q^m \exp\left(\frac{\mu_m N_m}{kT}\right) \right]_{\max}$$

$$\simeq \sum_{m=1}^{\alpha} Q_G{}^m$$

and

$$pV = kT \ln Q_G = kT \sum_{m=1}^{\alpha} \ln Q_G{}^m$$

It might at first seem that the methods of the grand canonical ensemble have given us nothing new over the results of Sec. 6.2. The important thing to remember here is that all our considerations at that point bore the restriction that particles were independent. While we have used G_{jk} for independent particles to *illustrate* how to evaluate Q_G, we have in no way restricted the preceding results for independent particles. We are now, therefore, in a position to describe a much broader class of equilibrium configurations than we were before.

It should be clear that the results in this section represent the macroscopic properties of equilibrium system "precisely," in the same sense that experiments measure properties precisely over times that greatly exceed the period of microscopic fluctuations. On the microscopic level our results are correct for the *average* behavior of the systems comprising the ensemble. Therefore, our attention turns naturally to the problem of understanding the nature of fluctuations of thermodynamic properties.

8.5 FLUCTUATIONS OF THERMODYNAMIC PROPERTIES

The properties of the systems within an ensemble vary about the mean or equilibrium values in the same way as a single system varies with the passage of time. We have so far emphasized that these fluctuations are generally quite small. We now wish to ask what the circumstances would be for fluctuations to become significant because there are indeed situations in which they can no longer be ignored. The Brownian movement of colloidal particles and other small-systems anomalies are examples; so, too, are the density fluctuations of gases near the critical point. The gross averaging character of statistical mechanics cannot predict exact fluctuations in a system, but it can predict the relative probabilities of the occurrence of fluctuations in the ensemble.

Another important reason for studying fluctuations is that the equivalence of the microcanonical, canonical, and grand canonical ensembles to one another depends upon fluctuations being very small. If, for example, energy fluctuations in the systems of a canonical ensemble are small, the ensemble is equivalent to a microcanonical ensemble. If the number of particles as well as the energy of the systems in a grand canonical ensemble fluctuate negligibly, all three ensembles are equivalent.

FLUCTUATIONS OF ENERGY OF A CANONICAL ENSEMBLE

We wish to consider the fluctuations in energy from system to system in a set of systems of equal particle number and temperature — in a canonical ensemble, in other words. The distribution of systems in energy U_j is given by Eq. (8.13). From this result we can compute the mean-square deviation $\sigma_U{}^2$ which is defined by

$$\sigma_U{}^2 \equiv \overline{(U_j - \overline{U})^2} = \overline{U^2} - \overline{U}^2 \tag{8.51}$$

Actually a more helpful measure of the magnitude of fluctuations is the "relative dispersion" $\sigma_U/\overline{U}$ or comparison of the root-mean-square energy fluctuation with the mean energy. To compute it we must first obtain the mean energy,

$$\overline{U} \equiv \sum_j \frac{\tilde{N}_j}{\tilde{N}} U_j = \frac{\sum_j U_j G_j \exp(-U_j/kT)}{\sum_j G_j \exp(-U_j/kT)} \tag{8.52}$$

Equation (8.52) can be rewritten as

$$\overline{U} \sum_j G_j \exp\left(-\frac{U_j}{kT}\right) = \sum_j U_j G_j \exp\left(-\frac{U_j}{kT}\right)$$

and differentiated with respect to T. The result is

$$\left.\frac{\partial \overline{U}}{\partial T}\right|_{V,N} \sum_j G_j \exp\left(-\frac{U_j}{kT}\right) + \frac{\overline{U}}{kT^2} \sum_j U_j G_j \exp\left(-\frac{U_j}{kT}\right) = \frac{1}{kT^2} \sum_j U_j{}^2 G_j \exp\left(-\frac{U_j}{kT}\right)$$

or, if we recognize in this expression the constant-volume specific heat, $(\partial U/\partial T)_{V,N}$, the average energy, and the average energy squared,

$$kT^2 c_v + \overline{U}^2 = \overline{U^2} \tag{8.53}$$

or

$$\frac{\sigma_U}{\overline{U}} = \frac{T\sqrt{kc_v}}{\overline{U}} \tag{8.53a}$$

For ordinary thermodynamic systems the relative dispersion is very small, owing to the presence of the Boltzmann constant in Eq. (8.53a). To be more specific, let us consider substances for which the equipartition principle holds. In this case $\overline{U} = DR^0T/2$ where D is the number of degrees of freedom, so

$$\frac{\sigma_U}{\overline{U}} = \sqrt{\frac{2}{D}}\frac{1}{\sqrt{N}} \tag{8.54}$$

For ideal monatomic gases, $D = 3$; for an Einstein solid that also obeys the equipartition principle, $D = 6$; and generally $\sqrt{2/D}$ is on the order of unity. Thus we find that in general

$$\frac{\sigma_U}{\overline{U}} \simeq N^{-1/2} \tag{8.54a}$$

The inverse-root-N result is well known for the relative dispersion of samples of size N from the mean value, and should be familiar to the student with a background in elementary statistics. It tells us that all but the tiniest molecular samples will yield negligible deviations from the mean. A 1-in.^{3} box filled with an ideal gas at room temperature will contain on the order of 10^{20} molecules and exhibit energy fluctuations of only about 0.00000001 percent of the total energy.

The pressure fluctuations of a monatomic ideal gas are also of interest. We find that Eq. (2.31),

$$p = \frac{\rho}{3}\overline{C^2} = \frac{2n}{3}\frac{m\overline{C^2}}{2} = \frac{2}{3}\frac{U}{V} \tag{2.31b}$$

can be used to replace U in Eq. (8.54),

$$\frac{\sigma_U}{U} = \sqrt{\frac{\overline{(U_j - \overline{U})^2}}{\overline{U}^2}} = \frac{\sigma_p}{\overline{p}} = \sqrt{\frac{2}{3}}\frac{1}{\sqrt{N}} \simeq \frac{1}{\sqrt{N}} \tag{8.55}$$

so that the inverse-root-N relation also characterizes the relative dispersion of pressure.

FLUCTUATIONS OF DENSITY IN A GRAND CANONICAL ENSEMBLE

We shall now shift our attention to the fluctuations of density that might occur as time passes in a system of constant T, V, and μ — or from system to system in a grand canonical ensemble at a given instant. Using the grand canonical distribution function Eq. (8.40) in the definition of an average, we obtain for $\overline{N}$,

$$\overline{N} = \sum\sum_{k} \frac{\tilde{N}_{jk}}{\tilde{N}} N_j = \frac{1}{Q_G}\sum_j\sum_k N_j G_{jk} \exp\left(\frac{\mu N_j - U_k}{kT}\right) \tag{8.56}$$

Cross-multiplying by Q_G and differentiating with respect to μ we obtain, as we did in Eq. (8.53),

$$\left.\frac{\partial \overline{N}}{\partial \mu}\right|_{T,V} + \frac{\overline{N}^2}{kT} = \frac{\overline{N^2}}{kT} \tag{8.57}$$

The relative dispersion of number density $\sigma_N/\overline{N}$ is then

$$\frac{\sigma_N}{\overline{N}} = \frac{(\overline{N^2} - \overline{N}^2)^{1/2}}{\overline{N}} = \frac{1}{\overline{N}}\left[kT\left(\frac{\partial \overline{N}}{\partial \mu}\right)_{T,V}\right]^{1/2} \tag{8.57a}$$

The derivative $(\partial \overline{N}/\partial \mu)_{T,V}$ can be reduced to more convenient form as follows.

We first employ the familiar chain-rule form to write

$$\left.\frac{\partial \overline{N}}{\partial \mu}\right|_{T,V} = -\left.\frac{\partial \overline{N}}{\partial V}\right|_{T,\mu} \left.\frac{\partial V}{\partial \mu}\right|_{T,\overline{N}}$$

We can consider p, instead of μ, to be held constant in the first derivative on the right-hand side because constant T and p implies constant μ. Then, after expanding the last derivative, we obtain

$$\left.\frac{\partial \overline{N}}{\partial \mu}\right|_{T,V} = -\left.\frac{\partial \overline{N}}{\partial V}\right|_{T,p} \left.\frac{\partial V}{\partial p}\right|_{T,\overline{N}} \left.\frac{\partial p}{\partial \mu}\right|_{T,\overline{N}} \tag{8.58}$$

Next, it is possible to construct a Maxwell relation by cross-differentiating the coefficients of dp and dN in the thermodynamic relation

$$dG = -S\,dT + V\,dp + \mu\,d\overline{N}$$

The result,

$$\left.\frac{\partial \mu}{\partial p}\right|_{T,\overline{N}} = \left.\frac{\partial V}{\partial \overline{N}}\right|_{T,p}$$

can be substituted into Eq. (8.58) to obtain

$$\left.\frac{\partial N}{\partial \mu}\right|_{T,V} = -\left[\left.\frac{\partial \overline{N}}{\partial V}\right|_{T,p}\right]^2 \left.\frac{\partial V}{\partial p}\right|_{T,\overline{N}} \tag{8.59}$$

The derivative in brackets reduces to

$$\left.\frac{\partial \overline{N}}{\partial V}\right|_{T,p} = \frac{\overline{N}}{V}$$

because $\overline{N}$ is directly proportional to V at constant T and p. Thus, Eq. (8.59) reduces to

$$\left.\frac{\partial N}{\partial \mu}\right|_{T,V} = -\left(\frac{\overline{N}}{V}\right)^2 \left.\frac{\partial V}{\partial p}\right|_{T,\overline{N}} \tag{8.60}$$

which simplifies Eq. (8.57a) to the convenient form

$$\frac{\sigma_N}{\bar{N}} = -\frac{1}{V}\left[kT\left(\frac{\partial V}{\partial p}\right)_{T,\bar{N}}\right]^{1/2} \tag{8.61}$$

It is no surprise at this point to discover that substitution of the ideal-gas law, $V = \bar{N}kT/p$, in Eq. (8.61) results in an inverse-root-N law again:

$$\frac{\sigma_N}{\bar{N}} = \frac{1}{\sqrt{\bar{N}}} \tag{8.62}$$

Thus density fluctuations in a system of constant T, V, and μ will normally be very small. Only when the number density of a system becomes very small will fluctuations have to be reckoned with. This result also justifies the use of the grand canonical ensemble to describe even systems possessed of a constant number of particles.

An interesting example of the role of density fluctuations in everyday experience is manifest in the appearance of blue sky. If there were no density fluctuations in the air, sunlight would not be scattered and the sky would be black except in the direct path to the sun. In actuality, very small adjacent regions of greatly different densities act as scattering centers, deflecting light from its straight path. To have appreciable scattering requires that a significant density fluctuation must occur over a region with dimensions of the order of the wavelength of the sunlight. Since red light has longer wavelength than blue, and significant fluctuations in density are more likely to occur in smaller regions, blue light is scattered more than red. On the same basis the glancing rays that penetrate through a very deep layer of air at sunrise and sunset tend to be red — the blues tend to be scattered out before they can be seen. This is why dawn and dusk are characteristically rich in red hues.

In addition to density fluctuations we can also calculate the energy fluctuations for a grand canonical ensemble. The result, however, will be a sum of two terms: One is the fluctuation in the energy of the *canonical* ensemble with a fixed number of particles; the other is the fluctuation in energy associated with the fluctuation of the number of particles.[10] The relative dispersion in energy computed for this case is of the same order of magnitude as given by Eq. (8.54) for the canonical ensemble.

[10]For a more complete discussion see N. Davidson, *Statistical Mechanics*, McGraw-Hill, Inc., New York, 1962, p. 270.

8.6 SUMMARY OF ENSEMBLES IN STATISTICAL MECHANICS

Our discussion has embraced only three ensemble methods, but it should be clear that others can be developed as the need dictates. Table 8.1 illustrates how additional ensembles can be proposed and their properties developed. The table paraphrases one presented by Knuth[11] in his excellent elementary discussion of ensemble methods.

TABLE 8.1 List of Ensembles That Can Be Formed

	Character of systems in the ensemble	Independent parameters of systems	Generating function for properties	Volume of systems
1	Isolated (microcanonical)	U, V, N	$S/k = \ln W$	Fixed
2[a]	Closed (canonical)	β, V, N	$-F/kT = \ln Q$	Fixed
3	Insulated to energy, but allowing mass transfer	$U, V, \mu/T$	H/kT	Fixed
4[a]	Open (grand canonical)	$\beta, V, \mu/T$	$pV/kT = q = \ln Q_G$	Fixed
5[b]	Isolated	U, π, N	$-pV/kT + S/K$	Variable
6[a,b]	Closed	β, π, N	$-G/kT$	Variable
7[b]	Insulated to energy but allowing mass transfer	$U, \pi, \mu/T$	U/kT	Variable
8[a,b]	Open	$\beta, \pi, \mu/T$	0	Variable

[a]The parameter $1/T$ is characterized as the Lagrangian multiplier, β.
[b]The parameter π is another Lagrangian multiplier, which enters by virtue of the variability of the volume.

[11]E. L. Knuth, *Introduction to Statistical Thermodynamics*, McGraw-Hill, Inc., New York, 1966, chap. 3.

Problems 8.1 Provide a descriptive illustration of some ensemble which might be observed in the world around you and which you think would be approximately ergodic. Is your example comparable with one of the three types described in Section 8.1?

8.2 Verify Eq. (8.30). Does the Lagrangian multiplier β appear in this result? Explain.

8.3 Develop Eq. (8.41) fully.

8.4 Use the Legendre transform of U to show that Eq. (8.46) is a proper fundamental equation. Then verify Eqs. (8.47), (8.48), (8.49), and (8.50).

8.5 Suppose the systems in an ensemble are impermeable but that they interact with constant-pressure and constant-temperature reservoirs. What are the governing macroscopic parameters? The characteristic microscopic expression? The relationship among them?

8.6 Obtain the derivatives $\partial\bar{U}/\partial V|_{\beta,N}$ and $\partial\bar{p}/\partial\beta|_{V,N}$ and show that

$$\left.\frac{\partial\bar{U}}{\partial V}\right|_{\beta,N} + \beta\left.\frac{\partial\bar{p}}{\partial\beta}\right|_{N,V} = -p$$

This result is interesting because comparison with the thermodynamic equation

$$\left.\frac{\partial\bar{U}}{\partial V}\right|_{T,N} = -\left[p + \frac{1}{T}\left.\frac{\partial p}{\partial(1/T)}\right|_{N,V}\right]$$

implies that $\beta = \text{constant}/T$, as of course it must be.

8.7 Use the method of Lagrangian multipliers to maximize $-\sum_i p_i \ln p_i$ subject to the constraint $\sum_i p_i = 1$. Show that when this quantity is maximum, $p_i = \text{constant}$.

8.8 For a microcanonical ensemble of single-component systems, show that

$$U\left(\frac{\partial \ln \Omega}{\partial U}\right)_{V,N} + V\left(\frac{\partial \ln \Omega}{\partial V}\right)_{U,N} + N\left(\frac{\partial \ln \Omega}{\partial N}\right)_{U,V} = \ln \Omega$$

8.9 Show how to evaluate the canonical partition function as a function of temperature, given $c_v(T)$ data.

8.10 At what approximate altitude do 1-in.^{3} samples of atmospheric air exhibit fluctuations in pressure that would appear in functional measurements?

8.11 In the study of nucleate boiling one is interested in the active nucleation sites or locations on a heating element from which bubbles are triggered. A common measure is the active site density, n sites/ft^2. Suppose you are studying boiling on a representative surface of area 10 in.^{2}. How high must n rise before the variation in observed n among different surfaces of the same type drops below ± 3 percent?

8.12 If we wished to do so, we could take $\overline{\Delta U^3} \equiv \overline{(U_j - \overline{U})^3}$ as a measure of energy fluctuations in the canonical ensemble. Show that

$$\overline{\Delta U^3} = -\frac{\partial^3 \ln Q}{\partial \beta^3}$$

8.13 Show that $c_v = N(\overline{\epsilon^2} - \bar{\epsilon}^2)/kT^2$ and decide whether or not it can ever be negative.

8.14 Given the result of Problem 8.13 show that

$$\frac{\sigma_\epsilon}{\bar{\epsilon}} = N^{1/2}\frac{\sigma_U}{U}$$

This result emphasizes that the distribution of particle energies is much broader than the distribution of system energies.

8.15 For a grand canonical ensemble of single-component ideal-gas systems of N average number of particles, show that the probability of finding a system with N_i particles is given by the Poisson distribution,

$$\mathcal{P}(N_i) = \frac{1}{N_i!}e^{-N}N^{N_i}$$

9 thermostatic properties of dense fluids

The ensemble formulation of statistical mechanics gives, as we indicated in chapter 8, a general basis upon which it is possible to predict the equilibrium behavior of nondilute gases. It is necessary to introduce an intermolecular potential-energy function into the description of molecular behavior, and the resulting computations can understandably become quite complicated. The description of such interacting or "dependent" particles requires an increasingly accurate model for the intermolecular potential function as the density increases. As the gas becomes more dilute, long-range forces are slight and can be characterized fairly crudely to give a correction to basically ideal-gas behavior. For more dense gases and for liquids, these computations must be done far less casually and computations are increasingly difficult.

Most of this chapter deals primarily with "moderately dense" gases, for which the theory is well developed and reasonably simple. Many useful applications exist in this range of behavior and the treatment also serves as a starting point for the more complex and general theories. In Sec. 9.4 we consider some applications of the law of corresponding states, which can be used to describe very nonideal behavior.

9.1 STATISTICAL-MECHANICAL DESCRIPTION OF MODERATELY DENSE GASES

CANONICAL PARTITION FUNCTION AND FUNDAMENTAL EQUATION

As we indicated in chapter 8, we have considerable latitude in choosing any particular ensemble approach. For practical purposes the canonical ensemble suits our present needs well. We have no need to consider chemical interactions (variable N) at the moment, and the canonical ensemble provides a statistical-mechanical description that allows particle interactions. The fundamental parameter with which we must deal is the canonical partition function Q,

$$Q = \sum_{i=0}^{\infty} G_j \exp(-\beta U_j) \tag{8.14}$$

The energy of the system U_j is expressed in hybrid form. We are well aware that the quantization of the translation of ordinary particles is unimportant in *all* cases of practical importance, while quantization of the internal modes frequently must be considered. Finally, U_j contains a potential energy $\phi(\mathbf{r}_1, \ldots, \mathbf{r}_N)$ of the N particles relative to one another. We defer discussion of ϕ until later. The resulting U_j is

$$U_j = \sum_{i=1}^{N} (\epsilon_i)_{\text{int}} + \sum_{i=1}^{N} \frac{|\mathbf{p}_i|^2}{2m} + \phi(\mathbf{r}_1, \ldots, \mathbf{r}_N) \tag{9.1}$$

where the $\mathbf{p}_i$'s are momenta of the individual particles and the $\mathbf{r}_i$'s are their positions. Because we wish to take a quantum view of only the internal modes in this hybrid expression, we factor their contribution out of Q and write[1]

$$Q = Q_{\text{int}}\left[\frac{1}{N!h^{3N}} \int \cdots \int \exp\left(-\sum_{i=1}^{N} \frac{|\mathbf{p}_i|^2}{2mkT}\right) \times \exp\left(-\frac{\phi}{kT}\right) d\mathbf{p}_1 \cdots d\mathbf{p}_N \, d\mathbf{r}_1 \cdots d\mathbf{r}_N\right] \tag{9.2}$$

[1]We use a notational device here that we have previously avoided: we let $d\mathbf{r}$ denote a volume element dV—the change of position in three-dimensional space.

in concordance with the phase integral, Eq. (5.37),[2] and the relation between Q and Z, Eq. (8.31) (see Problem 9.1). The internal partition function for the system Q_{int} is simply

$$Q_{\text{int}} = Z_{\text{int}}{}^{N} \tag{9.3}$$

because the *internal* energy state of each particle can be assumed independent of every other particle.

The integration over momenta is easily done, and it leads to a recognizable result,

$$\int_{-\infty}^{\infty} \cdots \int \exp\left(-\sum_{i=1}^{N} \frac{|\mathbf{p}_i|^2}{2mkT}\right) d\mathbf{p}_1 \cdots d\mathbf{p}_N = (2\pi mkT)^{3N/2} \tag{9.4}$$

The canonical partition function is, therefore,

$$Q = (Z_{\text{int}})^N \frac{1}{N!} \Lambda^{-3N} Z_\phi \tag{9.5}$$

where Λ, the *thermal de Broglie wavelength* (see Example 5.4) is

$$\Lambda \equiv \left(\frac{h^2}{2\pi mkT}\right)^{1/2}$$

and Z_ϕ, often called the *configuration integral*, is defined as

$$Z_\phi(T, V) \equiv \int \cdots \int \exp\left[-\frac{\phi(\mathbf{r}_1, \ldots, \mathbf{r}_N)}{kT}\right] d\mathbf{r}_1 \cdots d\mathbf{r}_N \tag{9.6}$$

When the configuration integral is evaluated, the fundamental equation can be written and used to evaluate all thermodynamic information:

$$F(T, V, N) = -kT \ln Q(T, V, N) \tag{8.25b}$$

In this case it takes the form

$$F = -NkT[\ln Z_{\text{int}} - (\ln N - 1) - 3 \ln \Lambda] - kT \ln Z_\phi \tag{9.7}$$

EVALUATION OF THE CONFIGURATION INTEGRAL

It should be clear from Eq. (9.7) that the most important element in the statistical mechanical description of non-ideal gases is the evaluation of the configuration integral. But this is not a simple matter. To achieve a simple workable formulation, a number of simplications and assumptions must be made. We consider here a gas for which the following restrictions are applicable:

[2]Although Eq. (5.37) was developed for independent particles, Eq. (9.2) is developed without need for this assumption.

1. The potential energy of interaction between any pair of molecules can be regarded as a function only of the distance between them. Their relative orientation can be ignored. This is valid for spherically symmetrical potentials, which are commonly exhibited by nonpolar, uncharged molecules with reasonably spherically symmetrical forms. A more general treatment must be employed for molecules with permanent dipole moments, grossly unspherical forms, or electrical charges.

2. The potential energy of the gas is the sum of the potential energies of all the pairs. This restriction is reasonable when gases are not too dense. It says in essence that the gas is sufficiently "thin" that only two particles interact at a time and additional energies of interaction can be ignored. Thus

$$\phi(\mathbf{r}_1, \ldots, \mathbf{r}_N) = \tfrac{1}{2} \sum_{i \neq j} \phi_{ij}(r_{ij}) \tag{9.8}$$

where $r_{ij} = |\mathbf{r}_j - \mathbf{r}_i|$ and the factor of $\frac{1}{2}$ is included to compensate for counting each interaction twice — once as ij and once as ji. Another way of expressing this summation is

$$\phi(\mathbf{r}_1, \ldots, \mathbf{r}_N) = \sum_{1 \leq i < j \leq N} \phi_{ij}(r_{ij}) \tag{9.9}$$

For ideal gases there is no potential energy of interaction: $\phi = \phi_{ij} = 0$ and the exponential terms, $\exp(-\phi_{ij}/kT)$, are equal to unity. Thus, for weakly interacting particles, it is reasonable to introduce a function, $f_{ij}(r_{ij})$,

$$f_{ij}(r_{ij}) \equiv \exp\left(\frac{-\phi_{ij}}{kT}\right) - 1 \tag{9.10}$$

Since the magnitude of f_{ij} is generally much less than unity, we can then rewrite Eq. (9.9) as

$$\exp\left(-\frac{\phi}{kT}\right) = \prod_{1 \leq i < j \leq N} \exp\left(-\frac{\phi_{ij}}{kT}\right) = \prod_{1 \leq i < j \leq N} (1 + f_{ij})$$

or

$$\exp\left(-\frac{\phi}{kT}\right) \simeq 1 + \sum_{1 \leq i < j \leq N} f_{ij} \tag{9.11}$$

The approximation in Eq. (9.11) results from dropping the second- and higher-order product terms, $\sum f_{ij} f_{i'j'}$, and so forth. A detailed treatment of such terms is the subject of the theory of cluster integrals.[3] It is rather complicated and we shall not take it up here.

[3] T. L. Hill, *Statistical Mechanics*, McGraw-Hill, Inc., New York, 1956, pp. 122–129, 136–144.

The product terms actually represent the interactions of more than two particles at once. Although we do not wish to treat these terms in any complete way, it will pay us to investigate their meaning in a little more detail. Figure 9.1 illustrates in a schematic way how a "moderately dense" and a "dense" gas might look. In Fig. 9.1(a), particle number 1 of a moderately dense gas interacts with all others. However, the effective range of interaction — or the distance beyond which ϕ becomes vanishingly small — is much less than its distance from any neighboring particles. Accordingly the sum over f_{1_j} gives no contribution. The sum over all f_{ij} picks up two important contributions from the 4-5 and 6-7 pairs, and in this case these two pairs represent the entire meaningful contribution to ϕ.

Figure 9.1(b) shows a gas that is dense enough so that, in addition to interacting *pairs*, there are some *clusters* of three and four interacting particles. The single or noninteracting particles in Fig. 9.1(b) as well as in 9.1(a) contribute *in toto* only the lead term of unity in Eq. (9.11). The pairs contribute through the $\sum f_{ij}$ term. The term $\sum f_{ij}f_{i'j'}$ would be roughly zero in all the interactions in Fig. 9.1(a), because the two factors are never both significant at once. This is

Fig. 9.1 Clustering in a gas whose effective range of intermolecular potential is *R*.

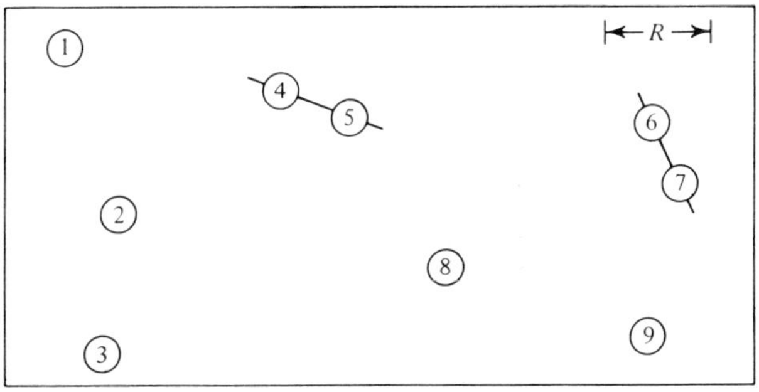

(a) A "moderately dense" gas

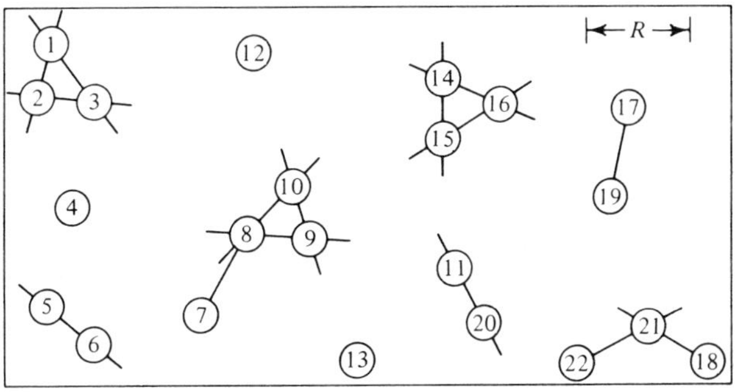

(b) A "dense" gas

not the case in Fig. 9.1(b), and the second term has to be included in the case of $f_{22,21}f_{21,18}$ or $f_{2,1}f_{2,3} + f_{1,3}f_{1,2} + f_{3,2}f_{3,1}$, and so on. Similarly, higher-order terms will arise as the complexity of the clusters of interacting particles becomes greater.

Returning to the configuration integral, Eq. (9.6), with Eq. (9.11) we obtain

$$Z_\phi \simeq \int_{(3N)} \cdots \int \left(1 + \sum_{i<j} f_{ij}\right) \prod_{k=1}^{N} d\mathbf{r}_k \tag{9.12}$$

for moderately dense gases. Integration of the first term gives simply V^N, so

$$Z_\phi = V^N + \int_{(3N)} \cdots \int \sum_{i<j} f_{ij} \prod_{k=1}^{N} d\mathbf{r}_k \tag{9.13}$$

Equation (9.13) is extremely interesting because it reveals an additive form for the configuration integral that yields the ideal-gas case from its lead term.

Since each individual f depends only upon $r_{ij} = |\mathbf{r}_i - \mathbf{r}_j|$, it is possible to rearrange the $3N$-fold integral in Eq. (9.13) without the summation as follows:

$$\begin{aligned}\int_{(3N)} \cdots \int f_{ij}(r_{ij})\, d\mathbf{r}_1 \cdots d\mathbf{r}_N &= \iint f_{ij}(r_{ij})\, d\mathbf{r}_i\, d\mathbf{r}_j \int_{3(N-2)} \cdots \int \prod_{\substack{k=1 \\ \neq i,j}}^{N} d\mathbf{r}_k \\ &= V^{N-2} \iint f_{ij}(r_{ij})\, d\mathbf{r}_i\, d\mathbf{r}_j \end{aligned} \tag{9.14}$$

The integral is broken down further with the help of a clever approximation. The position $\mathbf{r}_j$ is of no interest to us unless it is very close to $\mathbf{r}_i$ (unless, of course, $\mathbf{r}_j$ is near enough to a wall to interact with *it*). We therefore imagine that the jth particle is at the origin; we change $f_{ij}(r_{ij})$ to $f_{ij}(|\mathbf{r}_i|)$; and we factor out $\int d\mathbf{r}_j$. The result is

$$\int_{(3N)} \cdots \int f_{ij}(r_{ij})\, d\mathbf{r}_1 \cdots d\mathbf{r}_N = V^{N-2} \int d\mathbf{r}_j \int f_{ij}(r_i)\, d\mathbf{r}_i = V^{N-1} \int f_{ij}(r_i)\, d\mathbf{r}_i \tag{9.15}$$

But $\int f_{ij}(r_i)\, d\mathbf{r}_i = 4\pi \int_0^\infty f(r) r^2\, dr$ in a polar-coordinate representation, where we drop the subscripts because we have no further need of them. To carry out the summation in Eq. (9.13) we need simply to observe that for N particles there will be $N(N-1)/2 \simeq N^2/2$ identical terms in the sum. Thus Eq. (9.13) becomes

$$Z_\phi = V^N + V^{N-1}\, 2\pi N^2 \int_0^\infty f(r) r^2\, dr$$

The above equation can be written in the form

$$Z_\phi = V^N\left[1 - \frac{N^2B(T)}{N_A V}\right] \tag{9.16}$$

where

$$B(T) = -2\pi N_A \int_0^\infty f(r)r^2\,dr = 2\pi N_A \int_0^\infty \left[1 - \exp\left(-\frac{\phi}{kT}\right)\right] r^2\,dr \tag{9.17}$$

The parameter $B(T)$ is called the "second virial coefficient" and it plays a most important role in the equation of state for non-ideal gases as we shall indicate in the following sub-section.

It should be emphasized here that Eq. (9.16) is an approximate expression based on the approximation made in Eq. (9.11). If higher order product terms of f_{ij} are included, it will result in a series expression for Z_ϕ

$$Z_\phi = V^N\left[1 - \frac{N^2B(T)}{N_A V} + \cdots\right] \tag{9.16a}$$

Equation (9.16a), coupled with Eq. (9.7), provides the basis for developing a fundamental equation to describe gases for which ϕ is known. Clearly the knowledge of ϕ will be of crucial importance in developing realistic descriptions of dense-gas behavior. Before discussing potential functions for various molecular models, however, we should like to introduce the virial equation of state which has been widely used in calculations for non-ideal gases.

VIRIAL EQUATION OF STATE

Of the many "equations of state" obtainable from the fundamental equation, the pressure relation,

$$p = -\left.\frac{\partial F}{\partial V}\right|_{T,N} = kT \left.\frac{\partial \ln Z_\phi(T, V)}{\partial V}\right|_{T,N} \tag{9.18}$$

is of most interest to us because it gives the p-v-T surface directly. But Eq. (9.18) retains only the configuration integral contribution from the fundamental equation! This is not an entirely new result. It says merely that pressure depends upon action lodged wholly within the translational mode—that only this energy mode involves volume. Substituting Eq. (9.16a) in Eq. (9.18) and applying the power series expansion, $\ln(1 + x) = x - x^2/2 + \cdots$, gives

$$p = \frac{NkT}{V} + \frac{N^2kTB(T)}{N_A V^2} + \cdots$$

or

$$p = \frac{R^0T}{v} + \frac{R^0TB(T)}{v^2} + \cdots \tag{9.19}$$

where v is the molar volume ($v = N_A V/N$).

The importance of Eq. (9.19) is obvious. The lead term represents the contribution of the ideal-gas behavior. Anything beyond the lead term is called the *internal virial* (derived from the Latin *vires*, meaning forces) because it is the result of intermolecular forces. Likewise, the lead, or ideal-gas, term is sometimes called the *external virial* because it represents the external force imposed upon the system by the walls.[4]

Because ultimate accuracy in predicting the internal virial is not attainable, it is customary to express it in a power series, usually in inverse volume or density. Thus

$$\frac{pv}{R^0T} = 1 + \frac{B(T)}{v} + \frac{C(T)}{v^2} + \cdots = \sum_{n=1}^{\infty} \frac{A_n(T)}{v^{n-1}} \tag{9.19a}$$

Accordingly, the first virial coefficient would be unity, the second would be $B(T)$, and so on. Many means exists for developing virial equations of state. Often Eq. (9.19a) is simply used as a very effective formula for interpolating and extrapolating experimental data. We have just seen how the higher terms can be approximated formally using the methods of classical statistical mechanics.

9.2 VAN DER WAALS'S EQUATION AND OTHER RESULTS OF ELEMENTARY MOLECULAR MODELS

THREE SIMPLE MODELS FOR MOLECULAR INTERACTION

The three models for the intermolecular potential energy function, ϕ, that we wish to consider first are shown in Fig. 9.2. They can be characterized as the *noninteracting point particle* [Fig. 9.2(a)], the *noninteracting rigid sphere* of diameter σ [Fig. 9.2(b)], and the *weakly attracting rigid particle* [Fig. 9.2(c)]. In each case the intermolecular force, $\mathfrak{F} = -d\phi/dr$, is sketched in an inset. In the present considerations we consider attraction to be positive and repulsion to be negative in sign.

If we go to Eq. (9.18) with ϕ for noninteracting point particles, the resulting value of $B(T)$ is zero and we obtain (as we would expect to) from Eq. (9.19), the ideal-gas law.

The substitution of the potential for noninteracting rigid spherical particles results in a finite contribution from $B(T)$, because

$$B(T) = 2\pi N_A\left[\int_0^\sigma (1-0)r^2\,dr + \int_\sigma^\infty (1-1)r^2\,dr\right] \tag{9.20}$$

[4]These ideas are developed quantitatively by J. O. Hirschfelder, C. F. Curtiss and R. B. Bird, *Molecular Theory of Gases and Liquids*, John Wiley & Sons, Inc., New York, 1954, chap. 1.

Fig. 9.2 Intermolecular potential functions for three very simple molecular models. (Corresponding force versus distance behavior is shown in the insets.)

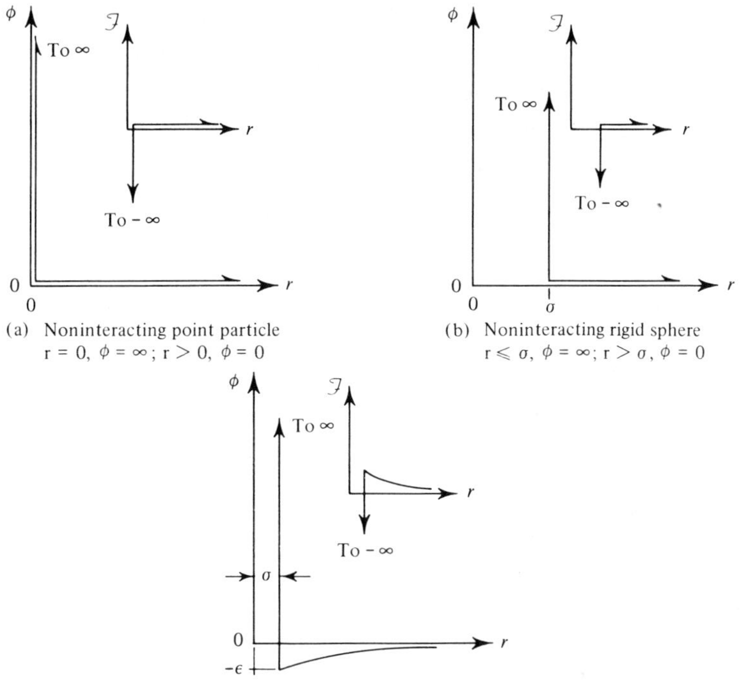

(a) Noninteracting point particle
$r = 0,\ \phi = \infty;\ r > 0,\ \phi = 0$

(b) Noninteracting rigid sphere
$r \leqslant \sigma,\ \phi = \infty;\ r > \sigma,\ \phi = 0$

(c) Weakly attracting rigid particle
$r \leqslant \sigma,\ \phi = \infty;\ r > \sigma,\ \phi = \dfrac{-c}{r^{\gamma}}$

where σ is usually taken as the closest approach of two particles. As such it is equal to an effective diameter of the particles. Thus

$$B(T) = 2\pi N_{\mathrm{A}}\left(\frac{\sigma^3}{3}\right) = 4\left[\frac{4\pi}{3}\left(\frac{\sigma}{2}\right)^3 N_{\mathrm{A}}\right] \qquad (9.21)$$

Equation (9.21) is an evocative result; it says that the second virial coefficient is four times the volume occupied by the molecules themselves. However, if the reader visualizes the collision of — say — two billiard balls, he can readily see that the presence of one billiard ball excludes the *center* of the other from a sphere whose diameter is *twice* σ. It excludes the other ball from a sphere of *eight* times its own volume. One half of this or four times its own volume is chargeable to each of the two spheres. Thus $B(T)$ is, in this case, the *covolume* b or the volume excluded to any one molecule by virtue of the presence of the other molecules. Calling $4(4\pi/3)(\sigma/2)^3 N_A \equiv b$, we then obtain the following equation of state[5]:

$$\frac{pv}{R^0T} = \frac{v+b}{v}$$

or since $b \ll v$ in our present considerations we can write the right-hand side as $[1-(b/v)]^{-1}$ and obtain an equivalent approximate equation of state,

$$p(v-b) = R^0T \tag{9.22}$$

When we use the weakly attracting rigid particle potential in Eq. (9.17),

$$B(T) = 2\pi N_A\left\{\int_0^\sigma r^2\,dr + \int_\sigma^\infty \left[1-\exp\left(\frac{c}{kT}r^{-\gamma}\right)\right]r^2\,dr\right\} \tag{9.23}$$

But the requirement that attraction must be weak means that $|\phi|$ is a small number, never more than ϵ. Thus $\exp[(c/kT)r^{-\gamma}] \simeq 1 + (c/kT)r^{-\gamma}$, so

$$B(T) = 2\pi N_A\left(\frac{\sigma^3}{3} + \frac{c}{kT}\frac{\sigma^{3-\gamma}}{3-\gamma}\right)$$

This expression has no validity for $\gamma \leq 3$, however γ is typically much larger than 3. This kind of limitation is discussed in Example 9.2. Since $c\sigma^{3-\gamma}/(3-\gamma)$ is a constant, this $B(T)$ can be written as

$$B(T) = b - \frac{a}{R^0T} \tag{9.24}$$

For weak attractions in general, a is of the form

$$a = -2\pi N_A^2 \int_\sigma^\infty r^2\phi(r)\,dr \tag{9.25}$$

If we consider weak attraction in a gas for which the covolume is negligible ($b \ll a/R^0T$), then

$$\frac{pv}{R^0T} = 1 - \frac{a}{R^0Tv}$$

or

$$p + \frac{a}{v^2} = \frac{R^0T}{v} \tag{9.26}$$

[5]Hirschfelder *et al.*, footnote 4, show that if cluster collisions are taken into account, this purely geometrical equation of state takes the following power-series form as the gas becomes more dense:

$$\frac{pv}{R^0T} = 1 + \frac{b}{v} + 0.6250\frac{b^2}{v^2} + 0.2869\frac{b^3}{v^3} + \cdots$$

Equations (9.22) and (9.26) represent two very important limiting cases of moderately dense gas behavior. Equation (9.22) shows the purely geometrical effect upon the ideal-gas law of molecules "consuming" part of the space that they occupy. Equation (9.26) shows that weak attraction accounts for part of the pressure that molecules might otherwise exert upon the walls of their container. These ideas provide the key to a very important early attempt to describe real-gas behavior.

VAN DER WAALS'S EQUATION

Van der Waals developed his famous equation of state in 1873 in an attempt to show the intrinsic continuity between the liquid and gas states. He achieved his objective and gave us the most remarkable equation of state yet conceived. It is utterly simple and plausible, and it provides at least a qualitative representation of every major feature of real gas behavior.

To make his prediction van der Waals suggested that real fluid behavior was the result of two molecular effects. The first was the occupancy of part of the free volume by the covolume; the second was the removal of part of the pressure imposed by the walls, by internal attraction. The corrections were made on the basis of intuitive arguments, and they took exactly the form of these corrections in Eqs. (9.22) and (9.26). But, whereas Eqs. (9.22) and (9.26) were subject to the "moderately dense" assumptions set down in Sec. 9.1, van der Waals's simple physical assumptions were not. He made both assumptions simultaneously and obtained the appropriate combination of Eqs. (9.22) and (9.26),

$$p = \frac{R^0 T}{v - b} - \frac{a}{v^2} \tag{9.27}$$

for use over the *entire* range of fluid behavior. Equations (9.21) and (9.25) give reasonable estimates of a and b for this equation as long as it is applied only to moderately dense gases. However, we shortly discover that much better estimates of a and b can be made from critical data for use over the entire range of conditions.

The first quality of real gas behavior that we find borne out in the van der Waals equation is the *law of corresponding states*. This is an approximate empirical law which says that all fluids (at least families of those that are thermodynamically "similar") can be described by the same p-v-T surface in the transformed coordinates

$$p_r \equiv \frac{p}{p_c} \qquad T_r \equiv \frac{T}{T_c} \qquad v_r \equiv \frac{v}{v_c} \tag{9.28}$$

where the subscript c denotes the thermodynamic critical point and the subscript r identifies the *reduced* variables. Thus the law suggests that there is a universal function

$$f(p_r, v_r, T_r) = 0 \tag{9.29}$$

which will represent all fluids fairly well. We have more to say about the microscopic basis for this law later. For the moment we wish only to view it as observed macroscopic behavior.

Now suppose that the one point in the van der Waals surface at which

$$\left.\frac{\partial p}{\partial v}\right|_T = 0 \quad \text{and} \quad \left.\frac{\partial^2 p}{\partial v^2}\right|_T = 0 \tag{9.30}$$

corresponds, as experimental results suggest it should, with the critical point. The simultaneous solution of conditions (9.30) and the van der Waals equation (9.27) gives

$$a = 3p_c v_c^2 \tag{9.31}$$

$$b = \frac{v_c}{3} \tag{9.32}$$

and

$$R^0 = \frac{8}{3}\frac{p_c v_c}{T_c} \tag{9.33}$$

These conditions are used to eliminate the constants from Eq. (9.27) so that it takes the dimensionless form

$$p_r = \frac{8T_r}{3v_r - 1} - \frac{3}{v_r^2} \tag{9.34}$$

This is in the form of Eq. (9.29); therefore the van der Waals equation is not only true to observation, but it also provides some basis for the law of corresponding states.

Equation (9.33) bears out another aspect of real-gas behavior by requiring that the compressibility factor $Z \equiv pv/R^0T$ has a universal value at the critical point

$$Z_c = \frac{p_c v_c}{R^0 T_c} = \frac{3}{8} \tag{9.33a}$$

Experiments generally place this constant below $\frac{3}{8}$. Table 9.1 gives typical values of critical data and Z_c which show that Z_c varies from about 0.3 down to about 0.22, depending on the complexity of the molecule.

TABLE 9.1 Critical Data for Several Types of Substance

		Observed values				Comparison with molecular values[a]		
Substance	Type	p_c, atm	T_c, °K	v_c, cm³/g mole	$\frac{p_c v_c}{R^\circ T_c}$	p_c/p_c^*	T_c/T_c^*	v_c/v_c^*
Argon	Almost spherical, nonpolar	48	151	75.2	0.291	0.243	1.26	1.51
H_2		12.8	33.3	65.0	0.304	0.134	0.90	2.05
Helium		2.26	5.3	57.8	0.300	0.057	0.52	2.75
N_2		33.5	126.1	90.1	0.292	0.274	1.33	1.41
O_2		49.7	154.4	74.4	0.292	0.297	1.31	1.28
Xenon		57.89	289.81	120.2	0.293	0.276	1.31	1.38
CH_4		45.8	190.7	99.0	0.290	0.264	1.29	1.41
Propane	Hydro-carbons	42.0	370	195.2	0.270			
n-Butane		37.5	425.2	248.5	0.257			
Ethane		48.2	305.5	144.5	0.267			
n-Hexane		29.9	507.9	359	0.260			
Benzene		48.6	562	254.5	0.265			
Ethylene		50.5	282.4	121.4	0.291			
Steam	Polar molecules	218.0	647.3	56.8	0.233	0.164	1.70	2.35
Ammonia		111.3	405.5	70.8	0.238	0.094	1.27	3.20
Methanol		78.5	513.2	115.1	0.220	0.027	0.82	6.59
Methyl chloride		65.9	416.3	139.8	0.258	0.107	1.10	2.75

[a]Molecular data for spherical molecules are based on the Lennard-Jones potential. Molecular data for polar molecules are based on the Stockmayer potential.

Now let us return to van der Waals's original thesis — that the liquid and gas states are intrinsically continuous. Figure 9.3 shows several lines of constant temperature on the van der Waals surface. When $T < T_c$, v becomes triple-valued in p but the conventional liquid-vapor transition is not in evidence. However, if the gas and liquid are to coexist in (stable) equilibrium at one temperature, the chemical potentials must be equal in the two phases,

$$\mu_g - \mu_f = \int_f^g d\mu|_T = 0 \tag{9.35}$$

or

$$\int_f^g (v\,dp - s\,dT)_T = \int_f^g v\,dp = 0 \tag{9.36}$$

Fig. 9.3 Isotherms of the van der Waals equation of state. (Not to scale — v_r coordinate stretched on the left.)

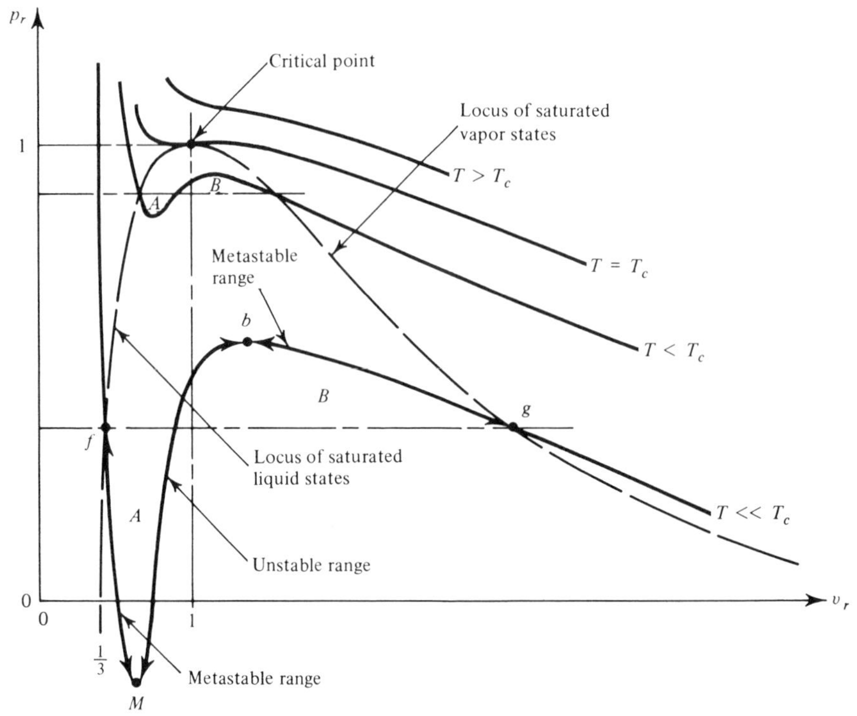

Equation (9.36) is satisfied only if the liquid–vapor equilibrium states are located such that areas A and B in Fig. 9.3 are equal.

The resulting locus of saturated liquid and vapor states is shown as a dashed line in Fig. 9.3. The corresponding vapor-pressure curve is similar in form to that of real fluids, although it predicts saturation temperatures that are low by as much as 35 percent.

It is in the region between f and g that the equation best displays its remarkable fidelity to real fluid behavior. The two regions of negative slope, f to M and b to g, are regions of metastable equilibrium. The region of positive slope, M to b, can be shown (Problem 9.4) to be unstable — no states can persist in it. The points M therefore represent the maximum superheat that a liquid can sustain (or the minimum pressure that a liquid of a given temperature can sustain).

The van der Waals equation predicts that at low temperatures liquids sustain enormous *tension* — a fact that has led some authors to take the equation lightly. In recent years measurements have been made that reveal this to be entirely correct.[6] Liquids that

[6]See, e.g., H. N. V. Temperley, "The Behavior of Water under Hydrostatic Tension: III," *Proc. Phys. Soc. (London)* **59,** 199 (1947).

are clean and free of dissolved gas can be subjected to tensions greater in magnitude than p_c. The reason is that, as the molecules become closely packed, the weak attraction forces become relatively very strong and serve to hold the liquid together. By the same token, it is possible to superheat liquids at constant pressure to temperatures hundreds of degrees in excess of their boiling point before enough thermal expansion has occurred that the relatively short-range attraction forces become small with respect to molecular momentum, and the fluid must come to a new equilibrium in the gas phase.

The equation also shows that, in the limit as either the temperature is lowered or the pressure is increased, the fluid will reach a minimum volume equal to b — on the order of the covolume — and that this is one third of the critical volume. For example, $v/v_c = 1/3.14$ for water at its triple point.

The van der Waals equation also provides some understanding of the "caloric," or heat-capacity, behavior of a real fluid. The classical expression for the effect of isothermal compression upon an ideal gas is

$$c_p - c_p^0 = \int_0^p T \left.\frac{\partial^2 v}{\partial T^2}\right|_p dp \tag{9.37}$$

where c_p^0 is the specific heat for very low pressures. But the integrand passes through infinity at points M and b and is negative in between. If it were possible to add heat in this range, the result would actually be to cool the gas, because the resulting expansion would return more potential energy of interaction than it would absorb kinetic energy.

Thus, if we remember that $c_p/T = (\partial s/\partial T)_p$ we can readily see that isobaric lines in the T-s plane must exhibit two points of zero slope connected by a region of negative slope. Real fluids accordingly behave as shown in Fig. 9.4 in the two-phase region in T-s coordinates. For lines of constant pressure, Eq. (9.36) takes the form

$$\int_f^g (v\,dp - s\,dT)_p = -\int_f^g s\,dT = 0 \tag{9.38}$$

which tells us that areas C and D must be equal in this figure.

These are only a few of the attributes of real fluid behavior described by the van der Waals equation. It can also be used to make a prediction of the variation of latent heat of vaporization with pressure, which is correct in shape but low in magnitude, to estimate surface tension, to show how pressure influences energy and enthalpy, and so forth.

As we now turn our attention back to the complications of statistical-mechanical methods, it is with the intention of developing accuracy that the van der Waals equation fails to provide. We do not

Fig. 9.4 Typical real-gas isobaric line in the $T - s$ plane.

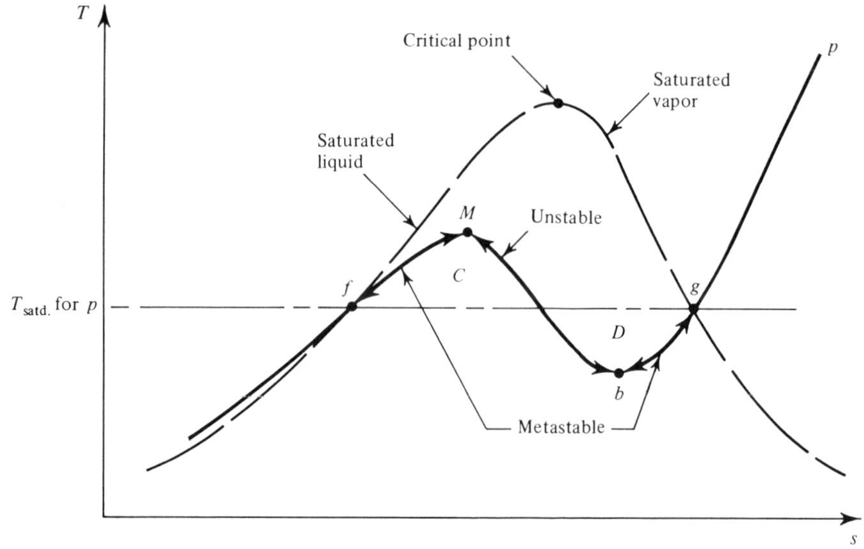

again achieve so total and simple a description of behavior, however.

EXAMPLE 9.1 Estimate the pressure exerted by intermolecular forces in water at 300°K and 1 atm.

This is a very "dense gas" and our best estimate is that given by the van der Waals equation,

$$p \text{ of intermolecular forces} = -\frac{a}{v^2}$$

But, from Eq. (9.31),

$$a = 3p_c v_c^2 = 3(218.3 \text{ atm})(3.135 \text{ cm}^3/\text{g})^2 = 6450 \text{ atm-cm}^6/\text{g}^2$$

and, from Eq. (9.32),

$$b = \frac{v_c}{3} = 1.044 \text{ cm}^3/\text{g}$$

Finally, for compatibility we must use Eq. (9.33) for R, instead of the actual value,

$$R^0 = \frac{8}{3}\frac{p_c v_c}{T_c} = 2.815 \text{ atm-cm}^3/{}^\circ\text{K-g}$$

Thus van der Waals's equation,

$$p = \frac{R^0 T}{v - b} - \frac{a}{v^2} = \left(2.815 \frac{\text{atm-cm}^3}{{}^\circ\text{K-g}}\right) \frac{300}{v - 1.044} \frac{{}^\circ\text{K-g}}{\text{cm}^3} - \frac{6450}{v^2} \text{ atm}$$

can be solved by trial and error for v at $p = 1$ atm. The smallest root is

$$v = 1.248 \text{ cm}^3/\text{g}$$

Therefore,

$$p \text{ of intermolecular forces} = -\frac{6450}{1.248^2} \text{ atm} = -4130 \text{ atm}$$

This might seem at first glance to be an incredible result. It says that the intermolecular pressure is thousands of times the pressure exerted on the walls. But it only serves to emphasize the enormous resistance that a liquid offers to being torn apart. This pressure is balanced by the tiny pressure exerted by the wall and the immense momentum of the densely packed molecules.

9.3 INTERMOLECULAR POTENTIAL FUNCTIONS

The prediction of a second virial coefficient, accurate enough to give a useful description of moderately dense gas behavior, requires an accurate knowledge of the molecular potential function. We first consider the character of the intermolecular forces that form the varied potentials of different molecules. It is then possible to show how actual ϕ functions evolve from these forces.

NATURE OF INTERMOLECULAR FORCES

A variety of kinds of force fields are at play between two particles as they undergo an encounter. These fields are widely different in character; some attract and some repel; some depend upon particle orientation during collision; and they are not always independent of one another. Because of the complexity of the forces, any attempt to put them together in a single potential function must necessarily result in some inaccuracy. We now attempt to catalog the basic kinds of forces.

1. *Electrostatic forces* are the interparticle forces arising because the particles are electrically charged. The elementary Coulomb force between two charged particles are proportional to the inverse square of the spacing, r, and negative (or repulsive) for like charges. Thus their contribution to ϕ is ϕ_{ab},

$$\phi_{ab} = \frac{C_a C_b}{r} \tag{9.39}$$

where C_a and C_b are the charges on the two interacting particles, a and b. This potential is negative, and the resulting force attractive, if the charges differ in sign.

When the particles have a finite dipole moment μ — that is to say, opposite sides of the particles are oppositely charged — the electrostatic forces are of higher order and varied in character. If the intrinsic angular dependence is averaged out,[7] we obtain for the interaction between a charged particle and a point dipole

$$\overline{\phi_{ab}^{(c,\mu)}} = -\frac{C^2\mu^2}{3kTr^4} \tag{9.40}$$

and for an interaction between two point dipoles,

$$\overline{\phi_{ab}^{(\mu,\mu)}} = -\frac{2\mu_a{}^2\mu_b{}^2}{3kTr^6} \tag{9.41}$$

When quadrapole particles are brought into the picture, other inverse power interactions (r^{-8} and r^{-10}) result. Examples of common dipolar molecules include H_2O, NCl, aniline, and the alcohols.

The simple Coulomb forces drop off slowly with distance and are relatively long range in character. The various dipole and quadrapole interactions, on the other hand, are only effective at very short range by virtue of the $r^{-4}, \ldots, r^{-10}$ factors, and they generally become negligible at high temperatures.

2. *Induction forces* occur when a permanently charged particle or a dipole is brought into the proximity of a neutral particle. The effect is that a dipole is induced in the neutral particle. If we introduce a quantity α, the *polarizability* of the neutral particle, then for the interaction between a charged particle and a neutral particle,

$$\phi_{ab} = -\frac{C^2\alpha}{2r^4} \tag{9.42}$$

and for the interaction between a dipole and a neutral particle,

$$\phi_{ab} = -\frac{\mu^2\alpha}{r^6} \tag{9.43}$$

The induction forces are always present and are uninfluenced by temperature. They are also usually fairly small in comparison with the dispersion and repulsive forces, forces 3 and 4. They do not fall off at high temperatures the way dipole–dipole and dipole–charge interactions do.

3. *Dispersion forces* are those that arise between two transient dipoles in neutral particles: one dipole is that induced in the electrons of one particle by the dipole in the other; the other dipole is the result of the instantaneous asymmetry of the electric field associated with the moving electrons of the approaching particle.

[7]See, for example, Hirschfelder *et al.*, *op. cit.*, chaps. 1 and 13.

This force is attractive and fairly strong. It is the basic "van der Waals" force of short-range attraction and it takes the form

$$\phi_{ab} = -\frac{3}{2}\frac{h\nu_a h\nu_b}{h\nu_a + h\nu_b}\frac{\alpha_a \alpha_b}{r^6} \tag{9.44}$$

where $h\nu_a$ and $h\nu_b$ are characteristic energies approximately equal to the ionization potentials of the particles.

4. *Repulsion* at very short range occurs as the result of the electronic orbits of two molecules coming into overlapping range. The forces can not be measured with much accuracy but are usually approximated by a repulsive potential of the form

$$\phi_{ab} = \frac{\text{constant}}{r^\kappa} \qquad 9 \leq \kappa \leq 15 \tag{9.45}$$

These electronic or *valence* repulsions are also sometimes approximated with an exponential function of the form

$$\phi = ae^{-br} \tag{9.46}$$

where a and b are constants related to ionization properties of the particles.

ANGLE-INDEPENDENT, SEMIEMPIRICAL POTENTIAL FUNCTIONS

The potential function for a given type of particle is a combination of the various components we have just described. But it is very important that the combination assume a simple enough analytical form to permit its use in the configuration integral. Figure 9.2 displayed three very simple approximations to real potential functions. Of these, only the weakly attracting rigid-particle potential, shown in Fig. 9.2(c), makes any attempt to include any of the attractive forces, and repulsion is approximated by a sudden infinite resistance. This is called the *Sutherland potential.*

A variety of refined potential functions have been developed to provide reasonable accuracy while retaining some simplicity. We discuss five of them here. They are (1) the Sutherland potential, which we have already used; (2) the *point center of repulsion;* (3) the *square-well potential;* (4) the *Lennard-Jones potential;* and (5) the *Buckingham potential.* Potentials 2, 3, 4, and 5 are shown graphically in Fig. 9.5. The analytical expressions for the potentials are also given in the figure.

The point center of repulsion is useful when a differentiable potential function is needed to express relatively rigid repulsion. It also leads to simpler mathematical descriptions of collisions in some respects than does the rigid noninteracting sphere. The second virial coefficient can be evaluated as follows for this model.

Fig. 9.5 Four comparatively sophisticated intermolecular potential functions.

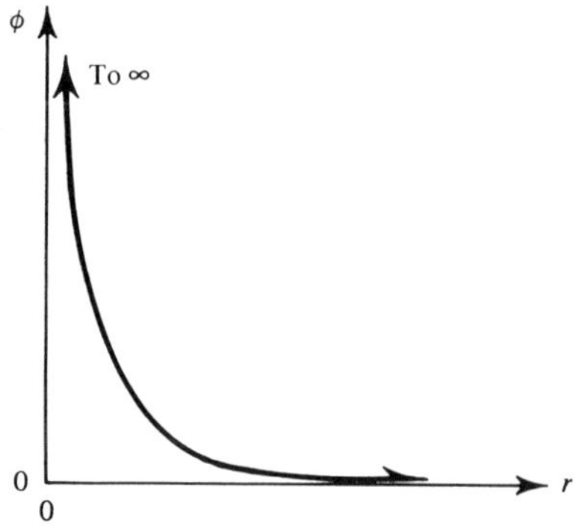

(a) Point center of repulsion

$$\phi(r) = \frac{\text{constant}}{r^{\delta}};\ 9 \lesssim \delta \lesssim 15$$

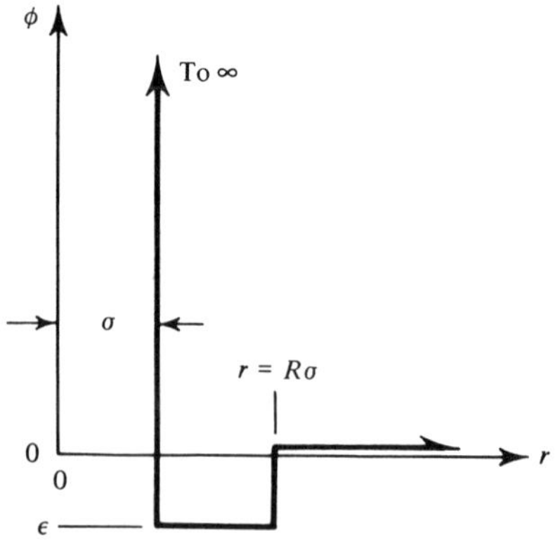

(b) Square well potential

$\phi(r) = \infty,\ r \leqslant \sigma;\ \phi(r) = -\epsilon,\ \sigma < r < R\sigma;$ and $\phi(r) = 0,\ r \geqslant R\sigma$

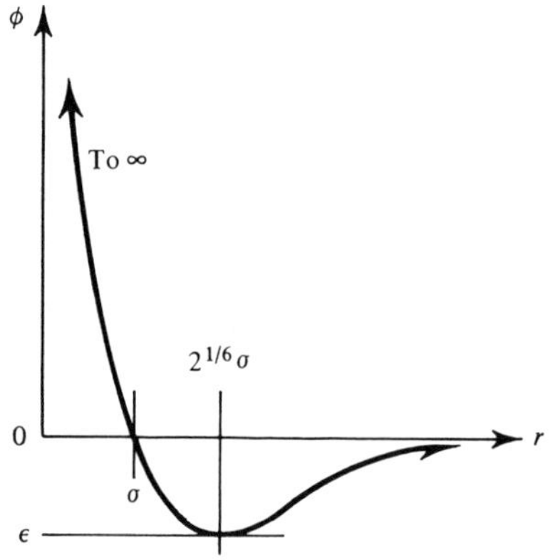

(c) Lennard–Jones (6–12) potential

$$\phi(r) = 4\epsilon\left[\left(\frac{\sigma}{r}\right)^{12} - \left(\frac{\sigma}{r}\right)^{6}\right]$$

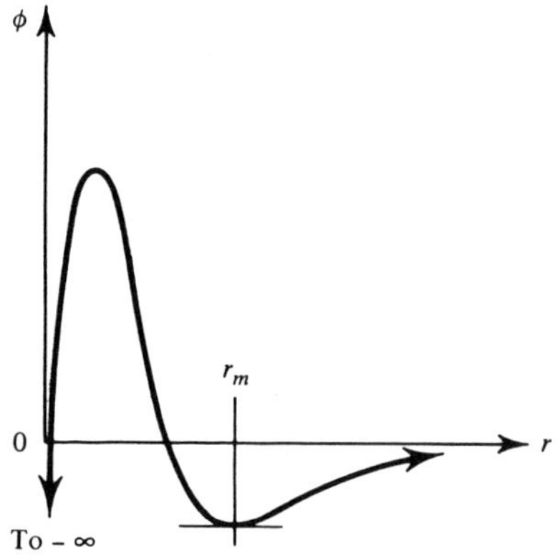

(d) Buckingham potential

$\phi(r) = b'\exp(-ar) - cr^{-6} - c'r^{-8}$

where a, b', c, and c' are constants

EXAMPLE 9.2 Evaluate the second virial coefficient for a gas composed of point centers of repulsion.

In this case, $\phi =$ constant $r^{-\delta}$. Then from Eq. (9.17) we obtain

$$B(T) = 2\pi N_{\mathrm{A}} \int_0^{\infty} [1 - \exp(-ar^{-\delta})] r^2\, dr \qquad a \equiv \frac{\text{constant}}{kT}$$

Now let us pick an r_0 for which $ar_0^{-\delta}$ is fairly big — say, on the order of 2 or more. Then $\exp(-ar^{-\delta})$ will be much less than unity for $0 \le r \le r_0$, and

$$B(T) \simeq 2\pi N_A \left\{ \int_0^{r_0} r^2\,dr + \int_{r_0}^{\infty} \left[1 - \sum_{i=0}^{\infty} \frac{(-ar^{-\delta})^i}{i!} \right] r^2\,dr \right\}$$

or

$$B(T) \simeq \tfrac{2}{3}\pi N_A r_0^3 \left[1 + 3 \sum_{i=1}^{\infty} \frac{(-ar_0^{-\delta})^i}{i!(3 - i\delta)} \right]$$

This result is limited to $\delta > 3$. If $\delta \le 3$, $B(T)$ is infinite. The physical meaning of this can be explained in the following way. The energy needed to bring particles from infinity to within r of the particle of interest is $cr^{-\delta}$, and the number of particles at a distance r is proportional to r^2. It follows that the energy required for a surrounding gas to be placed about the particle is proportional to

$$\int_0^{\infty} (r^{-\delta})(r^2)\,dr$$

and this is infinite for all $\delta \le 3$. This means that no walls could contain such a gas.

The expression that we have derived for $B(T)$ becomes increasingly accurate for larger and larger values of $ar^{-\delta}$, but more and more terms are needed in the summation. There is some value of $ar^{-\delta}$ on the order of magnitude of unity which optimize convergence and accuracy. The r_0 corresponding with the value would be an approximate effective σ for the particle; thus our expression for $B(T)$ reduces to

$$B(T) = \frac{2\pi N_A \sigma^3}{3} \begin{pmatrix}\text{number on the}\\ \text{order of unity}\end{pmatrix} \simeq \text{effective covolume, } b$$

which we might have anticipated.

The Sutherland potential [Fig. 9.2(c)] is a widely used and reasonably accurate approximation. We can show (Problem 9.8) that the second virial coefficient is in general

$$B(T) = b\left[1 - \sum_{j=1}^{\infty} \frac{1}{j!} \frac{3}{j\gamma - 3} \left(\frac{c}{\sigma^{\gamma} kT} \right)^j \right] \tag{9.47}$$

where b is the covolume, and γ, c, and σ are explained in Fig. 9.2(c). Equation (9.47) contains Eq. (9.24) as a special case.

We might do well to look at the square-well and Lennard-Jones potentials simultaneously. The square-well potential is a kind of blocky imitation of the more accurate Lennard-Jones potential, and the results of the two are somewhat similar. The second and third virial coefficients have been worked out for the square-well potential. The former is

$$B(T) = b\left\{1 - (R^3 - 1)\left[\exp\left(\frac{\epsilon}{kT}\right) - 1\right]\right\} \tag{9.48}$$

where R and ϵ are defined in Fig. 9.7(b).

The Lennard-Jones potential is given in its most familiar, "6-12," form in Fig. 9.7(c). That is, the exponents of the two terms in brackets are chosen as 12 and 6. It accordingly combines a strong repulsion with a dispersion attraction. The second virial coefficient for this case does not come out in a closed general form. However, it compares well with Eq. (9.48) when $\epsilon_{sw} = 0.56\epsilon_{LJ}$ and $R = 1.85$. Very broad use has been made of this function, and a third virial coefficient has been computed.

For purposes of comparing the second virial coefficient for the Lennard-Jones potential with other values, it is convenient to define

$$T^* \equiv \frac{kT}{\epsilon} \qquad \text{and} \qquad B^* \equiv \frac{B}{b} \tag{9.49}$$

Then the second virial coefficient for the Lennard-Jones potential can be put in the following dimensionless form:

$$B^* = B^*(T^*)$$

On the other hand, Eq. (9.48) can be divided by b to give B^* for the square-well potential in the form

$$B^* = 1 - (R^3 - 1)\left[\exp\left(\frac{1}{T^*}\right) - 1\right] \tag{9.48a}$$

The quantity R is very nearly 1.85 for many simple molecules and it remains on this order of magnitude for a large variety of substances. Thus B^* is of approximately the same functional form as B^* for the Lennard-Jones potential.

The second virial coefficient for the Lennard-Jones potential is plotted in the generalized coordinates, B^* versus T^*, in Fig. 9.6. Equation (9.48a), based on $R = 1.85$, is also included in this graph for comparison. The two curves are very similar to each other in form, and the Lennard-Jones plot matches experimental data for many substances, very closely.

The Buckingham potential is a refinement of the Lennard-Jones potential, intended to take better account of the short-range valence effects. Although it does this, it also introduces an incorrect singularity at $r = 0$. This singularity can be avoided in making many of the calculations that one normally uses potential functions for. Another more complicated modification, called the *Buckingham–Corner potential*, eliminates the singularity but does so at the cost of further complicating the equation for ϕ.

Fig. 9.6 Generalized second virial coefficients for the Lennard-Jones and square-well potentials.

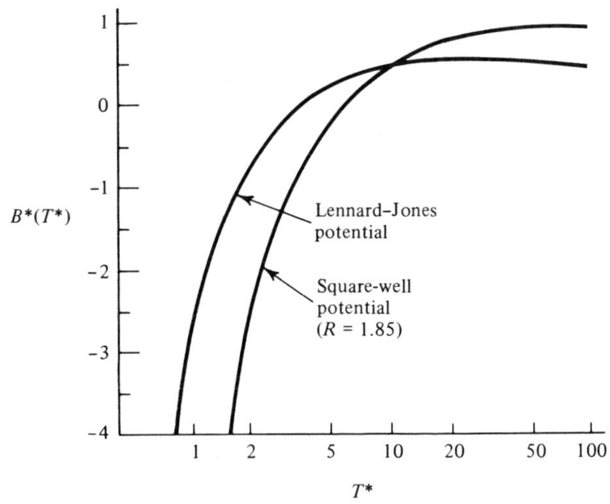

It would be well to understand at this point that these potentials are empirical in the sense that they include such parameters as σ, ϵ, and r_m. These parameters are often obtained by comparing the resulting equations of state with thermodynamic data rather than by proceeding directly from measured molecular parameters. The value of the resulting equations of state is judged by how well they represent data in the moderately dense gas region. They are not expected to provide good descriptions for the entire range of fluid behavior.

ANGLE-DEPENDENT POTENTIAL FUNCTIONS[8]

Intermolecular potential functions might be asymmetric by virtue of a nonuniformity of attractive forces, such as would arise from a dipole. They might also be asymmetric by virtue of nonuniformities of the very short range repulsive forces.

In the latter case the particle is effectively asymmetric in its geometrical form. Accordingly, a number of rigid noninteracting particle models have been formulated for molecules for which short-range repulsion dominates. Consideration has been given to cubes, tetrahedrons, octagons, cylinders, prolate and oblate ellipsoids, and to cylinders with spherical ends. Second virial coefficients have been computed and used to predict the properties of real gases at moderate densities.

[8]See Hirschfelder *et al.*, footnote 4, sec. 13.10a, for a relatively complete coverage of angle-dependent potential functions.

By the same token, a good deal of work has been done toward treating asymmetric attraction. Typical of these is the Keesom model. Keesom envisioned a rigid spherical molecule containing a dipole and suggested that

$$\phi(r, \theta_1, \theta_2, \phi_2 - \phi_1) = \begin{cases} \infty & \text{for } r < \sigma \\ -\dfrac{\mu^2}{r^3} g(\theta_1, \theta_2, \phi_2 - \phi_1) & \text{for } r \geq \sigma \end{cases} \tag{9.50}$$

where the four angles define the relative orientations of the two dipoles as shown in Fig. 9.7 and the factor g is

$$g = 2 \cos \theta_1 \cos \theta_2 - \sin \theta_1 \sin \theta_2 \cos (\phi_2 - \phi_1)$$

Almost 30 years after Keesom suggested this model in 1912, Stockmayer improved upon it by combining it with the relatively short range Lennard-Jones forces as follows:

$$\phi = 4\epsilon \left[\left(\frac{\sigma}{r}\right)^{12} - \left(\frac{\sigma}{r}\right)^{6} \right] - \frac{\mu^2}{r^3} \tag{9.51}$$

The μ^2/r^3 factor in the dipole term can be shown to be characteristic of a dipole–dipole interaction. It was the averaging of such a term as $\mu^2 g/r^3$ that led to Eq. (9.51).

Fig. 9.7 Schematic diagram showing angular coordinates of dipole orientation for two dipole particles.

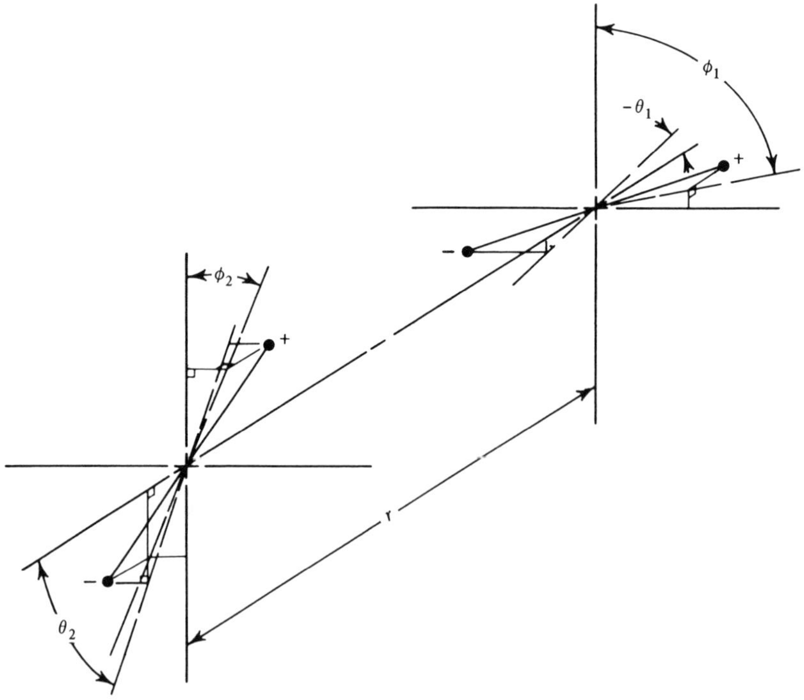

Second and third virial coefficients can be worked out for gases for which these potentials apply, if appropriate methods of averaging are used. For the Keesom potential the second virial coefficient is

$$B(T, \mu) = b\left[1 - \frac{1}{3}\left(\frac{\mu^2}{\sigma^3 kT}\right)^2 - \frac{1}{75}\left(\frac{\mu^2}{\sigma^3 kT}\right)^4 - \frac{29}{55{,}125}\left(\frac{\mu^2}{\sigma^3 kT}\right)^6 - \cdots\right] \tag{9.52}$$

For the Stockmayer potential, $B(T, \mu)$ and $C(T, \mu)$ are complicated, but they can be reduced to the following dimensionless forms:

$$B^*(T^*, \mu^*) = \frac{B(T, \mu)}{b} \quad \text{and} \quad C^*(T^*, \mu^*) = \frac{C(T, \mu)}{b} \tag{9.53}$$

where

$$\mu^* = \frac{\mu}{\sqrt{\epsilon\sigma^3}} \tag{9.54}$$

EXAMPLE 9.3 Express the Joule–Thompson coefficient, $\partial T/\partial p|_h$, in terms of the intermolecular potential, through the second virial coefficient.

The formula for the Joule–Thompson coefficient is

$$\left.\frac{\partial T}{\partial p}\right|_h = \frac{T(\partial v/\partial T)|_p - v}{c_p}$$

but

$$\frac{pv}{R^0 T} = 1 + \frac{B(T)}{v} + \cdots$$

or

$$v = \frac{R^0 T}{p} + \frac{B(T)}{pv/R^0 T} + \cdots \simeq \frac{R^0 T}{p} + B(T)$$

so

$$\left.\frac{\partial T}{\partial p}\right|_h = \frac{1}{c_p}\left[T\frac{\partial B(T)}{\partial T} - B(T)\right]$$

This can be put in terms of B^* and T^*,

$$\left.\frac{c_p}{b}\frac{\partial T}{\partial p}\right|_h = \frac{\partial B^*(T^*)}{\partial \ln T^*} - B^*$$

so the right-hand side can be evaluated directly from plots similar to Fig. 9.6.

As a point of interest, we note that this expression predicts that the Joule–Thompson inversion point, $(\partial T/\partial p)_h = 0$, will occur where

$$\frac{d \ln B^*}{d \ln T^*} = 1$$

DETERMINATION OF THE PARAMETERS OF ϕ

The parameters σ and ϵ in the Lennard-Jones potential are most commonly obtained by comparing the predicted second virial coefficients with second virial coefficients obtained from data. This may seem to be a circular procedure. However, many advantages acrue from it. Equations (9.52) and (9.53) give analytical form to $B(T)$ and make extrapolation possible. Furthermore, the potentials can be used in the prediction of transport as well as thermodynamic properties.

Table 9.2 gives values of σ and ϵ/k determined in this way, as well as the polarizability α for a variety of common gases. The latter is needed in the prediction of potentials for interactions between polar and nonpolar molecules. The numbers given for ϵ/k and σ are sometimes obtained from viscosity data instead of experimental $B(T)$'s. When this is done, the resulting values may differ somewhat from those given here.

TABLE 9.2 Constants for the Lennard-Jones Potential, and the Polarizability, of Common Gases

[computed from experimental values of $B(T)$]

Gas	ϵ/k, °K	σ, Å	α, 10^{-25} cm³
He	10.8	2.57	—
Ne	35.8	2.75	—
Ar	119.7	3.41	—
H_2	36.7	2.96	7.9
N_2	95.1	3.70	17.6
O_2	117.5	3.58	16.0
CO	100.2	3.76	19.5
CO_2	187.5	4.47	26.5
CH_4	148.1	3.81	26.0

Several rough empirical relationships have been developed to estimate ϵ/k and σ for those instances in which appropriate $B(T)$ and viscosity data are unavailable. One such pair of relationships (cf. Table 9.1) is

$$\frac{\epsilon}{k} = 0.77T_c \qquad \text{and} \qquad b = b(\sigma) = 18.4\,\frac{\text{cm}^3\text{-atm}}{°\text{K}}\,\frac{T_c}{p_c} \tag{9.55}$$

Table 9.3 gives the constants for the Stockmayer potential, along with the dimensionless dipole moment μ^* for five typical polar molecules.

TABLE 9.3 Constants for the Stockmayer Potential [computed from experimental values of $B(T)$]

Gas	ϵ/k, °K	σ, Å	μ^*
H_2O	380	2.65	1.85
NH_3	320	2.60	1.68
CH_3Cl	380	3.43	1.30
CH_3OH	630	2.40	1.51
n-C_3H_7OH	866	2.61	1.19

EMPIRICAL LAWS FOR INTERACTIONS BETWEEN TWO DISSIMILAR MOLECULES

Tables 9.2 and 9.3 — like most tabulations available in the literature — are only applicable to molecules of the same chemical species. To calculate the thermostatic properties of mixtures, however, we must know the potential function that describes the interaction between molecules of two different species. It turns out that the best way to infer this information is through a method involving the temperature dependence of measured binary diffusion coefficients D_{12}. Since such data are not generally available for all cases of interest, we shall look briefly at some reasonably adequate empirical relations for estimating the Lennard-Jones and Stockmayer parameters.

Designating the two species as 1 and 2, we can use arithmetic and geometrical means, respectively, for σ and ϵ:

$$\sigma_{12} = \frac{\sigma_1 + \sigma_2}{2} \quad \text{and} \quad \epsilon_{12} = \sqrt{\epsilon_1 \epsilon_2} \tag{9.56}$$

Similarly, the dimensionless dipole moment μ^* can be expressed as a geometric mean,

$$\mu^*_{12} = \sqrt{\mu^*_1 \mu^*_2} \tag{9.57}$$

When a polar molecule (p) and a nonpolar molecule (n) interact, comparable expressions can be used:

$$\sigma_{pn} = \frac{\sigma_p + \sigma_n}{2} \xi^{-1/6} \quad \text{and} \quad \epsilon_{pn} = \sqrt{\epsilon_p \epsilon_n} \tag{9.58}$$

where

$$\xi \equiv 1 + \frac{\alpha^*_n \mu^*_p}{4} \sqrt{\frac{\epsilon_p}{\epsilon_n}} \tag{9.59}$$

and where, in turn, α^*_n is the polarizability of the nonpolar molecule divided by σ^3.

SOME NUMERICAL VALUES OF VIRIAL COEFFICIENTS

Since the accuracy of the computation does not warrant it, theoretical evaluations of virial coefficients are seldom extended beyond the third one, $C(T)$. Experimental applications of the virial equation, on the other hand, often make use of many terms and achieve very high accuracy.

We have thus far laid fairly heavy stress upon the Lennard-Jones and Stockmayer potential functions because they have proved to give as good accuracy as can be expected from such simple expressions. The dimensionless second and third virial co-

TABLE 9.4 Functions Used in Virial-Coefficient Calculations[a]

T^*	$\frac{\mu^{*2}}{\sqrt{8}} = 0$	$\frac{\mu^{*2}}{\sqrt{8}} = 0.2$	$\frac{\mu^{*2}}{\sqrt{8}} = 0.4$	$\frac{\mu^{*2}}{\sqrt{8}} = 0.6$	$\frac{\mu^{*2}}{\sqrt{8}} = 0.8$	$\frac{\mu^{*2}}{\sqrt{8}} = 1.0$
			$B^*(T^*, \mu^*)$			
0.30	−27.88	−42.97				
0.50	−8.720	−10.40	−17.03	−36.36		
0.75	−4.176	−4.630	−6.163	−9.413	−16.05	−30.4
1.00	−2.538	−2.744	−3.402	−4.657	−6.820	−10.54
1.25	−1.704	−1.821	−2.187	−2.852	−3.915	−5.559
1.50	−1.201	−1.277	−1.511	−1.925	−2.561	−3.490
2.00	−0.6276	−0.6671	−0.7875	−0.9953	−1.302	−1.727
2.50	−0.3126	−0.3370	−0.4108	−0.5368	−0.7194	−0.9658
3.00	−0.1152	−0.1318	−0.1820	−0.2671	−0.3892	−0.5517
4.00	−0.1154	0.1062	0.0784	0.0316	−0.0349	−0.1221
5.00	0.2433	0.2374	0.2197	0.1898	0.1476	0.0926
10.00	0.4609	0.4593	0.4547	0.4469	0.4359	0.4218
50.00	0.5084	0.5083	0.5080	0.5076	0.5071	0.5064
100.00	0.4641	0.4641	0.4640	0.4639	0.4637	0.4635
400.00	0.3584	0.3584	0.3583	0.3583	0.3583	0.3583
			$C^*(T^*, \mu^*)$			
1.0	0.4297	0.5304				
2.0	0.4371	0.4826	0.6496	0.995	1.595	2.46
2.5	0.3811	0.4076	0.5195	0.6871	0.999	1.482
3.0	0.3523	0.3692	0.4275	0.5403	0.7248	1.002
4.0	0.3266	0.3350	0.3630	0.4156	0.4986	0.6194
10.0	0.2861	0.2871	0.2902	0.2957	0.3039	0.3151

[a]R. B. Bird, J. O. Hirschfelder, and C. F. Curtiss, "The Equation of State and Transport Properties of Gases and Liquids," *Handbook of Physics*, 2nd ed. (E. V. Condon and H. Odishaw, Eds.) McGraw-Hill, Inc., New York, 1967, chap. 4.

efficients for the Stockmayer potential [recall Eq. (9.53)] are general and can be applied to any gas for which the parameters μ, ϵ, and b are known or can be estimated. Furthermore, they reduce to Lennard-Jones values when μ vanishes. Accordingly, we give numerical values of these coefficients in Table 9.4. The $B(T, \mu)$ values calculated from Table 9.4 compare very well with experiment, but $C(T, \mu)$ values are only moderately accurate.

The numerical values in Table 9.4 can also be used to obtain B and C for mixtures. The idea is comparatively straightforward, but the method is not very simple in execution; we write

$$B_{\text{mix}} = \sum_{i=1}^{\nu} \sum_{j=1}^{\nu} x_i x_j B_{ij} \tag{9.60}$$

and

$$C_{\text{mix}} = \sum_{i=1}^{\nu} \sum_{j=1}^{\nu} \sum_{k=1}^{\nu} x_i x_j x_k C_{ijk} \tag{9.61}$$

where there are ν components in the mixture. The terms B_{jj} and C_{jjj} are the conventional second and third virial coefficients for a single (jth) species. The B_{ij}'s are calculated on the basis of ϵ_{ij} and σ_{ij} for each of the component binary interactions. The C_{ijk}'s have a similar meaning for three particle interactions. The x's are mole fractions of the various species. Equations (9.60) and (9.61), of course, represent simple weighted averages of the appropriate component values of B and C (see Problem 9.13).

9.4 LAW OF CORRESPONDING STATES—APPLICATIONS

CORRESPONDING p-v-T STATES

The law of corresponding states, which was explained briefly in Sec. 9.2, deserves further attention at this point. We first discuss the use of the p-v-T statement of the law given by Eq. (9.29) and then proceed to show how it can be extended to describe other properties. Although we took care to identify the law as a physical hypothesis, separate from the van der Waals equation, the discovery of the law is actually credited to van der Waals.

The really important feature of the law is that it provides the most utilitarian general description of dense-gas behavior that there is. It is frankly empirical in application and it makes up the accuracy that is lacking in the van der Waals equation. It turns out that p-v-T data correlate better on a Z-p_r-T_r surface than on a

p_r-v_r-T_r surface; therefore, Eq. (9.29) can be restated in the following way. First we write the compressibility factor Z

$$Z \equiv \frac{pv}{R^0T} = \frac{p_c v_c}{R^0 T_c}\frac{p_r v_r}{T_r} = Z_c \frac{p_r v_r}{T_r} \tag{9.62}$$

Then, since $p_r v_r / T_r = (p_r/T_r)[v_r(p_r, T_r)] = f(p_r, T_r)$, and Z_c is approximately a universal constant,

$$Z = f(p_r, T_r) \tag{9.63}$$

Equation (9.63) can be used directly as a means for correlating data. From p-v-T data for a given substance at a given T_r, Z values can be computed and plotted against p_r. This was first done by Su,[9] who found that data for 17 comparatively simple substances correlated within about 3 percent. Figures 9.8 and 9.9 are such charts as prepared by Nelson and Obert.[10] The data points in these figures have been removed, and what remains are families of Z versus p_r curves for different T_r. The curves give their best accuracy when they are used for substances that are thermodynamically

Fig. 9.8 Nelson–Obert generalized compressibility factor, $0 \leq p_r \leq 1.0$.

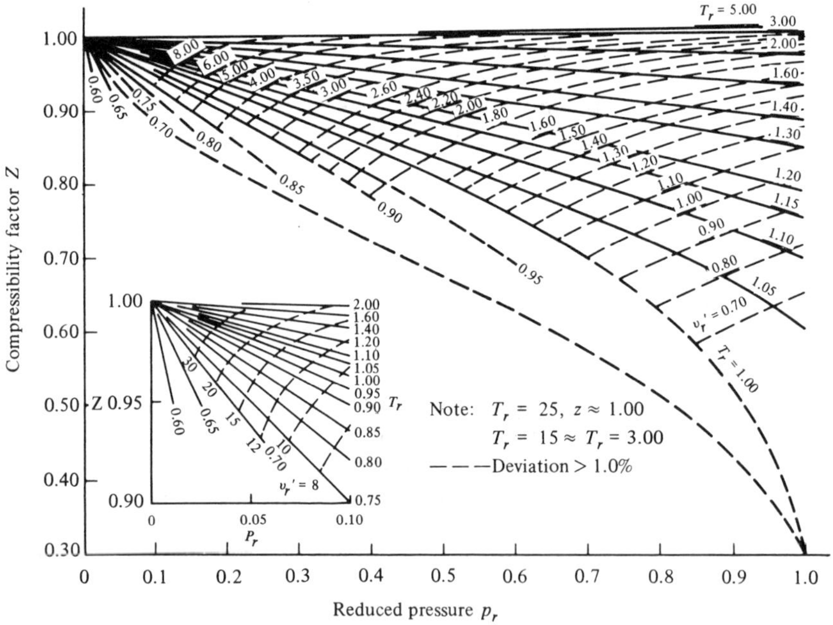

[9]Gouq-Jen Su, "Modified Laws of Corresponding States," *Ind. Eng. Chem.* **38,** 803–806 (1946).

[10]See, for example, E. F. Obert and R. A. Gaggioli, *Thermodynamics*, 2nd ed., McGraw-Hill, Inc., New York, 1963, chap. 10 and app. B.

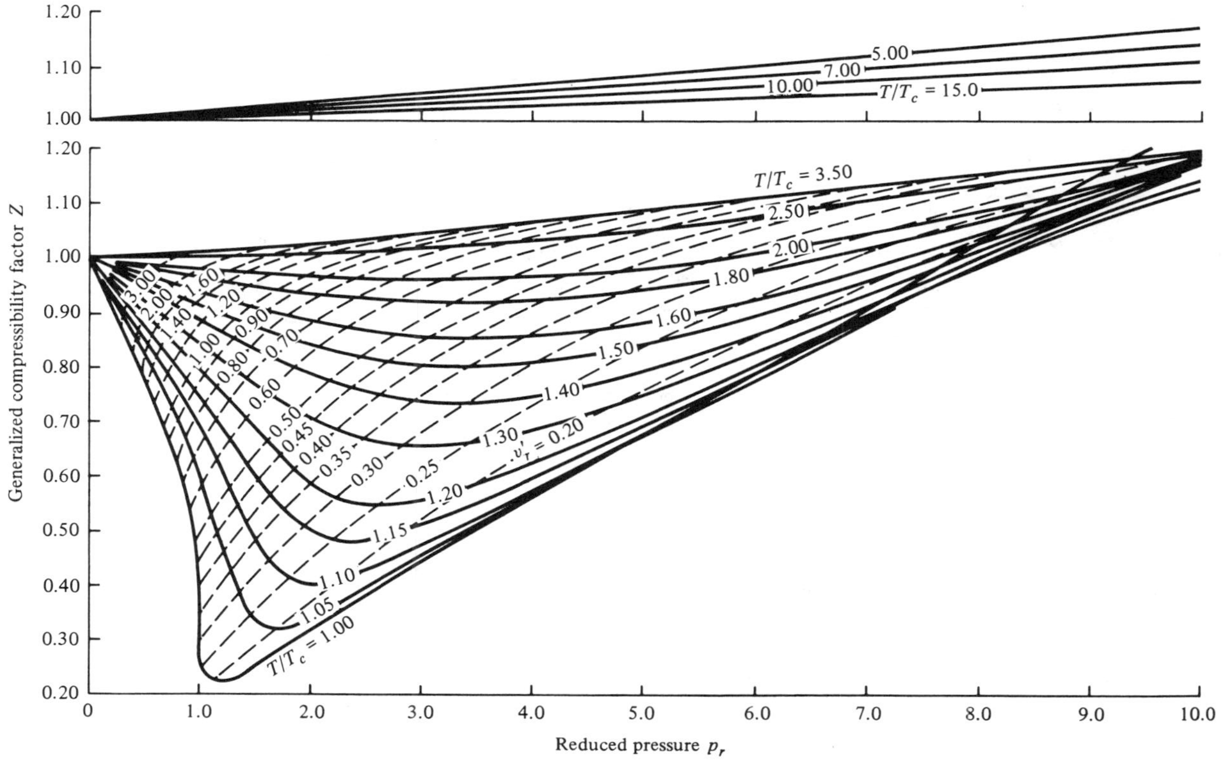

Fig. 9.9 Nelson–Obert generalized compressibility factor, $0 \leq p_r \leq 10.0$.

similar to those upon which the curve is based — in this case, molecules with fairly simple molecular structure.

Both Figs. 9.8 and 9.9 include another family of cross plots. These are lines of pseudo-reduced volume, v'_r, where

$$v_r'(T_r, p_r) \equiv \frac{v}{R^0 T_c/p_c} = Z_c v_r(T_r, p_r) \tag{9.64}$$

It turns out that this variable correlates more accurately upon Z versus p_r coordinates than does v_r. Although these lines are of some interest, the most accurate computation of v for a given T and p will be made with Eq. (9.63) and not Eq. (9.64).

Isotherms on the chart for the low-pressure range (Fig. 9.9) terminate to the left of $p_r = 1$ in what should be the saturated vapor line. Unfortunately, saturated vapor data correlate only rather coarsely in reduced coordinates, and this line is subject to fairly great error.

THERMODYNAMIC FUNCTIONS

All the thermodynamic functions — u, s, g, f, h, and so on — can be expressed solely in terms of p-v-T data and specific-heat data for low pressures. Enthalpy, for instance, can be computed above a reference datum at the liquid triple point by integrating the well-known relation

$$dh = c_p\,dT + \left(v - T\left.\frac{\partial v}{\partial T}\right|_p\right) dp$$

to give the enthalpy above some low-pressure reference or datum enthalpy, h_0:

$$h - h_0 = (h^* - h^*_0) + (h - h^*) \tag{9.65}$$

where the asterisks denote ideal-gas enthalpy changes computed by integrating $c_p\,dT$. The term $h - h^*$ represents a pressure correction at the temperature of interest,[11]

$$h - h^* = \int_{p_0}^{p} \left(v - T\left.\frac{\partial v}{\partial T}\right|_p\right)_T dp = T_c f(p_r, T_r) \tag{9.66}$$

which can be written

$$h - h^* = -\frac{T_c}{M}\,\frac{H^* - H}{T_c} \tag{9.66a}$$

[11]This step is a consequence of the law of corresponding states and involves some development. Details of this and other extensions of the law of corresponding states are given by O. A. Hougen, K. M. Watson, and R.A. Ragatz *Chemical Process Principles*, Part Two, Thermodynamics, John Wiley & Sons, Inc., New York, 1959, chap. 14.

In this case H is used to denote the molar enthalpy and M is the molecular weight.

The quantity $(H^* - H)/T_c$ is called the generalized *enthalpy defect*. It is obtained by graphical differentiation and other processing of the generalized compressibility factor, Z. The results are plotted on what is called an *enthalpy-defect chart*, shown in Fig. 9.10.

Many other applications of the law of corresponding states to the predictions of properties exist. *Entropy-defect charts* and the pressure corrections for c_p are among those very commonly presented, for example. To pursue these applications further would be beyond the scope of our present inquiry, however. We return to the law again in the context of *transport* properties at the end of chapter 11.

SOME FINAL OBSERVATIONS ON THE MEANING OF CORRESPONDING STATES

We have thus far taken the view that the law of corresponding states is simply an empirical result that happens to be borne out by van der Waals's equation. Actually we could have inverted the argument and offered the van der Waals equation as a rough molecular justification for the law.

A more careful attempt to make a molecular justification of the law shows that other forms of the law might also be admissible. If, for example, the potential takes the form

$$\frac{\phi}{\epsilon} = f\left(\frac{r}{\sigma}\right) \tag{9.67}$$

the resulting equation of state turns out (Problem 9.18) to be

$$p^* = p^*(v^*, T^*) \tag{9.68}$$

where the additional parameter, v^*, is

$$v^* = \frac{v}{b} \tag{9.69}$$

Equation (9.67) includes the utilitarian Lennard-Jones potential, the square-well potential with a particular value of R, and other forms of ϕ. The resulting equation of state, Eq. (9.68), is approximately the same for the different but similar potential functions. This implication of an approximate universality of Eq. (9.68) was borne out by Fig. 9.6, which showed that similar second virial coefficients appeared in this coordinate scheme for different approximations to ϕ.

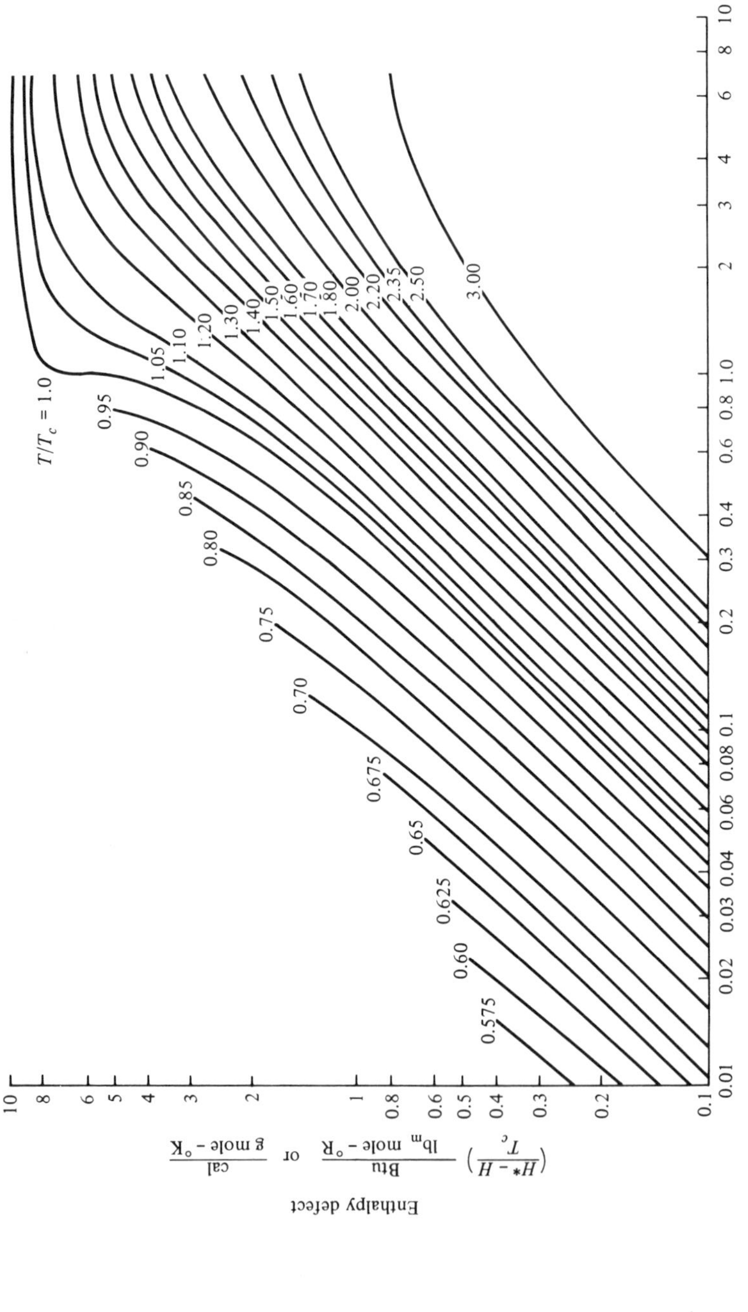

Fig. 9.10 Enthalpy-defect chart.

It is possible, in this connection, to identify *molecular* critical data:

$$p^*_c \equiv \frac{\epsilon N_A}{b} \qquad T^*_c \equiv \frac{\epsilon}{k} \qquad v^*_c \equiv b \tag{9.70}$$

We have already noted [Eq. (9.55)] that T^*_c, thus defined, is roughly 77 percent of measured T_c values. Similarly, v^*_c is $(\frac{1}{3}) \times$ (the van der Waals v_c). These values can be obtained from Table 9.1 for a variety of common substances. Table 9.1 reveals that, at least for almost-spherical molecules, the molecular critical data stand in reasonably constant proportion to p_c, T_c, and v_c.

Thus the law of corresponding states clearly has its origins in microscopic behavior and we are in a position to predict, for example, the Z charts for a Lennard-Jones gas. Such a prediction would not reach the accuracy of the existing Z charts so we do not do it. Nevertheless, with greater refinement, we have the means at hand for improving on the accuracy of the charts in special cases.

Problems 9.1 Verify that Eq. (9.2) is the proper expression for Q for the hybrid U_j under consideration. Where did the factor of $N!^{-1}$ come from?

9.2 Obtain the ideal-gas law directly from Eq. (9.6) using the appropriate potential, ϕ.

9.3 Verify Eqs. (9.31), (9.32), (9.33), and (9.34).

9.4 Prove that a system is unstable if for all processes that it can undergo, $(\partial p/\partial v)_T > 0$, or $(\partial T/\partial s)_p < 0$.

9.5 Show that the pressure at the point of maximum liquid superheat in a van der Waals fluid is given by $(3 - 2/v_{r_m})v_{r_m}^{-2}$, where v_{r_m} is the volume at that point.

9.6 Verify Eqs. (9.36) and (9.37).

9.7 Show that all virial coefficients are temperature-independent for a gas of hard spherical molecules.

9.8 Evaluate $B(T)$ for the Sutherland potential.

9.9 Compare b as given by Eq. (9.55) with the van der Waals value.

9.10 Determine the p-v-T relationship and c_v for a classical monatomic gas of rigid spheres in a gravitational field.

9.11 Sketch the following potential function and identify σ, ϵ, and $a\sigma$ on the sketch:

$$\phi(r) = \begin{cases} \infty & r < \sigma \\ \dfrac{\epsilon}{a-1}\left(\dfrac{r}{\sigma} - a\right) & \sigma < r < a\sigma \\ 0 & r > a\sigma \end{cases}$$

Compute the second virial coefficient for this potential.

9.12 Estimate the Joule–Thompson coefficient as a function of temperature for argon, using the square-well potential.

9.13 Obtain Eq. (9.60) formally, beginning with the canonical partition function for a two-component mixture.

9.14 A mixture of two gases undergoes an isentropic expansion. How does the entropy of each component behave if (a) one is a monatomic ideal gas and the other is a diatomic ideal gas, (b) both are rigid-sphere gases with $(\sigma_1/\sigma_2)^3 = \frac{1}{2}$, (c) both are Lennard-Jones gases with $\sigma_1 = \sigma_2$ and $2\epsilon_1/k = \epsilon_2/k$.

9.15 Compute the specific volume of air at 500 psia and 68°C by finding the "corresponding" point in the steam table and applying the law of corresponding states directly. Check this result in two ways: (a) use the generalized compressibility-factor charts, and (b) obtain a tabulated value of v for air.

9.16 Use van der Waals's molecular model to explain, qualitatively, why Z exceeds unity in certain ranges of the Z charts.

9.17 Steam at 1000 psia and 800°F is subjected to a Joule–Thompson expansion to atmospheric pressure. Predict the temperature change using the enthalpy-defect chart. Check your result with steam tables. ($c_p^0 \simeq [0.433 + 0.0000166T\,°\text{R}]$ Btu/lb$_m$-°R.)

9.18 Verify that Eq. (9.68) follows from Eq. (9.67).

9.19 An adsorbed surface layer of area A consists of N atoms that are free to move and interact with each other according to a potential $\phi(R)$ that depends only on their mutual separation, R. Find the fundamental equation of this surface layer.

10 thermostatic properties of solids and liquids

We have devoted considerable effort, thus far, to the treatment of ideal gases, but only in chapter 9 was the effort expanded to the treatment of weakly interacting, or moderately dense, gases. The complexity involved in the statistical-mechanical description of such gases increases as the density increases and molecular interaction becomes more pronounced. The molecules in a solid are subject to extremely strong intermolecular forces, but, because of these forces, they are generally well-ordered and easier to treat than the molecules of a liquid or dense gas. The positions of the molecules are fixed in a solid, instead of randomly distributed as they are in liquids and gases. The orderliness of the solid state allows simplification that makes possible an analysis of the average properties of the system. The liquid state, which lies between these two extremes, has been the most difficult to describe and to understand. As a result, theories of liquids are often based on solidlike or gaslike models.

The statistical-mechanical description of the solid or liquid state is distinctly different from that of the gaseous state in one essential point. The strong interactions among the molecules preclude a consideration of the dynamic behavior of each individual molecule as was possible in the theory of gaseous state. In solids, for example, we must consider the motions of the whole lattice and analyze these gross motions statistically. In this chapter we discuss the statistical-mechanical basis for such a description of the thermostatic or equilibrium properties of solids and liquids.

10.1 STRUCTURE OF SOLIDS

The structure of solids is often described in terms of a *lattice*, or of a *network*, of points defining the average locations of the molecules or atoms in a solid. The average distance between the neighboring points — on the order of 10^{-8} cm — is comparable with the dimensions of the molecules themselves. Despite this compactness, the lattice points can be arranged in many ways. Strictly speaking, at thermal equilibrium, there exists one unique arrangement of lattice points called the *ideal crystalline state*. The crystal lattice is in general highly ordered, periodic, and regular, and it possesses various forms of symmetry.

In actual situations most solids are seldom in the ideal crystalline state and, in particular, amorphous solids are characterized by randomly distributed lattice points. From the thermodynamic viewpoint, such solids are metastable and eventually crystallize. The relaxation time, or time required for reaching the equilibrium state, is so large in these cases that amorphous solids behave as though they were stable for almost unlimited lengths of time.

Needless to say, the regular and symmetric nature of crystal lattices presents ideal physical models for the study of the solid state, and most of the discussion in this chapter is based on such models. But we should bear in mind that most of the thermodynamic relations for crystal lattices apply fairly well to amorphous solids. One important difference is that, because noncrystal solids are not in equilibrium, the fourth postulate of thermodynamics (Sec. 1.4) does not apply, and their entropy assumes a nonzero constant value as the temperature approaches zero. This residual entropy, which characterizes the basic disorderliness in the structure of amorphous solids, must be added to the entropy expression for crystal lattices.

CLASSIFICATION OF CRYSTALLINE STRUCTURES

The particles composing a crystal are arranged, as we have observed, in a three-dimensional lattice. The lattice consists of a set of identical adjacent units, each of which by a suitable translation could be brought into coincidence with any other. These units are called *basic cells*. The configuration of these basic cells serves as the basis for classification of crystalline structures. According to theory there are 230 possible configurations of atoms in crystalline patterns. However, these 230 possible patterns may be grouped into 32 major categories, and these categories may be further simplified into six basic crystal systems as represented in Fig. 10.1.

Fig. 10.1 Basic crystal systems.

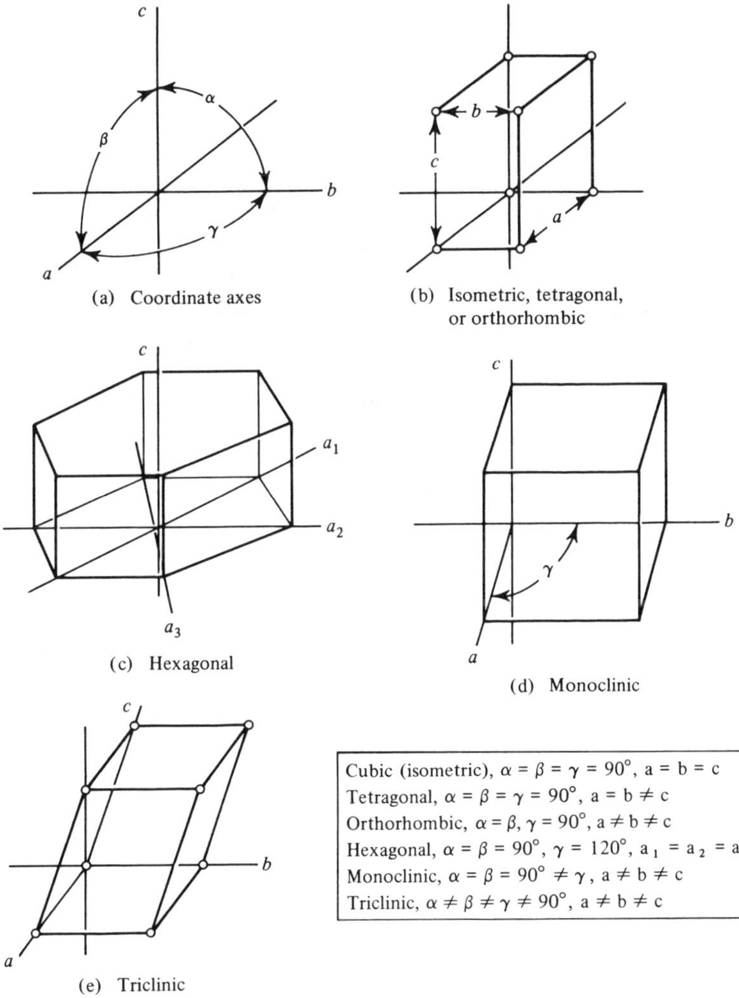

Cubic (isometric), $\alpha = \beta = \gamma = 90°$, $a = b = c$
Tetragonal, $\alpha = \beta = \gamma = 90°$, $a = b \neq c$
Orthorhombic, $\alpha = \beta, \gamma = 90°$, $a \neq b \neq c$
Hexagonal, $\alpha = \beta = 90°$, $\gamma = 120°$, $a_1 = a_2 = a_3 \neq c$
Monoclinic, $\alpha = \beta = 90° \neq \gamma$, $a \neq b \neq c$
Triclinic, $\alpha \neq \beta \neq \gamma \neq 90°$, $a \neq b \neq c$

The simplest of these crystal systems is the cubic, or isometric, system, which has three axes of symmetry, all mutually perpendicular with equal lengths. Although the cubic cell structure is quite common in metals, the actual matrix seldom exhibits the simple cubic cell structure shown in Fig. 10.1. Many metals are formed with a grouping of 8 symmetrically spaced particles about a single central particle. This is known as the body-centered cubic cell and is shown in Fig. 10.2(b).

Another common cubic structure shown in Fig. 10.2 is the face-centered cubic cell, in which each face consists of 4 corner particles and a central particle, resulting in a total of 14 particles in the cell. Perhaps the next most common crystal system, after the cubic system and its variations, is the close-packed hexagonal system. This hexagonal system consists of three axes of symmetry of equal lengths lying in a single plane and a fourth axis of a different length, perpendicular to the other three. In addition to the 12 corner particles, other particles may be arranged in the structure to form a close-packed hexagonal system as shown in Fig. 10.2(c).

About 70 percent of all elementary metals and alloys have either the face-centered cubic (fcc), the body centered cubic (bcc), or

Fig. 10.2 Three common cell structures.

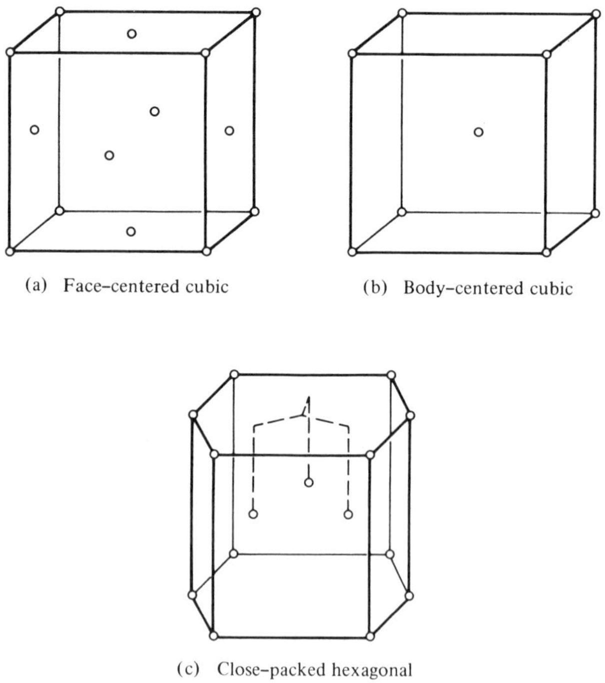

the close packed hexagonal (cph) structure. The large particle density and closeness in these structures is one reason for the high density and strength of metals in comparison with most nonmetallic solids. Table 10.1 lists the cell structures of some common metals.

TABLE 10.1 Cell Structure of Some Common Metals

Metal	Cell structure
Iron (delta)	bcc
Iron (gamma)	fcc
Iron (alpha)	bcc
Aluminum	fcc
Magnesium	cph
Copper	fcc
Zinc	cph
Nickel	fcc
Lead	fcc
Silver	fcc
Chromium	bcc
Tungsten	bcc
Indium	bct[a]
Tin	bct
Gold	fcc
Manganese	cubic[b]

[a]body-centered tetragonal
[b]simple

CLASSIFICATION OF CRYSTAL BINDING

The static forces binding atoms and molecules in solids are almost entirely electrostatic in nature. Differences in the distribution of electrons around the atoms and molecules result in various types of crystal binding. The four major types of crystal binding are

1. *Covalent bonds*, formed by the interlocking of the unfilled outer electron shells of the atoms. These bonds are associated with large binding energy, or the energy necessary to dissociate the solid into separated atoms. Covalent crystals such as diamond and quartz are very hard and have high heats of evaporation and low electrical and thermal conductivity.

2. *Ionic bonds*, representing the Coulombic attraction of the positive and negative ions, which in turn result from the transfer of one or more electrons from one atom to adjacent atoms in the

crystal structure. A typical ionic crystal is sodium chloride, common table salt.

3. *Metallic bonds*, representing the binding of an array of positive metallic ions in a uniform sea of negatively charged free electrons. The binding energy of metallic bonds is generally less than that of covalent or ionic bonds, because the attraction between ions and free electrons is relatively small. Because free electrons are available to participate in the conduction of electrical and thermal energy, they are called conduction electrons, and metallic crystals are characterized by high electrical and thermal conductivity.

4. *Molecular bonds* are formed by weak van der Waals forces, resulting, as we saw in Sec. 9.3, from induced dipole moments. The electronic structure of the atom or molecule is left essentially unchanged in the formation of molecular bonds. Such molecular crystals as solid helium and nitrogen are characterized by weak binding and thus have low melting and boiling points.

CRYSTAL AGGREGATES AND DEFECTS

Crystalline solids may consist of one or more single crystals. Most of the common materials of engineering interest are not single crystals but large polycrystalline aggregates with crystals bonded to one another along interfaces called grain boundaries. The average crystal size in the aggregate affects not only mechanical properties but, to a lesser degree, the transport properties—mass diffusivity, electrical conductivity, and thermal conductivity—as well. The effect of size on equilibrium thermodynamic properties, however, is comparatively small. In recent years, nearly perfect single crystals of microscopic size have been developed and have found significant applications in many electronic devices. These perfect single crystals seldom exist in nature, and they must be synthesized with extreme care.

Defects in crystals also exhibit certain effects on the physical properties of crystalline solids. Their effects upon optical, electrical, and thermal properties at low temperatures are particularly pronounced. In fact, many important properties of solids are controlled far more strongly by the nature of the defects than by the nature of the host crystal, which may be only a vehicle for the defects. Crystal defects may be classified as being either *point* or *line defects*. Point defects only affect the region surrounding a single particle site, whereas an entire series of particles are disturbed by a line defect.

There are three specific types of point defect: *vacancies*, *interstitialities*, and *impurities*. A vacancy represents the absence of a single atom or molecule from its expected location in the lattice. Lattice vacancies are often known as Schottky defects. Interstitial defects, which are also known as Frenkel defects, involve the transfer of an atom from its normal lattice position to an interstitial position. There always exist a certain number of lattice vacancies and interstitial defects in a crystal at thermal equilibrium, because the nonzero values of entropy at temperatures other than absolute zero require some disorderliness in the structure. Furthermore, the entropy is a monotonically increasing function of temperature, so the number of these defects increases with the rise of temperature. In the third type of point defect, impurity atoms may occupy the lattice positions that are normally reserved for the constituent atoms of the crystal, or they may lodge in interstitial positions. In either case they disturb the electronic states of their neighbors and affect the electrical behavior of the lattice.

Line defects include various kinds of dislocations. These may exert a strong influence on the mechanical properties, but they generally have little effect on the thermodynamic and transport properties.

10.2 STATISTICAL MECHANICS OF LATTICE VIBRATIONS

Thermal energy in a solid may exist in a variety of forms that correspond with various modes of motion in its fundamental particles. The common modes of motion include the vibration of lattice points around their equilibrium positions, internal vibration and rotation within the molecules, and the translation of free electrons. These modes, however, do not take place in all solids. For instance, there exists little or no free-electron contribution in nonmetals, and in monatomic crystals, such as occur in metals, there is no internal vibration or rotation. Furthermore, not all modes of motion are significant at all temperature levels. The free-electron contribution in metals, for instance, becomes appreciable as compared to the contribution of lattice vibration only at very low temperatures — below about 30°K. But, at room temperatures it is, as we saw in Sec. 6.4, almost negligible. One particular mode of motion, which always plays a significant role in the evaluation of the thermal properties of such solids, is the lattice vibration. The statistical-mechanical description of lattice vibrations will be developed in this section.

EINSTEIN MODEL

The first analysis of the thermodynamics of lattice vibrations was made by Einstein, who applied the Planck theory of quantized oscillators to the vibrations of a crystal. This analysis was described in Sec. 4.3. We now wish to reconsider Einstein's model from a more basic statistical-mechanical viewpoint. This model is not only a simple and useful approximation for crystalline solids, but it is also closely related to certain models or "cell theories" of the liquid state (Sec. 10.4).

In this highly idealized model, each molecule is assumed to be confined to a "cell" bounded by its immediate neighbors. It vibrates in this cell, independently of the vibrations of its neighbors. This idealization can not really be justified, because each atom in a solid is strongly coupled to its neighbors through binding forces, and any realistic model must be based on the vibration of the lattice as a whole. Despite this basic shortcoming, the Einstein theory gives a proper qualitative description of the thermodynamic behavior of crystalline solids and is worth pursuing.

We avoid the complications of internal molecular motion by limiting our discussion here to monatomic solids. The potential-energy function, ϕ, which results in a linear harmonic oscillation of the central atom in the "cell," is described by Eq. (7.40) and Fig. 7.1 or Fig. 5.1(b). For convenience of discussion here, the equilibrium position ξ of the central atom is taken as zero. Thus we have

$$\phi(\xi) = \phi(0) + \left(\frac{K}{2}\right)\xi^2 \tag{10.1}$$

where $K \equiv (d^2\phi/d\xi^2)_{\xi=0}$ is related to the frequency of the oscillation by

$$\nu = \frac{1}{2\pi}\sqrt{\frac{K}{m}} \tag{10.2}$$

as was indicated in the context of Eq. (5.10a). It is important to recognize that the two constants $\phi(0)$ and K depend not only on the structure of the interacting particles but on the size of the cell V/N as well. Furthermore, the motion of the central atom, described by Eq. (10.1), is actually three-dimensional. Thus we should view ξ as a polar coordinate that can be decomposed into a sum of three independent, one-dimensional, linear, harmonic motions in the x, y, and z directions,

$$\xi^2 = x^2 + y^2 + z^2$$

and the potential-energy function ϕ is

$$\phi(\xi) = \phi_x(x) + \phi_y(y) + \phi_z(z) = \sum_{w=x,y,z}\left[\phi_w(0) + \left(\frac{K}{2}\right)w^2\right] \tag{10.3}$$

Thus, according to the Einstein model, a crystalline solid of N atoms is described as an aggregation of $3N$ independent, one-dimensional, harmonic oscillators, all vibrating at the same frequency, ν.

We can now establish the canonical partition function for the Einstein solid. Using the canonical partition function for independent particles, given by Eq. (8.30), we have

$$\ln Q = \ln \left[e^{-N\phi(0,V/N)/2kT} \left(\sum_{i=0}^{\infty} g_i e^{-\epsilon_i/kT} \right)^{3N} \right] \tag{10.4}$$

where $N\phi(0, V/N)/2$ represents the total potential energy of all atoms at rest at their lattice points, and the factor of $\frac{1}{2}$ is introduced to avoid counting each intermolecular interaction twice. The summation term in Eq. (10.4) is the partition function Z for a one-dimensional harmonic oscillator, as given in Eq. (5.31):

$$Z = \frac{\exp(-\Theta_v/2T)}{1 - \exp(-\Theta_v/T)} = \left[2 \sinh\left(\frac{\Theta_v}{2T}\right) \right]^{-1}$$

where the characteristic temperature of vibration $\Theta_v = h\nu/k$ depends upon V/N, and where we drop the subscript v on Z_v since the only contributions to Z are vibrational. The fundamental equation for the Einstein solid can be then expressed as

$$F(T, V, N) = -kT \ln Q(T, V, N) = -kT \ln \left[e^{-N\phi(0,V/N)/2kT} Z^{3N} \right] \tag{10.5}$$

The only uncertainty involved in the above fundamental equation (10.5) is the unknown functional dependence of ϕ and Z (or Θ_v) on V/N. This dependence is complicated and varies from one kind of solid to another because of the change in the form of the intermolecular potential function. For certain thermodynamic properties, such as the constant-volume specific heat, this unknown factor does not cause any particular difficulty:

$$c_v \equiv \left.\frac{\partial U}{\partial T}\right|_{V,N} = -T \left.\frac{\partial^2 F}{\partial T^2}\right|_{V,N} = 3Nk\left(\frac{\Theta_v}{T}\right)^2 \frac{\exp(\Theta_v/T)}{[\exp(\Theta_v/T) - 1]^2} = f\left(T, \frac{V}{N}\right) \tag{10.6}$$

This is identical to Eq. (4.41), which we obtained directly, using Planck's quantized-oscillator concept. The limiting values given by Eq. (10.6) are

$$\lim_{T \to \infty} c_v = 3Nk \tag{10.7}$$

and

$$\lim_{T \to 0} c_v = 3Nk \left(\frac{\Theta_v}{T}\right)^2 \exp\left(-\frac{\Theta_v}{T}\right) \tag{10.8}$$

As we indicated in Sec. 4.3, the high-temperature limit agrees well with experiments and with the law of Dulong and Petit, but the predicted value of c_v approaches zero more rapidly in the low-temperature limit than the measured values do. Experiments show that $c_v \sim T^3$ when $T/\Theta_v \ll 1$.

FUNDAMENTAL EQUATION FOR LATTICE VIBRATIONS

The failure of the Einstein model to provide a good quantitative description of experimental data in the low-temperature range is largely the result of the assumption that the motion of each atom is independent of the motions of its neighbors. Improvement can only be achieved through an analysis of the vibration of the whole lattice. The total number of modes of vibration in a crystal composed of N atoms is equal, as we observed in Sec. 7.4, to half the total number of internal vibrational degrees of freedom, $2(3N - 6)$, or approximately $3N$. Each mode of vibration is distinguishable from the others, and in the harmonic-oscillator approximation (Fig. 7.1) the various vibrational modes do not interact.[1] Therefore, each vibrational mode can be regarded as an independent, distinguishable statistical element. It should be emphasized that these statistical elements represent the normal vibrations of the whole lattice — not just vibrations of the single atoms. The analysis of these independent elements, however, is almost identical to that of independent particles.

The canonical partition function Q for $3N$ normal modes of vibration is thus similar to Eq. (10.4),

$$\ln Q = \ln\left[e^{-N\phi(0,V/N)/2kT}\prod_{i=1}^{3N} Z(\Theta_{vi})\right] = -\frac{N\phi(0, V/N)}{2kT} + \sum_{i=1}^{3N} \ln Z(\Theta_{vi}) \tag{10.9}$$

where $Z(\Theta_{v_i})$ is given by Eq. (5.31), $\Theta_{v_i} = h\nu_i/k$, and the ν_i's are the vibrational frequencies and are functions of V/N. For convenience of mathematical manipulations, we replace the summation over a very large number ($3N$) of terms in Eq. (10.9) by an integral. This can be accomplished by introducing a continuous frequency distribution $g(\nu)$ such that $g(\nu)$ is the number of normal modes with frequencies between ν and $\nu + d\nu$. Since all separate ν's are functions of V/N, $g(\nu)$ is also a function of V/N. Therefore, we have from Eqs. (10.9) and (5.31),

$$\ln Q = -\frac{N\phi(0, V/N)}{2kT} + \int_0^\infty \left[\frac{h\nu}{2kT} - \ln\left(e^{-h\nu/kT} - 1\right)\right] g\left(\nu, \frac{V}{N}\right) d\nu \tag{10.10}$$

with the constraint

$$\int_0^\infty g(\nu, V/N)d\nu = 3N \tag{10.11}$$

[1] See, for instance, T. L. Hill, *Introduction to Statistical Thermodynamics*, Addison-Wesley Publishing Company, Inc., Reading, Mass. 1960, chap. 5, p. 2.

Based on the above equation, the fundamental equation for lattice vibration is simply

$$F(T, V, N) = -kT \ln Q(T, V, N) \tag{8.25b}$$

from which all thermodynamic information can be obtained.

This fundamental equation is not limited to any particular model and the only assumption invoked is that the vibrations are small and harmonic, so that a normal coordinate analysis can be carried out. Because of the large bonding forces in solids, the harmonic-oscillation assumption is generally very good, and the only major weakness is that it results in a zero thermal-expansion coefficient, as we see later. The frequency-distribution function $g(\nu, V/N)$ is in general a very complex function of ν, and very little is known as to the dependence of g on V/N. Fortunately, a simple approximation for $g(\nu)$ suggested by Debye yields a surprisingly good description of the thermodynamic behavior of solids.

DEBYE APPROXIMATION

The underlying reasoning behind the Debye approximation is rather simple and physically plausible. The fact that the discrepancies between Einstein's prediction and experimental measurements of c_v lie in the low-temperature region suggests the Debye correction. At low temperatures thermal excitations are relatively small, and lattice vibrations will occur primarily at low frequencies or long wavelengths. For wavelengths that are much longer than the atomic spacing in the lattice, lattice vibrations may actually be treated as elastic vibrations in a virtual continuum. This acoustic type of model is then suggested by the fact that the specific heats of both a photon gas (see Problem 6.14) and of a solid at low temperature increase as T^3. Accordingly, we are prompted to view a solid as though it were a *phonon gas* — a concept that was discussed briefly in Sec. 6.3.

For elastic vibrations it is known[2] that there are two different velocities of propagation — c_l for the longitudinal waves (or sound waves) in which one mode of vibration exists for a given wavelength in the direction of propagation, and c_t for transverse waves in which two modes of vibration exist for a given wavelength perpendicular to the direction of propagation:

$$c_l = \frac{3(1-\sigma)^{1/2}}{\kappa\rho(1+\sigma)} \tag{10.12}$$

$$c_t = \frac{3(1-2\sigma)^{1/2}}{2\kappa\rho(1+\sigma)} \tag{10.13}$$

[2]See, for example, A. E. H. Love, *A Treatise on Mathematical Theory of Elasticity*, Dover Publications, Inc., New York, 1944, pp. 297–302.

where κ is the volume compressibility, σ is Poisson's ratio, and ρ is the density. In a three-dimensional solid, the distribution function of vibrational modes is given by the following expression:[3]

$$g(\nu) = 4\pi V\left(\frac{1}{c_l^3} + \frac{2}{c_t^3}\right)\nu^2 \equiv \frac{12\pi V}{\bar{c}^3}\nu^2 \tag{10.14}$$

where $\bar{c}$ is defined through the last operation. If we recall from chapter 6 that $g(\nu) = dN_0/d\nu$, then Eq. (10.14) is three times the result we would obtain from Eq. (4.22). The factor of 3 accounts for the three modes of vibration.

The essence of the Debye approximation is to employ the limiting low-frequency expression for $g(\nu)$ as given in Eq. (10.14) for all frequencies and to terminate at a maximum frequency ν_m such that the total number of normal modes is $3N$, as prescribed by Eq. (10.11). This limiting frequency can be expressed in analytical form by writing

$$\int_0^{\nu_m} g(\nu)\,d\nu = \int_0^{\nu_m} \frac{12\pi V}{\bar{c}^3}\nu^2\,d\nu = 3N \tag{10.15}$$

which gives

$$\nu_m = \left(\frac{3N}{4\pi V}\right)^{1/3}\bar{c} \tag{10.16}$$

The Debye frequency spectrum is thus

$$g\left(\nu, \frac{V}{N}\right) = \begin{cases} \dfrac{9N\nu^2}{\nu_m^3} & 0 \le \nu \le \nu_m \\ 0 & \nu > \nu_m \end{cases} \tag{10.17}$$

where the dependence of g on V/N is carried through ν_m. Figure 10.3 shows the frequency spectra of three-dimensional lattice vibrations based on various types of approximations. It should be understood that the "Einstein approximation" refers to the frequency of the independent *normal modes* and not the independent particles that Einstein originally envisioned. The Debye approximation, as shown by comparison with more realistic calculations [Fig. 10.3(c) and (d)], is indeed a reasonable approximation for the two limiting-frequency regions: the low-frequency or large-wavelength region and the high-frequency cutoff region.

The fundamental equation for a Debye solid is obtained by combining Eqs. (10.10), (8.25b), and (10.17) to obtain

$$F(T, V, N) = \tfrac{1}{2}N\phi\left(0, \frac{V}{N}\right) + 9NkT\left(\frac{T}{\Theta_D}\right)^3 \int_0^{\Theta_D/T} u^2 \ln(1 - e^{-u})\,du \tag{10.18}$$

[3]See, for example, T. L. Hill, Reference 1, App. 6.

Fig. 10.3 Frequency spectra of three-dimensional lattice vibrations.

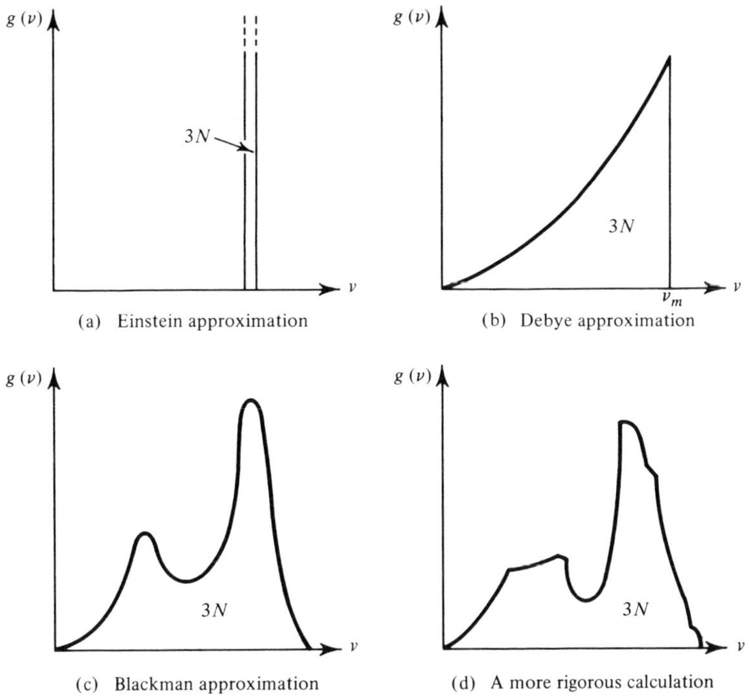

where the Debye temperature Θ_D is defined in terms of ν_m:

$$\Theta_D \equiv \frac{h\nu_m}{k} \tag{10.19}$$

and is a function of V/N. Table 10.2 gives average values of Θ_D for a few common substances based on various experimental investigations.[4] The integral in Eq. (10.18) can only be evaluated numerically. In principle, if the explicit expressions for $\phi(0, V/N)$ and $\Theta_D(V/N)$ are known for a particular solid, Eq. (10.18) contains all the thermodynamic information for that solid. For certain thermodynamics properties, this unknown dependence on V/N does not cause any real concern. One such property is the constant-volume specific heat, which can be obtained as

$$\begin{aligned} c_v = -T\left(\frac{\partial^2 F}{\partial T^2}\right)_{V,N} &= 9Nk\left(\frac{T}{\Theta_D}\right)^3 \int_0^{\Theta_D/T} \frac{u^4 e^u}{(e^u-1)^2}\,du \\ &= 3R^0\left[4D\left(\frac{\Theta_D}{T}\right) - \frac{3\Theta_D/T}{\exp(\Theta_D/T)-1}\right] \end{aligned} \tag{10.20}$$

[4]C. Kittel, *Introduction to Solid State Physics*, 3rd ed., John Wiley & Sons, Inc., New York, 1967, p. 180.

TABLE 10.2 Representative Values of Debye Temperature

Substance	Θ_D, °K	Substance	Θ_D, °K	Substance	Θ_D, °K
Be	1160	Fe	467	Al	418
Mg	406	Co	445	In	109
Ca	(219)	Ni	456	Tl	89
La	132	Pd	275	C (diamond)	(2000)
Ti	278	Pt	229	Si	658
Zr	270	Cu	339	Ge	366
V	273	Ag	225	Sn (gray)	212
Nb	252	Au	165	Sn (white)	189
Ta	231	Zn	308	Pb	94.5
Cr	402	Cd	300	Bi	117
Mo	425	Hg	(60–90)	W	(379)

The quantity D is the Debye function $D(\Theta_D/T)$ and is defined as

$$D\left(\frac{\Theta_D}{T}\right) \equiv 3\left(\frac{T}{\Theta_D}\right)^3 \int_0^{\Theta_D/T} \frac{u^3}{e^u - 1}\, du \tag{10.21}$$

Numerical values of D are given in Table 10.3. As $T \to \infty$ or $(\Theta_D/T) \to 0$, the integrand of D can be expanded to show that $D(0) \to 1$ and $c_v \to 3Nk$, the correct high-temperature limit. As $T \to 0$ or $(\Theta_D/T) \to \infty$, the definite integral in D with the upper limit now equal to infinity can be evaluated to give

$$c_v = 3Nk\,\frac{4\pi^4}{5}\left(\frac{T}{\Theta_D}\right)^3 \tag{10.22}$$

The above result, generally known as the T^3 law, is the most important result of the Debye theory. It is accurate to within 1 percent of experimental measurements, or better, for $T/\Theta_D < \frac{1}{12}$.

TABLE 10.3 Debye Function D

Θ_D/T	D
0	1
0.1	0.9630
0.2	0.9270
0.5	0.8250
1	0.6744
2	0.4411
3	0.2836
4	0.1817
5	0.1176
10	0.0193
20	0.0024
∞	0

We should remember that for metals at very low temperatures, the specific heat of the free electrons becomes appreciable, as we indicated in Sec. 6.4,

$$c_v \,|\, _{T \to 0°\mathrm{K}} = AT + BT^3 \tag{6.49}$$

and its contribution AT must be considered in addition to the lattice vibration contribution BT^3.

EXAMPLE 10.1 Determine the constant-area specific heat for a two-dimensional Einstein solid and for a two-dimensional Debye solid.

(a) *Einstein solid.* We begin by writing the partition function for a two-dimensional harmonic oscillator with constant ν:

$$Z = Z_x Z_y = Z_x^2 = \left[\frac{\exp(-h\nu/2kT)}{1 - \exp(-h\nu/kT)}\right]^2$$

but

$$U = NkT^2 \frac{\partial \ln Z}{\partial T} = Nh\nu + \frac{2Nh\nu}{\exp(h\nu/kT) - 1}$$

so

$$c_v = \left.\frac{\partial U}{\partial T}\right|_A = 2Nk\left(\frac{h\nu}{kT}\right)^2\left[1 - \exp\left(-\frac{h\nu}{kT}\right)\right]^{-2} \exp\left(-\frac{h\nu}{kT}\right)$$

As $h\nu/kT$ approaches zero this gives $c_v \to 2R^0$, which is what we would expect from the classical equipartition theory. As $h\nu/kT$ approaches infinity this result gives $c_v \to 0$, as it must, to satisfy the fourth postulate (or the third law) of thermodynamics.

(b) *Debye solid.* We must now consider a spectrum of phonon frequencies from zero to a maximum ν_m such that the number of vibrations is consistent with the total number of modes of energy storage in the solid $2N$. Thus

$$2N = \int_0^{\nu_m} g(\nu)\, d\nu$$

To integrate this we need $g(\nu)$. This can be obtained in the usual way,

$$dN_0 = d\left[2\left(\frac{\pi}{4}\,\frac{2L\nu}{c}\,\frac{2L\nu}{c}\right)\right] = g(\nu)\, d\nu$$

so

$$2N = \int_0^{\nu_m} 4\pi A \frac{\nu}{c^2}\, d\nu$$

where A is the area, equal to L^2. Integrating this we obtain

$$\nu_m = \frac{Nc^2}{\pi A}$$

To get the specific heat, we must integrate the Einstein expression for the energy of a single oscillator over all modes instead of just multiplying it by $2N$. Thus

$$c_v = k\int_0^{\nu_m} \frac{(h\nu/kT)^2 \exp(-h\nu/kT)}{[1-\exp(-h\nu/kT)]^2} g(\nu)\, d\nu$$

Introducing $x \equiv h\nu/kT$, we can transform this to

$$c_v = \frac{4Nk}{x_m^2}\int_0^{x_m} \frac{x^3 e^{-x}}{(1-e^{-x})^2}\, dx$$

In the high-temperature limit ($x_m \to 0$), c_v approaches $2R^0$, as we should expect, and as x_m approaches infinity, c_v approaches zero. It can also be shown that at low temperatures c_v varies as T^2 for a two-dimensional Debye solid. This is analogous to Eqs. (10.22) for the three-dimensional case.

10.3 EQUATION OF STATE OF SOLIDS

The Helmholtz-function form of the fundamental equation given in Eq. (8.25b) yields one equation of state of the form

$$S = -\left(\frac{\partial F}{\partial T}\right)_{V,N} = S(T, V, N)$$

This is often absorbed in

$$c_v = T\left(\frac{\partial S}{\partial T}\right)_{V,N} = c_v(T, V, N)$$

which is sometimes called the "thermal equation of state." In the Debye model of solids, the latter is given by Eq. (10.20).

What we commonly call the "equation of state" is the mechanical equation of state,

$$p = -\left(\frac{\partial F}{\partial V}\right)_{T,N} = p(T, V, N) \tag{10.23}$$

It is this equation and the related quantities that we wish to investigate next.

DEBYE EQUATION OF STATE

The Helmholtz function can be written as

$$F = U_0(V, N) + F_D(T, V, N) \tag{10.24}$$

where U_0 is the value of F (or $U - TS$) at 0°K and equals the total potential energy of all atoms in the lattice when the atoms are located at their equilibrium positions; F_D is the lattice-vibration contribution. Free-electron and other contributions are neglected here. For convenience we restrict our discussion to the Debye theory, and the two contributions U_0 and F_D can be readily identified by comparing Eqs. (10.18) and (10.24).

Now, defining a function $f(\Theta_D/T)$ such that

$$F_D(T, V, N) = Tf\left(\frac{\Theta_D}{T}\right) \tag{10.25}$$

we obtain

$$\left(\frac{\partial F_D}{\partial \Theta_D}\right)_T = \frac{df}{d(\Theta_D/T)} \tag{10.26}$$

The total derivative of f can be cast in a more convenient form by taking V, N, and consequently Θ_D, as constant, so

$$\frac{df}{d(\Theta_D/T)} = \frac{1}{\Theta_D}\left[\frac{\partial}{\partial(1/T)}\frac{F_D}{T}\right]_{V,N} = \frac{U_D}{\Theta_D} \tag{10.27}$$

where the partial derivative, by definition, gives an internal energy U_D, which is the internal energy of lattice vibrations in a Debye solid.

The pressure of a Debye solid is then given by Eq. (10.23) as

$$p = -\frac{\partial U_0}{\partial V} - \frac{\partial F_D}{\partial V} = -\frac{\partial U_0}{\partial V} - \frac{\partial F_D}{\partial \Theta_D}\frac{\partial \Theta_D}{\partial V} \tag{10.28}$$

or, with the help of Eqs. (10.26) and (10.27),

$$p = -\frac{\partial U_0}{\partial V} - \frac{\partial \Theta_D}{\partial V}\frac{U_D}{\Theta_D} \tag{10.29}$$

If we define a quantity called the Grüneisen constant γ,

$$\gamma \equiv -\frac{\partial \ln \Theta_D}{\partial \ln V} \tag{10.30}$$

we have, from Eq. (10.29), the Debye equation of state,

$$p = -\frac{\partial U_0}{\partial V} + \frac{\gamma U_D}{V} \tag{10.31}$$

which is a rather simple equation relating pressure to internal energy and volume. Moreover, it bears a remarkable resemblance to the ideal-gas equation of state, if U_0 is taken to be zero. For example, Θ_t is given for an ideal gas as $h^2/8mkV^{2/3}$ in Sec. 5.3. It follows that γ is equal to $\frac{2}{3}$ and

$$p = \frac{NkT}{V} = \frac{3}{2}\frac{U}{V} \tag{2.31b}$$

GRÜNEISEN RELATION[5]

An analytical determination of the Grüneisen constant γ is not very fruitful because we cannot make a quantitative prediction

[5]E. Grüneisen, "Zustand des Festen Körpers," *Handbuch der Physik*, ed. H. Geiger and K. Scheel, Springer Verlag, Berlin 1926, vol. 10 ("Thermische Eigenschaften den Stoff"), pp. 1–59.

of the explicit dependence of Θ_D on V. Although Eqs. (10.14), (10.16), and (10.19) do give the relationship between Θ_D and elastic properties of solids, it is not clear how the elastic properties, such as volume compressibility and Poisson's ratio, depend on volume. Thus γ generally has to be determined experimentally from measurements of related macroscopic properties. We derive here a simple relation that relates γ to other thermodynamic properties.

Differentiating Eq. (10.31) and assuming that γ is independent of T gives

$$\left(\frac{\partial p}{\partial T}\right)_V = \frac{\gamma C_v}{V} \tag{10.32}$$

In terms of the linear coefficient of thermal expansion β which is equal to one third of the volume-expansion coefficient, we have

$$\beta = \frac{1}{3V}\left(\frac{\partial V}{\partial T}\right)_p = \frac{1}{3V}\left(\frac{\partial p}{\partial T}\right)_V\left(\frac{\partial V}{\partial p}\right)_T = \frac{\kappa}{3}\left(\frac{\partial p}{\partial T}\right)_V \tag{10.33}$$

where κ is the volume compressibility. Combining Eq. (10.32) and (10.33) yields the so-called Grüneisen relation

$$\beta = \frac{\kappa\gamma C_v}{3V} \tag{10.34}$$

Grüneisen showed by examining experimental data that γ remains constant in many metals over a wide range of temperature and density. Some values of γ, calculated by Grüneisen with the help of Eq. (10.34), are given in Table 10.4.

TABLE 10.4 Some Values of the Grüneisen constant, γ

Substance	$\gamma = 3V\beta/\kappa C_v$
Na	1.25
K	1.34
Al	2.17
Mn	2.42
Fe	1.60
Co	1.87
Ni	1.88
Cu	1.96
Ag	2.40
Pt	2.54
NaCl	1.63
KF	1.45
KCl	1.60
KBr	1.68
KI	1.63

THERMAL-EXPANSION COEFFICIENT

We now consider the thermal-expansion coefficient from a microscopic viewpoint. This is particularly relevant here, because the harmonic (small-vibration) approximation used for the analysis of lattice vibrations gives a zero thermal-expansion coefficient, as is shown later. The thermal expansion of solids results from an increase in the average amplitude of atomic vibrations at the lattice points when the thermal energy increases. Let us therefore consider the potential energy ϕ of the vibrating atoms at a displacement ξ. If we proceed as we did in the context of Eq. (7.39) we can write

$$\phi(\xi) = \frac{1}{2!}\left.\frac{d^2\phi}{d\xi^2}\right|_{\xi_e}(\xi - \xi_e)^2 + \frac{1}{3!}\left.\frac{d^3\phi}{d\xi^3}\right|_{\xi_e}(\xi - \xi_e)^3 \tag{10.35}$$

where an additional anharmonic term is included for consideration and where the coefficients depend on V/N.

The anharmonic term is crucial to the phenomenon of thermal expansion. As the temperature is raised and vibration increases, it is the asymmetry of the potential function, as reflected in this term, that causes the average atomic displacement from equilibrium to be positive. By the definition of an average,

$$\bar{\xi} - \xi_e = \frac{\int_{-\infty}^{\infty} (\xi - \xi_e)\exp(-\phi/kT)\, d(\xi - \xi_e)}{\int_{-\infty}^{\infty} \exp(-\phi/kT)\, d(\xi - \xi_e)} \tag{10.36}$$

where the kinetic-energy contribution in the Boltzmann distribution function is not included, because it does not depend on ξ.

The anharmonicity effect can be obtained by substituting Eq. (10.35) into (10.36) and performing the integrations. The result for small displacements and correspondingly low anharmonic energies is obtained after making appropriate expansions and order-of-magnitude simplifications (the details of which are left as an exercise in Problem 10.8):

$$\bar{\xi} - \xi_e = -\frac{3kT}{4}\,\frac{\left.\frac{1}{3!}\frac{d^3\phi}{d\xi^3}\right|_{\xi_e}}{\left(\left.\frac{1}{2!}\frac{d^2\phi}{d\xi^2}\right|_{\xi_e}\right)^2} = (\text{constant})\,T \tag{10.37}$$

Thus the average increase of the spacing of molecules is directly proportional to the first anharmonic coefficient, $(d^3\phi/d\xi^3)_{\xi_e}/3!$. It also depends upon higher-order anharmonic coefficients—$(d^5\phi/d\xi^5)_{\xi_e}/5!$, and so on—if they are carried, but none of the harmonic terms contributes.

The common formula for the linear thermal expansion of solids,

$$\text{length} = \text{length}_{\text{ref}}[1 + \beta(T - T_{\text{ref}})] \tag{10.38}$$

is consistent with Eq. (10.37). They are, in fact, identical if T_{ref} is taken to be absolute zero and $\bar{\xi}/\xi_e$ is identified as length/(length at $T = 0$). Thus the coefficient of thermal expansion β is approximately $(\bar{\xi} - \xi_e)/T$ divided by ξ_e, which in turn is $\simeq (V/N)^{1/3}$. Thus

$$\beta \simeq \frac{k}{4}\left(\frac{N}{V}\right)^{1/3} \left(\frac{d^3\phi}{d\xi^3}\right)_{\xi_e} \Big/ \left(\frac{d^2\phi}{d\xi^2}\right)^2_{\xi_e} \tag{10.39}$$

EXAMPLE 10.2 Express β for the Lennard-Jones potential, $\phi = 4\epsilon[(\sigma/\xi)^{12} - (\sigma/\xi)^6]$.

Differentiating, and noting that $\xi_e = 2^{1/6}\sigma$ for the Lennard-Jones potential, we obtain

$$\left.\frac{d^2\phi}{d\xi^2}\right|_{\xi_e} = 4\epsilon\left(\frac{12\cdot 13}{2^{1/3}\sigma^2}\frac{1}{2^2} - \frac{6\cdot 7}{2^{1/3}\sigma^2}\frac{1}{2}\right)$$

and

$$\left.\frac{d^3\phi}{d\xi^3}\right|_{\xi_e} = 4\epsilon\left(-\frac{12\cdot 13\cdot 14}{1^{1/2}\sigma^3}\frac{1}{2^2} + \frac{6\cdot 7\cdot 8}{2^{1/2}\sigma^3}\frac{1}{2}\right)$$

Thus

$$\left(\frac{\bar{\xi}}{\xi_e} - 1\right)\frac{1}{T} = \beta \simeq \frac{(2^{1/6})(7)}{96}\frac{\sigma}{(N/V)^{1/3}}\frac{k}{\epsilon}$$

so β should vary directly as the relative size of the particles with respect to the lattice and inversely with the "critical temperature" $T^* = \epsilon/k$.

10.4 LATTICE THEORIES OF LIQUIDS

There are several general approaches in the study of the liquid state. The virial-expansion approach, which was used in chapter 9 to describe moderately dense gases, is quite general and exact, but problems associated with the convergence of the virial expansions and the complex numerical evaluation of many virial coefficients severely limit the utility of this approach. In practice, virial expansions can be applied only to dilute or moderately dense gases. The most significant advances in the statistical-mechanical description of the liquid state have been along two other avenues: the distribution-function approach and the lattice-theory approach.

In the fundamental sense, the former approach is more rigorous than the latter. The approximate lattice theories of liquids, although semiempirical in nature, are attractive in the simplicity of their physical models and in their strong analogy to the solid-state theory. In view of this analogy, the lattice theories should generally be appropriate for liquids at high pressure and density. After the preceding descriptions of the solid state, it is only natural to proceed to the lattice theories of liquids. A brief description of the distribution-function approach to the theory of liquids, however, is given at the end.

CELL THEORIES

In a cell theory of liquids, the volume V occupied by the N molecules is divided into a lattice of N cells, each of volume $v = V/N$ and each occupied by one molecule. The cell volume is sometimes called the *free volume*, or the volume free for a single molecule, and cell theories are often referred to as free-volume theories. The motion of each molecule within its cell and in the potential field of its neighbors is assumed to be independent of the motion of the other molecules.

Up to this point, the cell model of liquids is indeed identical to the Einstein model of solids. However, there are a number of basic differences. The molecule in a liquid is not as strongly influenced by its neighbors as that in a solid, so its motion is not restricted in a region near an equilibrium position of minimum potential energy. Thus the potential field experienced by the molecule in a liquid is not necessarily the near-parabolic one, which results in an almost harmonic oscillation, as is the case in solids. Moreover, we do not usually have to consider the quantum-mechanical description of the vibrations, as we did for solids, because the temperature level of the liquid state is usually not very low.[6]

Perhaps the most profound difference is the introduction of the "communal-entropy" concept in the cell model of liquids. The "communal entropy," which is absent in a crystal but present in a gas, is added to compensate for the decrease in entropy due to the ordered confinement of each molecule in a cell in the regular lattice. This concept can be made clearer through the following example.

Consider a monatomic ideal gas of N molecules in a volume V. Using Eq. (9.7), we can note that $Z_{\text{int}} = 1$ and $Z_\phi = V^N$, and obtain

$$F = -kT \ln Q = -kT \ln \left(\frac{V^N}{N!\Lambda^{3N}}\right) = -NkT \ln \left(\frac{ve}{\Lambda^3}\right) \tag{10.40}$$

For the cell model in which N molecules are equally distributed in N cells of volume v, the factor $kT \ln (1/N!)$ must be added to Eq.

[6]Liquid helium is a definite exception which displays a variety of quantum peculiarities.

(10.40) to account for the distinguishability of particles by virtue of their locations. Moreover, with no intermolecular forces, the configuration integral Z_ϕ becomes v^N in the cell model. Thus

$$F = -NkT \ln \left(\frac{v}{\Lambda^3}\right) \tag{10.40a}$$

This indicates that when we arbitrarily divide the system into cells, the free energy F of the system increases by NkT, and the entropy ($S = -\partial F/\partial T$) decreases by Nk. It therefore appears necessary to add a certain amount of "communal entropy" to the entropy obtained from a cell model. One of the difficult problems in cell theories of liquids is that of determining how much "communal entropy" should be added. For convenience of the present discussion, we take the value of the "communal entropy" to be $\Delta S = Nk$, as suggested by the preceding example.

Now let us consider the effect of the intermolecular potential between a given molecule and all other neighboring molecules in the cell-model formulation. We realize that the potential energy becomes very large when the confined molecule approaches its neighbors near the edges of its cell. Consequently, the probability of having the confined molecule in a given element of volume is no longer uniform throughout the cell but is proportional to the Boltzmann factor, $\exp\{-[\phi(\mathbf{r}) - \phi(0)]/kT\}$. The "effective" free volume v_f through which the confined molecule can move, is thus

$$v_f = \iiint \exp\left[-\frac{\phi(\mathbf{r}) - \phi(0)}{kT}\right] d\mathbf{r} \tag{10.41}$$

where the integration is carried out over the whole cell space. The effective free volume must clearly be less than the cell volume (V/N), and both $\phi(\mathbf{r})$ and $\phi(0)$ are functions of V/N; thus we can write $v_f(T, v)$ and $\phi(0, v)$. Substituting Eq. (10.41) into (10.40a) and adding the "communal-entropy" contribution, we arrive at the following fundamental equation:

$$F = -NkT \ln\left[\frac{v_f(T, v)e}{\Lambda^3}\right] + \frac{N\phi(0, v)}{2} \tag{10.42}$$

Now the problem is simply to obtain appropriate forms of $v_f(T, v)$ and $\phi(0, v)$.

One of the simplest fundamental equations for the liquid state is for the van der Waals liquid. It can be readily shown that to obtain the van der Waals equation, Eq. (9.27), we must have

$$v_f(v) = v - b \qquad \phi(0, v) = -\frac{2a}{v} \tag{10.43}$$

where a and b are constants related to intermolecular potential parameters by, for example, Eqs. (9.21) and Eq. (9.25).

The basic prototype for all cell theories of liquids is originally due to Lennard-Jones and Devonshire.[7] In the LJD model, as it is called, each molecule moves within its cell in the resultant potential field of its immediate neighbors, which are assumed to be fixed at the centers of their respective cells. The resultant potential field $\phi(\mathbf{r})$ is obtained approximately as a spherically symmetric function on the basis of the Lennard-Jones intermolecular potential. The fundamental equation can then be determined from Eqs. (10.41) and (10.42). More refined calculations have been made for such a model and the results for pv/kT are shown in Table 10.5.[8] Here, pv/kT, which is akin to the compressibility factor, is expressed as a function of the reduced variables T^* and v^*, given in Eq. (9.68). Similar tables for the various thermodynamic functions are also available. As we might expect, these calculations agree well with measurements at high pressures and densities.

TABLE 10.5 Function $pv/kT = f(T^*, v^*)$ Based on the LJD Theory

Reduced temperature	Reduced volume, $v^* = v/\sigma^3$				
$T^* = kT/\epsilon$	1.131	1.414	2.121	3.536	4.243
0.80	−1.442	−2.413	−0.885	−0.0661	0.0990
1.00	0.1881	−0.8515	−0.0547	0.3985	0.4897
1.30	1.6128	0.5151	0.6933	0.8230	0.8340
1.60	2.4417	1.3168	1.1453	1.084	1.030
2.50	3.570	2.460	1.806	1.471	1.266
5.00	4.253	3.185	2.291	1.700	1.311
400.00	2.543	2.082	1.296	1.026	1.007

OTHER APPROXIMATE LATTICE THEORIES

A number of attempts have been made to improve the cell theories. One kind of modification allows not just one molecule, but 0, 1, 2,··· molecules in a cell. Such theories are generally called "hole theories," because they allow the existence of an empty cell or a "hole." Unfortunately, these refinements do not provide much improvement over the LJD theory. Another variation of these treatments is the theory of significant structures,[9] in which the

[7] J. E. Lennard-Jones and A. F. Devonshire, *Proc. Roy. Soc.* (*London*) **A163,** 53 (1937); **A165,** 1 (1938).

[8] J. O. Hirschfelder, C. F. Curtiss, and R. B. Bird, *Molecular Theory of Gases and Liquids*, John Wiley & Sons, Inc., New York, 1964, pp. 1122–1125.

[9] H. Eyring and collaborators, *Proc. Natl. Acad. Sci. U.S.* **44,** 683 (1958); **45,** 1594 (1959); **46** 333, 336, 639 (1960); **47,** 526 (1961); **48,** 501 (1962).

liquid is regarded as a mixture of crystallike and gaslike regions. The physical and theoretical basis of this theory is not entirely clear; however, it has led to rather successful semiempirical equations of state for a variety of liquids. Other extended cell theories include the "tunnel" model[10] and the "worm" model,[11] which have found some success in rationalizing experimental data.

All these lattice theories of liquids are semiempirical in nature and are to be viewed as successful only if they agree with experimental results. The fundamental basis of these theories is usually not strong.

MORE EXACT CALCULATIONS

The distribution-function approach is more rigorous, but it incurs formidable mathematical complexity. The starting point is to define the radial distribution function $g(r)$ which characterizes, on the average, the number of molecules in an infinitesimal element of volume at a distance r from a particular molecule. It is theoretically possible to set up an equation for $g(r)$, and the thermodynamic properties of the liquid can be expressed in terms of it. In practice, one must make approximations before he can solve the equation for $g(r)$. It is possible to obtain $g(r)$ directly from x-ray diffraction experiments. It can also be evaluated on the basis of the Lennard-Jones potential, and resulting values of pv/kT, similar to those given by the LJD theory, are given in Table 10.6.[12]

TABLE 10.6 Function $pv/kT = f(T^*, v^*)$ Based on the Radial Distribution Function Theory

Reduced temperature $T^* = kT/\epsilon$	Reduced volume, $v^* = v/\sigma^3$				
	1.222	1.483	2.260	3.632	13.82
0.833	−2.829	−2.433	−1.445	−0.594	
1.000	−1.382	−1.268	−0.734	−0.156	0.629
1.250	0.052	−0.115	−0.038	0.264	0.768
1.677	1.467	1.018	0.649	0.670	0.883
2.500	2.856	2.139	1.326	1.064	
5.000	4.223	3.242	1.998	1.456	
∞	5.567	4.333	2.667	1.833	1.167

[10] J. A. Barker, *Proc. Roy. Soc.* (*London*) **A259,** 442 (1961); *J. Chem. Phys.* **37,** 631 (1962).
[11] H. S. Chung and J. S. Dahler, *J. Chem. Phys.* **40,** 2868 (1964); **42,** 2374 (1965).
[12] J. O. Hirschfelder, C. F. Curtiss, and R. B. Bird, Reference 8.

For a relatively small number of particles, it is also possible to perform an exact numerical calculation of liquid properties with high-speed computers. The numerical techniques employed in these calculations are the Monte Carlo method[13] and the molecular-dynamics method, in which the detailed molecular trajectories are computed.[14]

EXAMPLE 10.3 Calculate the specific volume of liquid nitrogen at its normal boiling point (77.2°K) (a) by assuming it to be a van der Waals liquid, and (b) using the LJD hole theory.

In both of these computations we can compare the results with the observed value of 0.811 cm³/g.

(a) The van der Waals equation [Eq. (9.34)] for $T_r = 77.2/126.2$, or 0.611, where $T_c = 126.2$°K, and $p_r = 1/33.5 = 0.02983$, where $p_c = 33.5$ atm, is

$$0.02983 = \frac{4.88}{3v_r - 1} - \frac{3}{v_r^2}$$

The solution, by trial and error, is $v_r = 0.437$, but $v_c = 3.22$ cm³/g, so $v = 0.437 \times 3.22 = 1.409$ cm³/g, which is high by 74 percent.

(b) We now must use values of $f(T^*, v^*)$ given in Table 10.5. By definition,

$$v = \frac{kT}{p} f(T^*, v^*) = 1.05 \times 10^{-20} \text{ cm}^3/\text{molecule} \cdot f(T^*, v^*)$$

or, using $v^* = v/\sigma^3$, where Table 9.2 gives $\sigma = 3.70$ Å as the Lennard-Jones radius for nitrogen molecules,

$$v^* = \frac{1.05 \times 10^{-20} \text{ cm}^3}{(3.70 \times 10^{-8})^3 \text{ cm}^3} f(T^*, v^*) = 208 \cdot f(T^*, v^*)$$

and T^* is $77.2°\text{K}/(\epsilon/k)$, where Table 9.2 gives $T_c^* = \epsilon/k = 95.1$°K for water. Thus $T^* = 0.812$. By interpolating a table similar to 10.5, we can obtain $f(0.812, v^*)$ values. The value of v^* for which $f(0.812, v^*)$ is equal to $v^*/208$ is 1.07. Thus

$$v = [1.07 \times (3.70)^3 \times 10^{-24} \text{ cm}^3/\text{molecule}] \times [(6.02 \times 10^{23}/28) \text{ molecules/g}]$$

or

$$v = 1.17 \text{ cm}^3/\text{g}$$

which is high by only 44 percent.

[13] W. W. Wood, F. R. Parker, and J. D. Jacobson, *J. Chem. Phys.* **27,** 720, 1207 (1957); Z. W. Salsburg, *J. Chem. Phys.* **37,** 798 (1962).
[14] B. J. Alder and T. E. Wainwright, *J. Chem. Phys.* **33,** 1439 (1960), **40,** 2724 (1964).

Problems 10.1 Write down all the steps in Sec. 10.2 leading to Eqs. (10.6), (10.7), and (10.8).

10.2 Some of the thermal energy of a gas will consist of sound waves having a wavelength longer than the mean free path, l. Determine the total energy of a monatomic gas with such waves.

10.3 Derive an equation for the vapor pressure $p_0(T)$ of an Einstein crystal, assuming that the vapor is an ideal gas and noting that μ_{solid} must equal μ_{vapor} in equilibrium.

10.4 Complete all the steps leading to Eq. (10.22).

10.5 Derive the fundamental equation for two-dimensional Debye solids. (Recall Example 4.1 and Problem 4.6.)

10.6 Determine the relative dispersion of energy fluctuations [recall Eq. (8.53a) and context] in a Debye solid at the high- and low-temperature limits, respectively.

10.7 The total blackbody radiation energy of a dielectric solid with a frequency-independent refractive index n is n^2 times the value in hohlraum at the same temperature. Obtain expressions for pressure and c_v in a Debye solid of this kind, including both radiation and the lattice-vibration contribution. Compare the order of magnitude of these contributions when the temperature is 300°K.

10.8 Verify Eq. (10.37).

10.9 Express the Joule–Thompson coefficient for liquids obeying the cell model, in terms of the functions $v_f(T, v)$ and $\phi(0, v)$.

10.10 Determine the speed of sound in a liquid obeying the cell model.

10.11 Determine the coefficient of thermal expansion of a liquid that obeys the cell model.

10.12 Obtain the fundamental equation of a two-dimensional liquid, based on the cell model.

10.13 What is c_v for a van der Waals liquid? Can you explain this rather odd result?

11 elementary kinetic theory of transport processes

11.1 INTRODUCTION

The methods of statistical mechanics have proved to be a powerful tool for treating equilibrium behavior. Now we wish to enlarge upon our capability for deducing the macroscopic behavior of gaseous[1] groups of microscopic particles. We wish to develop predictions for such nonequilibrium systems as shear flows and thermally conducting gases.

Kinetic theory provides the means by which we can do this. Its methods are generally less abstract and more explicit than those of statistical mechanics, but they are also more complicated in that they take closer account of what is happening to particles. In particular, kinetic theory focuses upon collisions and the way in which they change the distribution function $f(\mathbf{c}, \mathbf{r}, t)$.

The reason that we have been permitted to ignore the fre-

[1]Transport in liquids is considerably more complicated. See, for example, J. Frenkel, *Kinetic Theory of Liquids*, Dover Publications, Inc., New York, 1955; or J. O. Hirschfelder, C. F. Curtiss, and R. B. Bird, *Molecular Theory of Gases and Liquids*, John Wiley & Sons, Inc., New York, 1954, chap. 9.

quency of collisions is that it has nothing to do with the equilibrium state. Collisions provide the means by which a system changes its state. But once an equilibrium state has been achieved in a system, we can learn all that we need to know by considering the system during a single instant in which almost no collisions occur.

Suppose, for example, that a wall, maintained at a constant high temperature, is separated from a low-temperature gas by an adiabatic partition. When the partition is removed, cold molecules strike the wall. They are excited by the high-temperature molecules comprising the wall, and leave with a higher average energy. In each subsequent collision the hotter molecules will, on the average, share their greater energies of translation, rotation, and so on, with the cooler molecules.

The conduction of heat in gases is the gross effect of this process. Energy is transported from one position to the next by collisions. If a cold wall were held at constant temperature some distance from the hot wall, the macroscopic system would eventually become steady. Although we might observe, in a gross way, that the system has reached steady state, it should be clear that the gas is in a nonequilibrium state. The system is causing entropy to be generated in its surroundings even though its macroscopic properties are not changing. More specifically, although the distribution function $f(\mathbf{c}, x)$ is time-independent, it is still position-dependent, and thus it does not describe an equilibrium gas.

The derivation of the energy distribution function in chapter 3 was an elementary example of the way in which the methods of statistical mechanics lead to a description of equilibrium behavior. No consideration at all was given to the mechanics of collision, and the kinetic hypothesis was not expanded beyond the minimal statement given at the end of Sec. 2.1.

Secs. 11.2 and 11.3 use the methods of kinetic theory in an elementary description of the processes of heat, momentum, and mass diffusion. The description centers on the mechanics of collision and it does not use the equilibrium assumption of maximum probability. It employs an expanded form of the kinetic hypothesis as stated in Sec. 2.1, and it deals with gases that are *classical* in the sense that they are neither relativistic nor subject to quantum limitations.

11.2 MEAN FREE PATH

To describe any process that depends upon molecular collisions, we must first characterize the frequency with which particles collide. This is generally done by giving the average distance that particles travel between collisions — the "mean free path," l. Although a more precise derivation is delayed until chapter 12, it is possible to show the essential features of the mean free path in a simple approximate derivation.

The kinetic hypothesis must be expanded to include an idealization of molecular size. The molecules are considered as spherical and of diameter σ. If σ is large, collisions are frequent, and, as σ is shrunk to zero, collisions become impossible.[2]

If all but one of the molecules are considered to be stationary, and that one molecule is considered to move at the average speed $\bar{c}$ of the molecules in the gas, then it collides with particles within a volume swept through at the rate of $(\pi/4)(\sigma + \sigma)^2\bar{c}$ cm^3/sec. But this space contains n molecules/cm^3. Accordingly, the frequency of collisions is

$$\text{frequency} = (\pi\sigma^2\bar{c})(n) \text{ collisions per second}$$

and the mean free path l is

$$l = \frac{\text{velocity}}{\text{frequency}} = \frac{1}{n\pi\sigma^2} \tag{11.1}$$

The mean free path of ideal-gas molecules is thus inversely proportional to the density, $\rho = mn$, and the cross-sectional area of the molecules.

Two kinds of approximation have been made in developing Eq. (11.1). When the other molecules are in motion, the likelihood of collision is improved. Conversely, a slow-moving molecule generally has a shorter free path than a fast molecule. We discover, later, that when the velocity of the other particles is considered, l is reduced by a factor of $1.051/\sqrt{2}$.

The second approximation lies in the use of a constant molecular diameter. A collision occurs when one molecule (or other particle) moves into the local force field of another molecule — not when one hard sphere bumps into another hard sphere. In an actual collision, the effective diameter of the molecule is less when the relative velocity between particles is high, because forces at a distance have less time to act. This effect becomes especially important when very high energy particles move through a medium. For example, high-energy neutrons in a nuclear reactor must be slowed down (or "moderated") before they have a large enough "collision cross section" to hit other particles and continue the fission process.

Finally, it is instructive to compare the mean free path with the

[2]The interesting corollary to this is the fact that changes in state — which result from collisions — become increasingly slow in very dilute gases.

average spacing l_0 between molecules. The latter is equal to $n^{-1/3}$. Thus we obtain from Eq. (11.1),

$$\frac{l}{l_0} = \frac{1.051}{\sqrt{2}\pi}\left(\frac{l_0}{\sigma}\right)^2 \tag{11.2}$$

from which we learn that l is a great deal larger than l_0. For molecular hydrogen at atmospheric pressure and 0°C, n is 2.7×10^{19} particles/cm^3, and σ is about 2.7×10^{-8} cm. Then l is found to be 1.15×10^{-5} cm and Eq. (11.1) gives $l_0 = 3 \times 10^{-7}$ cm, or $\frac{1}{38}\,l$.

EXAMPLE 11.1 Consider an electron moving in a gas whose molecules have a diameter σ. It can be assumed to have negligible size and a far greater speed than the molecules do. What will its mean free path l_e be?

In this case the assumption used in the derivation of Eq. (11.1) is satisfied with good accuracy; the molecules are essentially stationary with respect to the electron. Thus we need only to use the correct collision cross section in Eq. (11.1). This would be $(\pi/4)(\sigma_{\text{molecules}} + 0)^2$ instead of $(\pi/4)(\sigma + \sigma)^2$. Thus

$$l_e = \frac{4}{\pi n(\sigma_{\text{molecules}})^2}$$

and, unlike Eq. (11.1), this result is quite accurate.

EXAMPLE 11.2 Obtain the distribution function for the individual free paths, whose length is designated by ξ.

Let $\mathcal{P}(\xi)$ be the probability that a molecule has a free path of at least ξ. If a particle has already traveled a distance ξ, then the likelihood of its going a little bit ($d\xi$) further without collision is $d\xi/l$. Thus

$$\mathcal{P}(\xi + d\xi) = \mathcal{P}(\xi)\left(1 - \frac{d\xi}{l}\right)$$

but we also can write

$$\mathcal{P}(\xi + d\xi) = \mathcal{P}(\xi) + \frac{d\mathcal{P}}{d\xi}\,d\xi$$

Comparing these expressions, we obtain

$$\frac{d\mathcal{P}}{d\xi} = -\frac{\mathcal{P}}{l}$$

Integrating this expression and noting that $\mathcal{P}(\xi = 0) = 1$, we get

$$\mathcal{P}(\xi) = \exp\left(\frac{-\xi}{l}\right)$$

Thus the probability that a molecule has a free path less than ξ is $1 - \exp(-\xi/l)$, and, in accordance with Eq. (2.14), the normalized distribution function for ξ is

$$F(\xi) = \frac{d[1 - \exp(-\xi/l)]}{d\xi} = \frac{\exp(-\xi/l)}{l}$$

This result is sketched in Fig. 11.1.

The fraction of free paths between 0 and ξ can be expressed as

$$\int_0^\xi F(\xi)\, d\xi = 1 - \exp\left(-\frac{\xi}{l}\right)$$

and this result is also plotted in Fig. 11.1.

11.3 RELATION BETWEEN MEAN FREE PATH AND TRANSPORT PROPERTIES

Once the mean free path and the number density are known, we can estimate the coefficients of momentum, heat, and mass transport[3] resulting from a known gradient of velocity, tempera-

Fig. 11.1 Distribution of free paths.

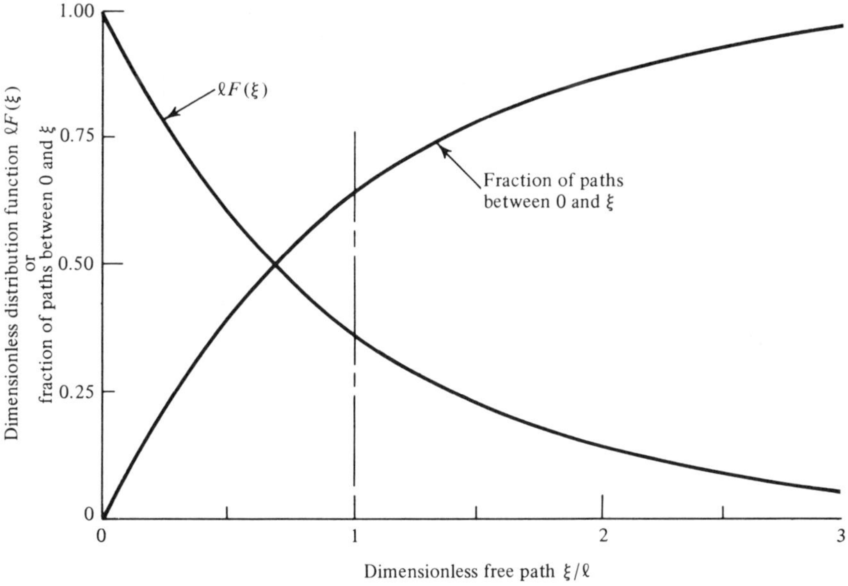

[3]That is, the transport properties: viscosity, thermal conductivity, and mass diffusivity.

ture, or concentration. Figure 11.2, for example, shows a particle moving past a reference plane, $y = y_r$, in a fluid with a velocity gradient, $\partial u/\partial y$, where u is the x component of the gross velocity c_0. The average particle that crosses the plane $y = y_r$ will travel a distance of a little more than one mean free path — say $2\alpha l$ — where α is a constant that is approximately equal to unity.

FACTORS INFLUENCING THE PENETRATION OF MOLECULAR TRANSPORT

Why, one might ask, should α exceed $\frac{1}{2}$? There are several different effects that bear upon this value. The depth of penetration of molecules is of great importance in the estimation of transport properties, so a qualitative explanation of these effects is in order.

Suppose, by way of explanation of the first contribution to α, that a line were to be drawn across the floor and a handful of straws cut to various lengths were strewn about the line. The average straw touching the line would be longer than the average straw in the original handful because the longer straws would have a greater chance of touching the line.

Although the free paths of molecules are distributed in three dimensions, the principle is the same. We can estimate how much longer than l the free paths are in the following way. The presence of certain molecules at $y = y_r$ is an accomplished fact. In the same way that a coin that has been tossed "heads" nine times in succession still has a 50 percent chance of coming out "heads" in the next toss, these particles have a full mean free path of l ahead of them. And by symmetry we expect that they have a full mean free path

Fig. 11.2 Motion of a particle in a velocity gradient.

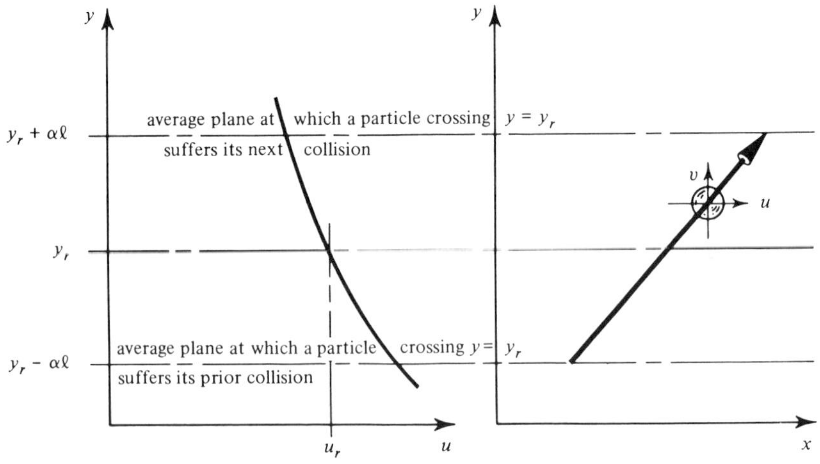

behind them as well. Consequently, the mean free path *of those particles that cross* $y = y_r$ is about *twice* l.

We can also show by a straightforward geometrical argument that the average vertical travel of randomly scattered paths is two thirds of the average path (Problem 11.1). Consequently, α is $\frac{2}{3}[2(\frac{1}{2})]$, or $\frac{2}{3}$.

As it happens, this value is still too low. The effective penetration of momentum continues a little beyond the next collision because scattering is not perfect in any collision. When this correction for the "persistence of velocities" is made, α comes out close to unity. The actual value depends slightly on the character of the molecules.[4]

VISCOSITY COEFFICIENT

The x-direction momentum of a particle at $y_r + \alpha l$ is $mu|_{y_r+\alpha l}$. This can be expressed in terms of the momentum at $y = y_r$ with the help of a Taylor-series expansion,

$$mu|_{y_r+\alpha l} = mu|_{y_r} + m\left.\frac{du}{dy}\right|_{y_r}(\alpha l) + \cdots \tag{11.3}$$

Higher-order terms should be negligible, because l is usually small with respect to macroscopic changes. The molecule flux from the top to the bottom is given by Eq. (2.52) as $n\bar{C}/4$. We multiply this by $mu|_{y_r+\alpha l}$ to get the momentum flux leaving the top, J^{+}_{mom},

$$J^{+}_{\text{mom}} = \frac{nm\bar{C}}{4}\left(u + \alpha l\left.\frac{du}{dy}\right|_{y_r}\right) \tag{11.4}$$

Similarly, for the momentum flux leaving the bottom we obtain

$$J^{-}_{\text{mom}} = \frac{nm\bar{C}}{4}\left(u - \alpha l\left.\frac{du}{dy}\right|_{y_r}\right) \tag{11.4a}$$

The net momentum flux, J_{mom}, crossing the surface, $y = y_r$, from bottom to top is, accordingly,

$$J_{\text{mom}} = -\tfrac{1}{2}\alpha\rho\bar{C}l\left.\frac{du}{dy}\right|_{y_r} \tag{11.5}$$

[4]More complete discussions of the persistence of velocities are given by Sir James Jeans, *The Dynamical Theory of Gases*, 4th ed., Dover Publications, Inc., New York, 1954, chaps. X and XI, and by E. H. Kennard, *Kinetic Theory of Gases*, McGraw-Hill, Inc., New York, 1938, chap. IV. The precise value of α is not of major importance in the approximate mean-free-path theory. The actual value of α is obtained directly from the more sophisticated calculations that we discuss in chapter 12.

But J_{mom} [recall Eq. (1.49)] is the same as τ_{yx} and is known from elementary fluid mechanics to be

$$\tau_{yx} = -\mu \left.\frac{du}{dy}\right|_{y_r} \tag{11.6}$$

for this unidimensional flow configuration. It follows that the viscosity coefficient μ is

$$\mu = \tfrac{1}{2}\,\alpha\rho\bar{C}l \tag{11.7}$$

When the appropriate corrections are all applied to α and l, we obtain at last

$$\mu \simeq 0.499\,\frac{m\bar{C}}{\sqrt{2}\pi\sigma^2} \tag{11.8}$$

The temperature dependence of this μ is proportional to the square root and it arises strictly in C.

Equation (11.8) is remarkable in that it shows that the viscosity of a dilute gas is *pressure-independent*. Maxwell first predicted this result and later offered experimental justification for it. Considerable experimental justification was required because the result jarred people's intuition. An isothermal increase of pressure results in a more dense gas, and one feels that a more dense gas should somehow be more viscous. However, the mean free path is shorter for the higher-pressure gas. Thus, although more particles cross the plane $y = y_r$ each one passes through a smaller change in average velocity and is therefore less effective in retarding the flow. The increase in number and decrease in l are effects that just balance each other out. The pressure independence of μ breaks down at high pressure, but so, too, do the dilute-gas assumptions upon which Eq. (11.1) is based.

THERMAL CONDUCTIVITY

We are now interested, not in the transport of momentum but in the average thermal energy of particles $\bar{\epsilon}$. The penetration of energy turns out to be generally a little greater than the penetration of momentum. The reason is that the energy a molecule carries in translational kinetic form "persists," not as the velocity but as the *square* of the velocity. The persistence effect accordingly contributes more heavily to energy transport than to momentum transport and α must be replaced with a somewhat larger constant β. Another effect contributing to the greater magnitude of β is that faster molecules have larger free paths, and they can carry considerably more energy over the longer paths than do slow molecules over shorter paths.

Equation (11.5) can then be paraphrased as

$$J_{\text{energy}} = -\tfrac{1}{2}\,\beta n\bar{C}l\,\frac{\partial\bar{\epsilon}}{\partial y}\bigg|_{y_r} \tag{11.9}$$

However, the specific heat at constant specific volume c_v is such that

$$mc_v = \frac{\partial\bar{\epsilon}}{\partial T}\bigg|_v = \frac{\partial\bar{\epsilon}}{\partial y}\bigg|_{y_r}\frac{dy}{dT}\bigg|_{y_r} \tag{11.10}$$

where constant v and $y = y_r$ are equivalent restrictions on the partial derivatives. Thus, we obtain

$$J_{\text{energy}} = -\tfrac{1}{2}\,\beta\rho\bar{C}lc_v\,\frac{dT}{dy}\bigg|_{y_r} \tag{11.11}$$

But J_{energy} is the unidimensional heat flux, q_y. From Fourier's law [Eq. (1.50)] we have

$$q_y = -\lambda\,\frac{dT}{dy}\bigg|_{y_r} \tag{11.12}$$

Combining Eqs. (11.11) and (11.12), we obtain for the thermal conductivity, λ,

$$\lambda = \tfrac{1}{2}\beta\rho\bar{C}lc_v \tag{11.13}$$

or

$$\lambda = \left(\frac{\beta}{2\sqrt{2}\pi}\right)\frac{mc_v\bar{C}}{\sigma^2} \tag{11.14}$$

where the temperature dependence of λ is carried entirely in $\bar{C}$.

Equations (11.13) and (11.14) bear a strong similarity to Eqs. (11.7) and (11.8) for the viscosity. Both μ and λ vary as $\bar{C}/\sigma^2$ and both are pressure-independent. Before we go on to consider the diffusion of mass, let us explore the similarities between heat and momentum transfer in more detail.

PRANDTL NUMBER AND EUCKEN'S FORMULA

The Prandtl number, Pr, is the well-known dimensionless group that characterizes the relative influences of the thermal and viscous diffusion processes in a fluid. It is defined as

$$\text{Pr} \equiv \frac{c_p\mu}{\lambda} = \frac{\nu}{a} \tag{11.15}$$

where ν is the kinematic viscosity μ/ρ and a is the thermal diffusivity $\lambda/\rho c_p$. We have just seen that thermal and viscous diffusion in a gas are very similar processes, and we can anticipate that Pr will not be very far from unity in a gas. The elimination of $\bar{C}$ from Eqs. (11.7) and (11.13) gives

$$\frac{\alpha}{\beta} = \frac{c_v \mu}{\lambda} \tag{11.16}$$

and the introduction of $\gamma = c_p/c_v$ in Eq. (11.15) yields

$$\text{Pr} = \left(\frac{\alpha}{\beta}\right)\gamma \tag{11.17}$$

For an ideal monatomic gas γ is $\frac{5}{3}$ (see Sec. 3.7) and the experimental value of Pr for argon in the range $0°\text{F} \leq T \leq 2300°\text{F}$ is 0.662 ± 0.008. (Other typical values of Pr, ν, and a are given in Table 11.1). It follows that

$$\left(\frac{\beta}{\alpha}\right)_{\text{monatomic}} \simeq \beta_{\text{monatomic}} \simeq 2.5 \tag{11.18}$$

TABLE 11.1 Some Transport Properties for Selected Gases at Atmospheric Pressure and for Saturated Liquids

Fluid	Temperature, °K	Kinematic viscosity, ν (m²/s) $\times 10^5$	Thermal diffusivity, a (m²/s) $\times 10^5$	Prandtl No., Pr $= \nu/a$
Gases				
Argon	273	11.75	17.47	0.67
Helium	255	9.55	13.68	0.70
Hydrogen	300	10.95	15.54	0.70
Nitrogen	300	1.56	2.20	0.71
Oxygen	300	1.59	2.24	0.71
Air	300	1.57	2.22	0.71
CO_2	300	0.83	1.06	0.77
Ammonia	273	1.18	1.31	0.90
Steam	373	2.17	2.25	0.96
Liquids				
Water	273	0.179	0.0131	13.7
Water	373	0.0294	0.0168	1.75
Ammonia	303	0.0349	0.0174	2.01
Glycerin	303	50.0	0.00929	5382.
Mercury	293	0.0114	0.43	0.027

A more precise theoretical description than we discuss here[5] also leads to this value.

[5] A more detailed discussion and references are given by Sir James Jeans, *op. cit.*

Diatomic gases will transport rotational as well as translational energy unless the temperature is very low, and other kinds of energy as well, if the gas is very hot. For molecules that carry energy in rotational and translational form we found (Sec. 3.7) that $\gamma = \frac{7}{5}$. The Prandtl numbers for such gases — air at room temperature, for example — is about 0.72. Thus

$$\left(\frac{\beta}{\alpha}\right)_{\text{diatomic}} \simeq \beta_{\text{diatomic}} \simeq 1.94 \tag{11.19}$$

In 1913 Eucken used these simple ideas to relate viscosity to thermal conductivity in a more complete way. He began by dividing c_v into two parts: c_{v_t} was the specific heat related to just the energy of translation, and c_{v_i} was the specific heat related to all other modes of storage. He divided λ into a component λ_t the conductivity of translational energy and a component λ_i the conductivity of the remaining molecular energies. He could then write Eq. (11.16) in the form

$$\lambda = \lambda_t + \lambda_i = \left[\left(\frac{\beta}{\alpha}\right)_t c_{vt} + \left(\frac{\beta}{\alpha}\right)_i c_{vi}\right]\mu \tag{11.16a}$$

where $(\beta/\alpha)_t$ is the value of β/α appropriate to pure translational motion and $(\beta/\alpha)_i$ is the value appropriate to the internal motion.

Equation (11.18) gives the value of β/α for purely translational motion as about $\frac{5}{2}$. We can assume that $(\beta/\alpha)_i$ is unity, on the other hand, because the internal modes of molecular energy are transported in exactly the same way as is momentum.

Furthermore, we obtained, in Sec. 3.7, the elementary relationship $c_{v_{\text{monatomic}}} = 3R^0/2$, where R^0 is equal to $(c_p - c_v)$ or $(\gamma - 1)c_v$. The following relations can then be written easily:

$$\left(\frac{\beta}{\alpha}\right)_t c_{v_t} = \tfrac{5}{2}[\tfrac{3}{2}(\gamma - 1)c_v]$$

and

$$\left(\frac{\beta}{\alpha}\right)_i c_{v_i} = 1(c_v - \tfrac{3}{2}R^0) = c_v[1 - \tfrac{3}{2}(\gamma - 1)]$$

Substitution of these results in Eq. (11.16a) gives Eucken's formula,

$$\lambda = \tfrac{1}{4}(9\gamma - 5)\mu c_v \tag{11.20}$$

For most gases at low pressure Eucken's relationship is accurate to within just a few percent, although, under some conditions, some gases exhibit serious deviations from it (see Problem 11.9).

EXAMPLE 11.3 How is the Prandtl number affected by the complexity of molecules in a gas?

An immediate consequence of Eucken's formula is that

$$\mathrm{Pr}|_{\text{Eucken}} = \frac{\mu c_v \gamma}{\lambda} = \frac{4\gamma}{9\gamma - 5} = \frac{\alpha}{\beta}\gamma$$

The complexity of molecules influences the ratio of specific heats, γ, which appears in this result. The substitution of

$$\gamma = \frac{D+2}{D}$$

where D is the number of degrees of freedom of the molecule, in this expression gives

$$\mathrm{Pr}|_{\text{Eucken}} = \frac{2D+4}{2D+9} = \frac{\alpha}{\beta}\frac{D+2}{D}$$

and

$$\frac{\beta}{\alpha} = \frac{2D+9}{2D}$$

For the simplest (monatomic) gas there are three degrees of freedom and

$$\mathrm{Pr}|_{\text{Eucken}} = \tfrac{2}{3} \qquad \text{and} \qquad \frac{\beta}{\alpha} = \frac{15}{6} = \frac{5}{2}$$

This compares favorably with Pr = 0.67 for Ar and He, as given in Table 11.1.

Of the gases listed in Table 11.1, ammonia and steam are most complicated. For them, $D > 6$, because there will be three modes of translation, three modes of rotation, and some contribution due to partial excitation of vibration. The Prandtl number is given as 0.90 and 0.96, respectively, as compared with the value 0.76 that is predicted for $D = 6$.

In the limit as $D \to \infty$ the equation gives

$$\lim_{D\to\infty} \mathrm{Pr}|_{\text{Eucken}} = 1 \qquad \text{and} \qquad \lim_{D\to\infty} \frac{\beta}{\alpha} = 1$$

This result reflects the fact that the internal modes of energy storage are solely responsible for energy transport in extremely complex molecules, and that the internal modes of energy are transported in the same way as momentum [cf. Eq. (11.16a)].

MASS DIFFUSIVITY

Suppose now that two stationary gases are distributed, at constant temperature and pressure, along a conduit. The number densities of the two components n_1 and n_2 might be distributed as shown in Fig. 11.3. Since the number density, n, of the mixture must equal the sum of the components,

$$n_1 + n_2 = n, \text{ a constant} \tag{11.21}$$

it follows that the concentration gradients must obey the relation

$$\frac{dn_1}{dx} = -\frac{dn_2}{dx} \tag{11.22}$$

at each point. The molecule fluxes must also balance each other,

$$J_{n1} = -J_{n2} \tag{11.23}$$

Once more we can write by analogy with Eq. (11.5) that

$$J_{n1} = -\tfrac{1}{2}\eta\bar{C}_1 l_1 \frac{dn_1}{dx} \tag{11.24}$$

or

$$J_{n2} = -\tfrac{1}{2}\eta\bar{C}_2 l_2 \frac{dn_2}{dx} \tag{11.24a}$$

where η is a constant like α or β appropriate to the transport of the molecules themselves.

It follows from Eqs. (11.22), (11.23), (11.24), and (11.24a) that

$$\tfrac{1}{2}\eta\bar{C}_1 l_1 = \tfrac{1}{2}\eta\bar{C}_2 l_2 \tag{11.25}$$

and it follows from a comparison of Fick's law, Eq. (1.51), with Eqs. (11.24) and (11.24a), that

$$D_{12} = \tfrac{1}{2}\eta\bar{C}_1 l_1 = D_{21} = \tfrac{1}{2}\eta\bar{C}_2 l_2 \tag{11.26}$$

The subscript 12 on D_{12} denotes the diffusion of component 1 into component 2; the subscript 21 denotes the converse. However,

Fig. 11.3 Number-density distribution in a two-component mixture.

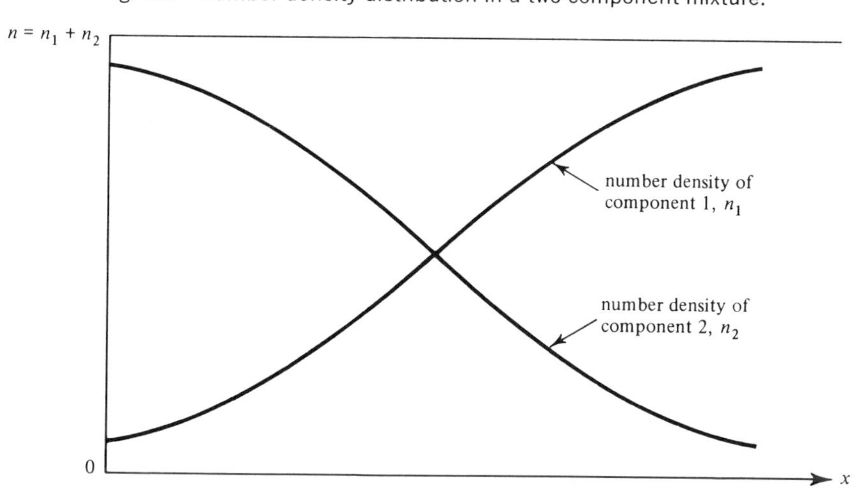

the elementary mean-free-path theory used to develop Eq. (11.26) becomes accurate only when two components are nearly identical. When this happens, D_{12} and D_{21} approach a common value, often called the coefficient of *self-diffusion*, D_{11}. Strictly speaking, D_{11} has little physical meaning because the diffusion of indistinguishable particles is a fictitious process. However, D_{11} is useful in describing the diffusion of very similar gases — an isotope of oxygen into normal oxygen, for example. Theoretical and experimental studies of self-diffusion show that η ranges between 1.20 and 1.55, depending upon the nature of the molecule.

Coefficients of *mutual* diffusion D_{12} or D_{21} are often presented for traces of different gases in a common gas. Table 11.2 lists

TABLE 11.2 Diffusion Properties for Low-Concentration Traces of Gases in Air[a] at Atmospheric Pressure and 20°C

Trace Gas	Molecular weight of trace, lb_m/lb_m-mole	Diffusion coefficient D, 10^5 ft²/sec	Schmidt No. Sc, ν/D	Lewis No. Le, D/a
Hydrogen (H_2)	2.02	71.9	0.22	3.24
Helium (He)	4.00	57.5	0.22	2.59
Methane (CH_4)	16.04	18.8	0.84	0.85
Ammonia (NH_3)	17.03	25.9	0.61	1.16
Steam (H_20)	18.02	26.4	0.60	1.18
Nitrogen (N_2)	28.02	16.1	0.98	0.73
Ethane (C_2H_6)	30.07	13.0	1.22	0.58
Oxygen (O_2)	32.00	21.4	0.74	0.96
Methanol (CH_3OH)	32.04	15.8	1.00	0.71
Argon (Ar)	39.95	21.5	0.73	0.97
Carbon Dioxide (CO_2)	44.01	16.5	0.96	0.74
Propane (C_3H_8)	44.09	10.5	1.51	0.47
Ethyl alcohol (C_2H_5OH)	46.07	12.2	1.30	0.55
Acetone (C_2H_6CO)	58.08	9.9	1.60	0.44
Butane (C_4H_{10})	58.12	9.0	1.77	0.40
n-Propyl alcohol (C_3H_7OH)	60.09	10.2	1.55	0.46
Sulfur dioxide (SO_2)	64.06	12.3	1.28	0.55
Chlorine (Cl_2)	70.90	11.1	1.42	0.50
Pentane (C_5H_{12})	72.15	8.0	1.97	0.36
n-Butyl alcohol (C_4H_9OH)	74.12	8.4	1.88	0.38
Benzene (C_6H_6)	78.11	9.25	1.71	0.42
n-Octane (C_8H_{18})	114.22	6.0	2.62	0.27
Naphthalene ($C_{10}H_8$)	128.16	6.15	2.57	0.28
Carbon tetrachloride (CCl_4)	153.84	7.4	2.13	0.34

[a]For dilute mixtures the properties of air can be considered unaltered by trace components.

diffusion coefficients D of this type for traces of various gases in air. Such a table is only made up for trace mixtures, because D generally depends upon the concentration of the diffusing species when it becomes large. As a first approximation, Eq. (11.26) may be employed to estimate the mutual diffusion coefficient of the trace gas, but the evaluation of the mean free path for the trace must be based on collisions between it and the surrounding foreign molecules.

Just as the Prandtl number compares ν with a, we have two additional dimensionless properties of a gas that compare D with both ν and a. They are

$$\textit{Schmidt}\text{ number, } \mathrm{Sc} \equiv \frac{\nu}{D} \tag{11.27}$$

$$\textit{Lewis}\text{ number, } \quad \mathrm{Le} \equiv \frac{D}{a} = \frac{\mathrm{Pr}}{\mathrm{Sc}} \tag{11.28}$$

and their values for traces of common gases in air are also included in Table 11.2.

SUMMARY OF SOME WORKING EQUATIONS

This section sets down a brief compilation of some working relations for the transport properties of a dilute gas. The basic determinant of μ, λ, and D is the mean free path l. With the factor $1.051/\sqrt{2}$ included, this is

$$l = \frac{1.051}{\sqrt{2}\pi n\sigma^2} \tag{11.29}$$

but $n = p/kT$, so the temperature and pressure dependence of l are reflected as follows:

$$l = \frac{1.051kT}{\sqrt{2}\pi\sigma^2 p} \tag{11.30}$$

and in terms of the mean speed, $\bar{C} = \sqrt{8kT/\pi m}$, l is

$$l = \frac{1.051m}{8\sqrt{2}\sigma^2 p}\,\bar{C}^2 \tag{11.31}$$

The viscosity equation resulting from this mean free path is

$$\mu = 0.499\frac{m\bar{C}}{\sqrt{2}\pi\sigma^2} = \frac{0.998}{\pi\sigma^2}\sqrt{\frac{mkT}{\pi}} \tag{11.32}$$

Equation (11.32) represents a gas composed of smooth elastic spheres. The advanced methods of kinetic theory[6] show that for

[6]See, for example, S. Chapman and T. G. Cowling, *The Mathematical Theory of Non-Uniform Gases*, Cambridge University Press, New York, 1960, chaps. 9 through 14.

more complex molecular descriptions the resulting viscosity expression is the quotient of the preceding expression and a correction factor. One of the most widely used results of this kind is "Sutherland's formula," which is the viscosity based on the Sutherland potential [Fig. 9.2(c)]:

$$\mu = 0.998 \frac{\sqrt{\pi m k T}}{\pi^2 \sigma^2} \frac{1}{1 + S/T} \tag{11.33}$$

The constant S characterizes the relative strength of attraction of the molecules of a given gas.

TABLE 11.3 Values of S for Sutherland's Formula[a]

Gas	S	Temperature range, °C
Hydrogen	83	−60.2 to 185.3
	71.7	−20.6 to 302
Helium	78.2	−60.9 to 183.7
	80.3	15.3 to 184.6
Methane (CH_4)	198	17 to 100
Ammonia	377	15 to 183.8
Neon	61	20 to 100
Carbon monoxide	118	15 to 100
Ethylene (C_2H_4)	225.9	−21.2 to 302
Nitrogen	118	15 to 100
	102.7	−76.3 to 250.1
Air	114	0 to 300+
Nitric oxide	128	20 to 200
Oxygen	138	16.75 to 185.8
Sulfuretted hydrogen	331	17 to 100
Hydrochloric acid	357	12.5 to 100.3
Argon	169.9	14.7 to 183.7
	147	20 to 100
Carbon dioxide	239.7	−20.7 to 302
	274	15 to 100
Nitrous oxide	274	15 to 100
	260	28.1 to 278
Methyl chloride	454	−15.3 to 302
Sulfur dioxide	416	18 to 100
Chlorine	325	12.7 to 99.1
Krypton	188	16.3 to 100
Xenon	252	15.3 to 100.1

[a]compiled by S. Chapman and T. G. Cowling, *The Mathematical Theory of Non-Uniform Gases*, Cambridge University Press, New York, 1960, p. 225. Reprinted by permission.

A very convenient form of Sutherland's formula is obtained by dividing this μ by a reference value μ_{ref}. The result,

$$\mu = \mu_{\text{ref}}\left(\frac{T}{T_{\text{ref}}}\right)^{3/2} \frac{T_{\text{ref}} + S}{T + S} \tag{11.34}$$

is useful and fairly accurate for interpolating data as long as two values of μ are known. Table 11.3 presents values of S for many gases.

Another general formula, suggested by Bromley and Wilke,[7] represents a useful semiempirical modification of the viscosity predicted on the basis of the Lennard-Jones potential. It is

$$\mu = 3.33 \times 10^{-5} \frac{(MT_c)^{1/2}}{v_c^{2/3}} f(T^*) \tag{11.35}$$

where $T^* = T(k/\epsilon)$, the Lennard-Jones reduced temperature, M is the molecular weight, T_c and v_c are the critical temperature and volume, and the units are T_c in °K, v_c in cm³/g-mole, and μ in poises. The function $f(T^*)$ is given graphically in Fig. 11.4.

Fig. 11.4 Empirical viscosity function, $f(T^*)$, for the Lennard-Jones potential

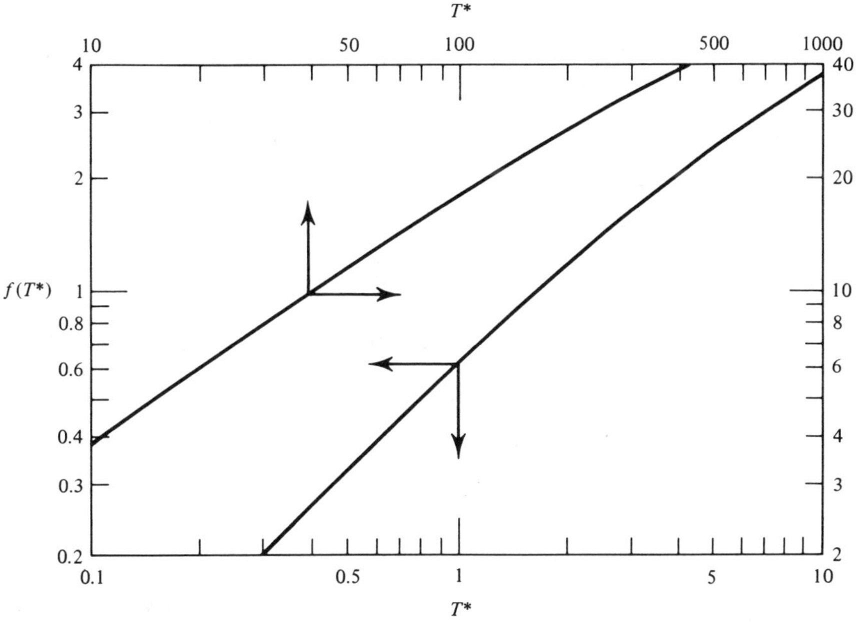

[7]L. A. Bromley and C. R. Wilke, *Ind. Eng. Chem.* **43,** 1641 (1951).

The viscosity of mixtures is also predictable, but the results are exceedingly cumbersome. The following approximate relation[8] for a binary mixture gives reasonable estimates of $\mu_{1,2}$:

$$\mu_{1,2} = \frac{\mu_1}{1 + (x_2/x_1)\phi_{1,2}} + \frac{\mu_2}{1 + (x_1/x_2)\phi_{2,1}} \tag{11.36}$$

where x_1 and x_2 are the mole fractions of components 1 and 2, respectively, and

$$\phi_{1,2} = \frac{[1 + (\mu_1/\mu_2)^{1/2}(M_2/M_1)^{1/4}]^2}{\sqrt{8}(1 + M_1/M_2)^{1/2}}$$
$$\phi_{2,1} = \frac{[1 + (\mu_2/\mu_1)^{1/2}(M_1/M_2)^{1/4}]^2}{\sqrt{8}(1 + M_2/M_1)^{1/2}} \tag{11.37}$$

The thermal conductivity based on mean-free-path theory generally takes the form

$$\lambda = \frac{\beta}{\alpha} c_v \mu \tag{11.16b}$$

so for rigid spherical molecules $\beta/\alpha \simeq \frac{5}{2}$ and

$$\lambda = \tfrac{5}{2} c_v\left(0.499 \frac{m\bar{C}}{\sqrt{2}\pi\sigma^2}\right) = \tfrac{5}{2} c_v\left(0.998 \frac{\sqrt{\pi m k T}}{(\pi\sigma)^2}\right) \tag{11.38}$$

The Sutherland model can also be employed to correct Eq. (11.38). The result is simply

$$\lambda = \tfrac{5}{2} c_v\left(0.998 \frac{\sqrt{\pi m k T}}{(\pi\sigma)^2}\right) \Big/ \left(1 + \frac{S}{T}\right) \tag{11.39}$$

or

$$\lambda = \lambda_{\text{ref}}\left(\frac{T}{T_{\text{ref}}}\right)^{3/2} \frac{T_{\text{ref}} + S}{T + S} \tag{11.40}$$

where S has the same values as given in Table 11.3.

Generally, the problem of predicting λ by statistical means reduces to the problem of approximating β/α. Eucken's formula (recall Example 11.3),

$$\frac{\beta}{\alpha} = \frac{9\gamma - 5}{4} = \frac{2D + 9}{2D} \tag{11.20a}$$

gives the basis for obtaining λ from μ. If experimental values of γ are available, this result will usually be correct within a few percent.

The prediction of λ for mixtures is again a complicated affair.

[8] Due to C. R. Wilke, *J. Chem. Phys.* **18,** 517 (1950).

Brokaw[9] suggests that λ_{mix} might be approximated as the average of a simple mean and a geometric mean. Thus

$$\lambda_{mix} = \frac{1}{2}\left[(x_1\lambda_1 + x_2\lambda_2 + \cdots) + \left(\frac{x_1}{\lambda_1} + \frac{x_2}{\lambda_2} + \cdots\right)^{-1}\right] \quad (11.41)$$

We have given a result for the coefficient of diffusion which is really only accurate for self-diffusion:

$$D_{11} = \tfrac{1}{2}\eta\bar{C}_1 l_1 = \frac{1.051\eta}{0.998\rho}\left(\frac{0.998}{\pi\sigma^2}\sqrt{\frac{mkT}{\pi}}\right)$$

or

$$D_{11} = \frac{1.053\eta}{\rho}\,\mu_{\text{rigid sphere}} \quad (11.42)$$

The constant, η, will range from $\frac{6}{5}$ for rigid spheres up to about 1.55. For dissimilar gases the equivalent expression can be shown to be

$$D_{12} = \frac{1.053\eta}{n\sqrt{m_1 m_2}}\left[\frac{0.998}{\pi\sigma_{12}^2}\sqrt{\frac{kT(m_1+m_2)}{\pi}}\right] \quad (11.43)$$

The Sutherland model gives this same result divided by $1 + S_{12}/T$, where S_{12} is a constant characterizing the interaction of species 1 with species 2. From this we can write

$$D_{12} = D_{12\text{ref}}\frac{n_{\text{ref}}}{n}\left(\frac{T}{T_{\text{ref}}}\right)^{3/2}\frac{T_{\text{ref}} + S_{12}}{T + S_{12}}$$

or

$$D_{12} = D_{12\text{ref}}\frac{p}{p_{\text{ref}}}\left(\frac{T}{T_{\text{ref}}}\right)^{5/2}\frac{T_{\text{ref}} + S_{12}}{T + S_{12}} \quad (11.44)$$

In this section we have tried to show, in a summary way, the kind of results that can be obtained for transport properties using mean-free-path methods, and we have sketched without proof some of the extensions of these results that can be made. Actually, the law of corresponding states once again provides the method by which we can circumvent the limitation of these predictions to dense gases. We consider these methods next.

11.4 LAW OF CORRESPONDING STATES FOR TRANSPORT PROPERTIES OF DENSE GASES

The linear phenomenological transport equations described in Sec. 1.6 have now been developed with the help of a neat and ele-

[9]R. S. Brokaw, *Ind. Eng. Chem.* **47,** 2398 (1955).

mentary theory. The mean-free-path method not only shows how these equations arise, but it provides considerable insight into transport phenomena as well. It also gives expressions for the transport properties, but these have some weaknesses. The first is their limitation to dilute gases. A second is the requirement that we know the effective radius of the molecule — a parameter that is often obtained by the circular process of comparing Eq. (11.8) with viscosity data.

Another shortcoming of the theory as a whole is that it is limited to small gradients of T, n, or $\mathbf{c}_0$. Thus, for the theory to be valid,

$$\frac{\partial u}{\partial y}\left(\frac{l}{\bar{u}}\right), \frac{\partial T}{\partial y}\left(\frac{l}{\bar{T}}\right), \text{ or } \frac{\partial n}{\partial y}\left(\frac{l}{\bar{n}}\right) \text{ must be} \ll 1 \tag{11.45}$$

In chapter 12 we take up the more general strategy developed by Boltzmann for treating nonequilibrium problems. Here we shall find that the linear transport equations in chapter 1 appear as first approximations when such conditions as (11.45) are no longer valid.

But the Boltzmann theory is also limited by the diluteness assumption and the need for a knowledge of molecular detail; therefore, we wish to return briefly to the law of corresponding states. Once more, for dense gases, the law of corresponding states provides the basis for an empirical method that is effective and simple. As we saw in chapter 9, the existence of a potential of the form

$$\frac{\phi(r)}{\epsilon} = f\left(\frac{r}{\sigma}\right) \tag{9.67}$$

(which includes the Lennard-Jones potential) results in a dimensionless equation of state. It also results in dimensionless transport properties of the form[10]

$$D^* = \frac{D\sigma^3}{\sqrt{m\epsilon}}\frac{m}{\sigma^3} = D^*(v^*, T^*) \tag{11.46}$$

$$\mu^* = \frac{\mu\sigma^2}{\sqrt{m\epsilon}} = \mu^*(v^*, T^*) \tag{11.47}$$

and

$$\lambda^* = \frac{\lambda\sigma^2}{\sqrt{m\epsilon}}\frac{m}{k} = \lambda^*(v^*, T^*) \tag{11.48}$$

[10]For a brief discussion, and bibliography, see R. B. Bird, J. O. Hirschfelder, and C. F. Curtiss, "The Equation of State and Transport Properties of Gases and Liquids," *Handbook of Physics*, 2nd ed. (E. V. Condon and H. Odishaw, eds.), McGraw-Hill, Inc., New York, 1967, chap. 4.

where the terminology is the same as was used in chapter 9. After rationalization of critical data in terms of the molecular properties σ, m, k, and ϵ, it is possible to write

$$\frac{D}{D_c} = D_r(p_r, T_r) \tag{11.49}$$

$$\frac{\mu}{\mu_c} = \mu_r(p_r, T_r) \tag{11.50}$$

and

$$\frac{\lambda}{\lambda_c} = \lambda_r(p_r, T_r) \tag{11.51}$$

Correlations of this kind have been carried out with success. Figure 11.5, for example, shows Hougen and Watson's[11] reduced viscosity curve. Bird *et al.*[12] cite a number of recent works that have carried correlations of this kind on to greater refinement.

11.5 TRANSPORT PROPERTIES OF SOLIDS

The simple mean-free-path approach to the calculation of transport properties of gases can be applied to solids as well. Two important transport properties of solids are the thermal conductivity and the electrical conductivity, which we shall consider here.

As we noted in Sec. 10.2, the thermal energy of a solid may exist in various forms. In the transport of thermal energy, however, the predominant energy modes are the lattice vibration and the translation of free electrons. Because lattice vibrations can be treated as phonons, thermal transport in solids can be treated as energy transport in phonon and electron gases, and the mean-free-path theory of molecular gases is indeed directly applicable.

In most solids, one of the two energy modes — phonons or the translation of free electrons — usually dominates the other in energy transport. Just as we saw in Sec. 10.2 that there was little or no free-electron contribution to the thermostatic properties of solids, this contribution also proves to be negligible in the transport properties of dielectric solids. The situation, however, is quite different in metals. Although the free-electron contribution to the thermostatic properties of metals is negligible as compared to the phonon contribution except at very low temperatures, we shall see later that it is dominant in the transport properties.

[11]J. O. Hirschfelder, C. F. Curtiss, and R. B. Bird, *Molecular Theory of Gases and Liquids*, John Wiley & Sons, Inc., New York, 1954, chap. 9.
[12]Bird *et al., op. cit.*

Fig. 11.5 Generalized viscosity chart.

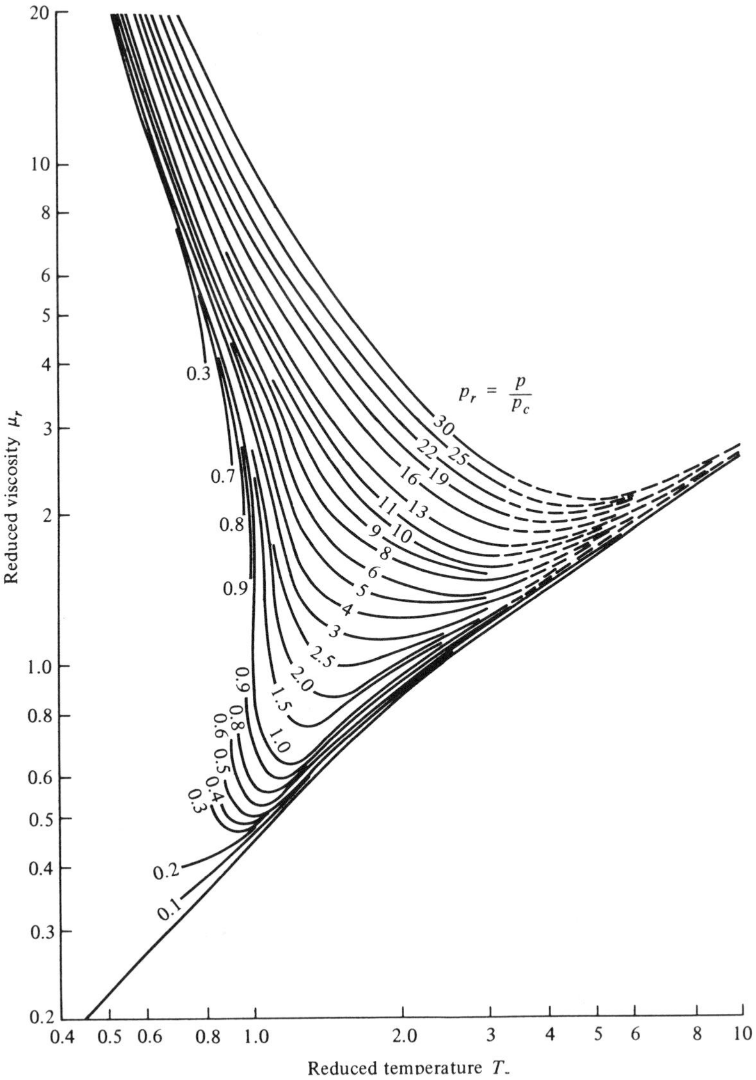

THERMAL CONDUCTIVITY OF A DIELECTRIC

Following an argument similar to that which gave Eq. (11.13), we may express the thermal conductivity of a dielectric in the approximate form as

$$\lambda = \tfrac{1}{3}(\rho c_v)\bar{C}l \tag{11.52}$$

where ρc_v is the phonon heat capacity per unit volume, $\bar{C}$ is the phonon velocity, and l is the phonon mean free path. The constant β

in Eq. (11.3) has been taken as two thirds on the basis of simple kinetic arguments (Problem 11.1), because little is known about the exact value of β and, moreover, the formula is in any case an approximate qualitative one. As given in Eqs. (10.12) to (10.14), $\bar{C}$ can be easily calculated from tabulated properties of solids, and it is relatively insensitive to temperature. For most crystalline solids, $\bar{C}$ is approximately 5×10^5 cm/sec. The phonon heat capacity ρc_v, however, is a function of temperature, as indicated in Eq. (10.20). At low temperatures in particular, $c_v \propto T^3$.

The mean free path l of a phonon is determined primarily by two processes, the collisions of a phonon with other phonons and with the crystal or grain boundary and lattice imperfections. These two collisional or scattering[13] effects can be combined accordingly to give a geometrical mean l:

$$\frac{1}{l} = \frac{1}{l_p} + \frac{1}{l_g} \tag{11.53}$$

where l_p and l_g are the mean free paths between phonon–phonon scattering and between phonon–geometric scattering, respectively. It is clear physically that l_p and l_g cannot be simply added to get l, but the inverse of the mean free paths can be superposed linearly, at least in an approximate sense, because the inverse directly characterizes the contribution of the particular scattering.

The geometrical scattering effect is almost constant over a wide range of temperatures but begins to decrease with decreasing temperature in the range below about 100°K. When at very low temperatures, say below 30°K, l_g becomes comparable with the width of the test specimen. The value of l is then limited by the width, and the thermal conductivity becomes a function of the specimen dimension. The phonon–phonon (or thermal) *scattering factor* $1/l_p$ is found experimentally to be almost directly proportional to the absolute temperature or $l_p \propto 1/T$. Because of the increase of thermal excitation, we naturally would expect this to increase with increasing temperature. Some typical values of l_p are shown in Table 11.4.

THERMAL AND ELECTRICAL CONDUCTIVITY OF METALS

For the thermal conductivity of metals, the first thing is to decide whether the electrons or the phonons carry the larger share of the thermal-energy transport. It is found that in pure metals the

[13]The term "scattering" is commonly used to designate a molecular collision, because no contact actually occurs between the two molecules — they are simply scattered by the interaction of force fields. The terms "scattering" and "collisional" are synonymous for our purposes.

TABLE 11.4 Representative Values of the Mean Free Path for Phonon–Phonon Scattering [calculated from Eq. (11.52) with $\bar{C} = 5 \times 10^5$ cm/sec]

Crystal	T, °K	ρc_v, cal/cm³-°K	λ, cal/cm-°K-sec	l_p, Å
Quartz[a]	273	0.48	0.03	40
	83	0.13	0.12	540
NaCl	273	0.45	0.17	23
	83	0.24	0.064	100

[a]Along the optical axis.

electron contribution is one or two orders of magnitude higher than the phonon contribution, whereas in very impure metals or in disordered metals, the phonon contribution may be comparable with the electron contribution. This point is made evident by comparing values of the thermal conductivity for pure metals, dielectrics, and alloys. In all cases the electron contribution is predominately that of electron–phonon scattering and not electron–electron scattering.

Equation (11.52) is directly applicable here, but the various terms now refer to the free electrons. In the kinetic description of free electrons, it is convenient to introduce a parameter called the *relaxation time*, τ, which is the period between collisions:

$$\tau \equiv \frac{l}{\bar{C}} \tag{11.54}$$

The electron velocity, $\bar{C}$, can be estimated from the Fermi level, μ_0 [as given in Eq. (6.44) and Table 6.1], which represents the kinetic energy of free electrons at 0°K. The use of μ_0 is justified because in the normal temperature range, as can be shown from Eq. (6.45), μ differs only slightly from μ_0. Therefore,

$$\mu_0 \simeq \tfrac{1}{2} m_e \bar{C}^2 \tag{11.55}$$

or, from Eq. (6.44),

$$\frac{h^2}{2m_e}\left(\frac{3}{8\pi}\frac{N_0}{V}\right)^{2/3} = \tfrac{1}{2} m_e \bar{C}^2 \tag{11.56}$$

where N_0/V is the number density of free electrons at 0°K. Since in general the number density n is relatively insensitive to temperature, $n \simeq N_0/V$. The values of n at 0°C, given for some metals in Table 6.1, can thus be used to estimate $\bar{C}$.

Equation (11.52) can be then expressed as

$$\lambda = \tfrac{1}{3}\rho c_v \bar{C} l = \frac{1}{3}\frac{\rho c_v l^2}{\tau} \tag{11.57}$$

Using Eq. (11.55) and the approximate expression for c_v given by Eq. (6.48), we have

$$\lambda = \frac{\pi^2}{6}\frac{nk^2Tl^2}{\mu_0\tau} = \frac{\pi^2}{3}\frac{nk^2T\tau}{m_e} \tag{11.58}$$

This result is used later in deriving a relationship between the thermal and electrical conductivity of metals.

For electrical conductivity we must first consider what the physical effect of an imposed voltage **E** is. A voltage difference accelerates a negatively charged particle toward the higher voltage. Without any collisions the electrons would accelerate without limit, but in a metal lattice they suffer collisions and have to be reaccelerated. We can calculate the average drift velocity $\mathbf{u}_d$ that is developed between collisions in a voltage gradient by using a heuristic argument. Newton's law gives for the force, on an electron, $\mathbf{F} = e\mathbf{E}$,

$$e\mathbf{E} = m_c \frac{d\mathbf{u}}{dt} \simeq m_e \frac{\mathbf{u}_d}{\tau} \tag{11.59}$$

where e is the charge on an electron. Thus

$$\mathbf{u}_d = \frac{e\tau}{m_e}\mathbf{E} \tag{11.60}$$

The flux of charge **i** subject to the gradient is simply

$$\mathbf{i} = ne\mathbf{u}_d \tag{11.61}$$

Combining Eqs. (11.60) and (11.61), we obtain

$$\mathbf{i} = n\frac{e^2\tau}{m_e}\mathbf{E} \tag{11.62}$$

This has to be equivalent to the phenomenological relation given by Ohm's law,

$$\mathbf{i} = \sigma\mathbf{E} \tag{11.63}$$

The important formula relating the electrical conductivity to the microscopic electronic parameters follows from Eqs. (11.62) and (11.63)

$$\sigma = n\frac{e^2\tau}{m_e} \tag{11.64}$$

As σ can be measured easily, Eq. (11.64) can be used to determine τ. From the value of τ, we can then calculate l using Eqs. (11.54) and (11.56). With the basic microscopic electronic parameters known, all thermodynamic and transport properties of metals can be calculated. The thermal conductivity, for example, is given by Eq. (11.57). The values of these electronic parameters for some common metals are tabulated in Table 11.5.

TABLE 11.5 Values of Electronic Parameters for Metals at 0°C[a]

Metal	$n \times 10^{-22}$,[b] cm^{-3}	$\sigma \times 10^{-17}$,[c] esu	$\bar{C} \times 10^{-8}$,[d] cm/sec	$\tau \times 10^{14}$,[e] sec	$l \times 10^{8}$,[f] cm
Li	4.6	1.06	1.31	0.84	110
Na	2.5	2.09	1.07	3.27	350
K	1.3	1.47	0.85	4.35	370
Cu	8.5	5.76	1.58	2.27	420
Ag	5.8	6.12	1.40	4.07	570
Au	5.9	4.37	1.40	2.93	410

[a]Reference should be also made to Table 6.1.
[b]n refers to the values at 0°C.
[c]σ refers to the measured values at 0°C.
[d]$\bar{C}$ refers to the calculated values from Eq. (11.56).
[e]τ refers to the calculated values from Eq. (11.64).
[f]l refers to the calculated values from Eq. (11.54).

A simple useful relationship between the electrical and thermal conductivity of metals can also be established. Division of Eq. (11.58) by Eq. (11.64) gives

$$\frac{\lambda}{\sigma} = \frac{\pi^2}{3}\left(\frac{k}{e}\right)^2 T \tag{11.65}$$

which is called the law of Wiedemann and Franz.

In conformity with Eq. (11.65) we can define the Lorenz number L

$$L \equiv \frac{\lambda}{\sigma T} = \frac{\pi^2}{3}\left(\frac{k}{e}\right)^2 = 2.45 \times 10^{-8} \frac{\text{Watt-ohm}}{°\text{C}^2} \tag{11.66}$$

Table 11.6 provides a comparison of this result with observed values of L for several elementary metals at moderate temperatures. The comparison is really very good — within 6 percent in most cases. At low temperatures, however, it can be shown that the Wiedemann–Franz law is no longer applicable when electron–phonon scattering is dominant. At very low temperatures, where the electron scattering due to defects and impurities is the primary scattering mechanism, Eq. (11.66) again seems to hold.

TABLE 11.6 Experimental Values of the Lorenz Number, $L = \lambda/\sigma T$

Metal	$L \times 10^{-8}$ Watt-ohm/°C²		
	Observed		Predicted
	0°C	100°C	$(\pi k)^2/3e^2$
Ag	2.31	2.37	2.45
Au	2.35	2.40	2.45
Cd	2.42	2.43	2.45
Cu	2.23	2.33	2.45
Ir	2.49	2.49	2.45
Mo	2.61	2.79	2.45
Pb	2.47	2.56	2.45
Pt	2.51	2.60	2.45
Sn	2.52	2.49	2.45
W	3.04	3.20	2.45
Zn	2.31	2.33	2.45

A rigorous treatment of electron-transport processes due only to electron–phonon scattering (as would be the case for pure metals) gives[14]

$$\frac{1}{\sigma} = 4A\left(\frac{T}{\Theta}\right)^5 J_5 \tag{11.67}$$

and

$$\frac{1}{\lambda} = \frac{4A}{LT}\left(\frac{T}{\Theta}\right)^5\left\{\left[1 + \frac{3}{\pi^2}\left(\frac{N_e}{2}\right)^{2/3}\left(\frac{\Theta}{T}\right)^2\right]J_5 - \frac{1}{2\pi^2}J_7\right\} \tag{11.68}$$

where

$$J_n\left(\frac{\Theta}{T}\right) \equiv \int_0^{\Theta/T} \frac{x^n\,dx}{(e^x - 1)(1 - e^{-x})} \tag{11.69}$$

In these expressions A is a constant characteristic of the electron–phonon interaction in the particular metal, Θ a characteristic temperature of the metal very close to the Debye temperature (Table 10.2), and N_e the effective number of conduction electrons per atom. Equation (11.69) can be evaluated analytically at the high-

[14]A. H. Wilson, *The Theory of Metals*, Cambridge University Press, New York, 1953.

and low-temperature limits, $\Theta/T \ll 1$ and $\Theta/T \gg 1$. Thus it follows (Problem 11.14) from Eqs. (11.67) and (11.68) that, for $T \ll \Theta$,

$$\sigma \propto T^{-5} \qquad \lambda \propto T^{-2} \tag{11.70}$$

and, for $T \gg \Theta$,

$$\sigma \propto T^{-1} \qquad \lambda \simeq \text{constant} \tag{11.71}$$

The above results indicate that the Wiedemann–Franz law breaks down at low temperatures but is indeed correct at the high-temperature limit.

EXAMPLE 11.4 A simple approximate expression for the total emissivity (the ratio of the total emissive power of a surface to that of a "black" surface at the same temperature) of a metal at moderate temperatures is

$$\epsilon = 4\left(\frac{kT}{\sigma h}\right)^{1/2}$$

How would the total emissivity vary with temperature at high temperatures? Determine the total emissivity of gold at 300°K on the basis of the given value of the thermal conductivity for gold at that temperature, 2.93Watt-cm/cm^2-°C.

At high temperatures $\Theta/T \ll 1$, we can approximate in the integrand of Eq. (11.54) $e^x \simeq 1 + x$, and obtain

$$J_5\left(\frac{\Theta}{T}\right) = \int_0^{\Theta/T} x^3\,dx = \frac{1}{4}\left(\frac{\Theta}{T}\right)^4$$

Equation (11.67) thus becomes

$$\frac{1}{\sigma} = A\left(\frac{T}{\Theta}\right)$$

[cf. Eq. (11.71)]. Consequently, the total emissivity can be expressed as

$$\epsilon = 4\left(\frac{Ak}{\Theta h}\right)^{1/2} T$$

which indicates that ϵ increases linearly with the increase of T. This temperature dependence of ϵ agrees with experimental observation.

For gold, $\Theta \simeq \Theta_D - 165$°K from Table 10.2 and $\Theta/T \simeq 0.55$, so we may use the Wiedemann–Franz law with good accuracy. The expression for ϵ given above can be rewritten in terms of λ as

$$\epsilon = 4\left(\frac{kL}{\lambda h}\right)^{1/2} T$$

or

$$\epsilon = 4\left[\frac{(1.38 \times 10^{-16})(2.45 \times 10^{-8})}{(2.93)(6.63 \times 10^{-27})}\right]^{1/2} 300 = 15{,}830\ \sqrt{\text{Ohm-cm/sec}}$$

But the conversion factor for ohms into electrostatic units is 1 Ohm = $(1/8.9876 \times 10^{11})$ sec/cm, so

$$\epsilon = 15830 \times 10^{-5}/\sqrt{89.876} = 0.0167$$

This predicted value is very nearly equal to the experimental value for pure gold.

Problems 11.1 Find what the average vertical component of travel of molecules across a plane $y = y_r$ is in terms of the mean free path.

11.2 A bundle of straws has lengths varying uniformly between 0 and L_0. If the bundle is scattered randomly over a line, what is the ratio of the average length of straws crossing the line to the average length of straws in the bundle?

11.3 Find the root-mean-square free path in terms of the mean free path. What is the most probable free path? What is the probability of finding a free path that is 0.001 mm or more in length, in air at standard conditions.

11.4 Obtain for air at room temperature and pressure: l, l_0, and the frequency of collisions. What are these values 200 miles above the surface of the earth? How small a volume could still be called "macroscopic" at sea level and at 200 miles elevation?

$$\sigma_{\text{air}} \simeq 3.7 \times 10^{-8} \text{ cm}$$

11.5 Combine Fourier's law with the first law of thermodynamics in such a way as to eliminate the heat flux q and obtain a second-order differential equation in T.

11.6 Does pressure have the same influence upon λ, μ, and D? Explain.

11.7 Derive an expression for μ for a two-dimensional Maxwellian gas.

11.8 Combine Fick's law with the principle of conservation of the ith component in such a way as to eliminate J_{n_i} and obtain a second-order differential equation in n_i.

11.9 Obtain λ, μ, c_v, and c_p data for several gases. Present in tabular form the following results: $\lambda/\mu c_p$, $\frac{9}{4}[(c_p/c_v) - \frac{5}{9}]$, the percent of deviation of the data from Eucken's formula, the pressure and temperature at which data were obtained, $4\gamma/(9\gamma - 5)$, and the observed Prandtl number. Can you explain any of the more serious deviations of theory from experiment?

11.10 Derive Eucken's formula and $\text{Pr}|_{\text{Eucken}}$ for a two-dimensional polyatomic gas.

11.11 Calculate μ, λ, and Pr for water vapor at 227°C and 1 atm. For H_2O, $\sigma \simeq 4.6 \times 10^{-8}$ cm and α is somewhere between 0.998 and 1.14. Elementary statistical mechanics (Sec. 3.7) indicates that $\gamma = \frac{4}{3}$ and $c_v = 3R^0$, in this case.

11.12 Calculate μ for a gaseous mixture of 40 percent H_2O and 60 percent CO_2 at 1000°R and 1 atm. In this case $\alpha = 1.14$, $\sigma_{H_2O} = 4.6 \times 10^{-8}$ cm, and $\sigma_{CO_2} = 4.59 \times 10^{-8}$ cm.

11.13 Plot p_r lines for steam on μ_r versus T_r coordinates and compare them with Fig. 11.5. Note that to do this you must first try to locate an approximate μ_c by graphical interpolation of existing data. Comment on the prediction for H_2O.

11.14 Verify Eqs. (11.70) and (11.71).

12 a more detailed kinetic theory of dilute gases

Chapter 11 provided an introduction to the methods of kinetic theory in which the major analytical difficulties of the subject were avoided. The use of approximate mean-free-path methods, for instance, made it possible to obtain many of the important features of molecular transport. However, kinetic theory can also be used to obtain more precise results. It can, for example, be used to expose aspects of transport behavior that only become really important when gradients of properties become extremely strong.

This chapter is therefore devoted to an elementary discussion of the methods of advanced kinetic theory. The starting point is Boltzmann's century-old description of molecular behavior, which has remained the foundation of the subject. The methods are more heavily analytical than anything we have considered previously, but they proceed in a very direct way from concrete considerations of particle dynamics. They also result in a surprisingly complete picture of nonequilibrium behavior.

12.1 BOLTZMANN INTEGRODIFFERENTIAL EQUATION

Boltzmann[1] developed the basic tool for treating nonequilibrium ideal gases in about 1872. This was an integrodifferential equation in the distribution function. It was formulated on consideration of collisions, subject to appropriate methods of averaging, and it provides a remarkably complete description of gaseous behavior.

ASSUMPTIONS

An important result of the diluteness assumption is that we need only consider binary collisions. Collisions of three or more particles become impossibly difficult to treat, but they very seldom occur in a dilute gas and can properly be ignored.

We again need an assumption akin to the principle of equal a priori probabilities that was so basic in the study of statistical mechanics. A related form of this idea is the "ergodic hypothesis," which says that the time-average behavior of individual particles is the same as their ensemble average behavior (recall Sec. 8.1). These statements are clearly restricted to spatially and temporally uniform gases. In the nonequilibrium situation we must state the assumption in the following restrictive form: *At any instant there is no correlation among the locations in phase space of particles occupying a small volume element* δV. This is the so-called "principle of molecular chaos."[2]

A particle can, in general, be acted upon by a body force or force that is proportional to the mass of the particle. Body forces are generally the result of such external fields as gravity or magnetism. The body force per unit mass of a particle is designated in this chapter as $\mathbf{F}(\mathbf{r}, t)$. In multicomponent systems, particles of the ith kind are acted upon by a force $F_i(\mathbf{r}, t)$. We take these forces to be independent of velocity $\mathbf{c}$. In a simple gravity field acting along the z axis, $\mathbf{F}$ would be

$$\mathbf{F} = (0)\mathbf{i} + (0)\mathbf{j} + g\mathbf{k}$$

The units of $\mathbf{F}$ are dynes/g or cm/sec^2.

[1]Boltzmann summarized this work in *Vorlesungen über Gas Theorie*, vol. 1, Leipzig, 1896; *Lectures on Gas Theory*, English translation by S. G. Brush, University of California Press, Berkeley, Calif., 1964.

[2]When we use the principle of molecular chaos to derive Eq. (12.12), we do so in the slightly different form which ter Haar identified by the name *Stosszahlansatz*. D. ter Haar, *Elements of Thermostatistics*, 2nd ed., Holt, Rinehart and Winston, Inc., New York, 1966, p. 17.

The additional assumption that translational energy is conserved — that collisions are elastic — is also introduced at an early stage of the development.

DERIVATION

Consider a volume element in μ space as shown schematically in Fig. 12.1. Within the time interval t to $(t + \delta t)$, a particle of the ith species moves from $\mathbf{r}$ to $(\mathbf{r} + \mathbf{c}_i\,\delta t)$ and its velocity changes from $\mathbf{c}_i$ to $(\mathbf{c}_i + \mathbf{F}_i\,\delta t)$, if we ignore collisions for the present. Newton's second law of motion has been used to determine the effect of the body force upon the velocity, and in turn upon the position, of the particle. We should note carefully that the body force is the only force acting upon the particle; intermolecular forces have yet to be considered.

The particles have a distribution function $f_i(\mathbf{r}, \mathbf{c}_i, t)$, which we wish to determine. As long as there are no collisions, we can write

$$\underbrace{f_i(\mathbf{r}, \mathbf{c}_i, t)\,\delta V\,\delta\Omega}_{\substack{\text{number of}\\ \text{particles}\\ \text{under con-}\\ \text{sideration at}\\ \text{time } t}} = \underbrace{f_i(\mathbf{r} + \mathbf{c}_i\,\delta t, \mathbf{c}_i + \mathbf{F}_i\,\delta t, t + \delta t)\,\delta V\,\delta\Omega}_{\substack{\text{same group of particles that we}\\ \text{considered at time } t\text{, as they}\\ \text{appear } \delta t \text{ later}}} \tag{12.1}$$

Equation (12.1) says that all particles which began at point $(\mathbf{r}, \mathbf{c}_i)$ arrive at point $(\mathbf{r} + \delta\mathbf{r}, \mathbf{c}_i + \delta\mathbf{c}_i)$. This would be true in a collisionless gas, but collisions *do* occur. Their effect is to deflect certain of the particles so that they end up in some other location. We can repre-

Fig. 12.1 Volume element in μ space.

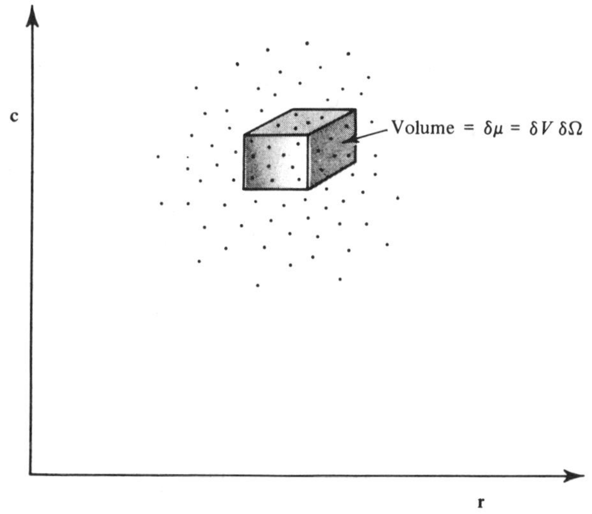

sent the effect of these removals (and of additional particles deflected into the element by external collisions) with the following schematic equation:

$$f_i(\mathbf{r} + \mathbf{c}_i\,\delta t,\ \mathbf{c}_i + \mathbf{F}_i\,\delta t,\ t + \delta t)\,\delta V\,\delta\Omega - f_i(\mathbf{r}, \mathbf{c}_i, t)\,\delta V\,\delta\Omega = \sum_{j=1}^{r} (\Gamma_{ij}^{(+)} - \Gamma_{ij}^{(-)})\,\delta V\,\delta\Omega\,\delta t \qquad (12.2)$$

The terms $\Gamma_{ij}^{(+)}$ and $\Gamma_{ij}^{(-)}$ are called the collision integrals, and we develop analytical expressions for them shortly. They have the following physical meaning:

$\Gamma_{ij}^{(+)}\,\delta V\,\delta\Omega\,\delta t$ = number of molecules of the ith kind that join the group in the position and velocity ranges $[\mathbf{c}_i, \mathbf{c}_i + \delta\mathbf{c}_i)$ and $[\mathbf{r}, \mathbf{r} + \delta\mathbf{r})$ as a result of collisions with molecules of the jth kind

$\Gamma_{ij}^{(-)}\,\delta V\,\delta\Omega\,\delta t$ = number of particles lost to the group as a result of such collisions

The first term of Eq. (12.2) can be expanded in a Taylor series in δt. If we restrict our attention to time increments that are short in the sense that they subtend relatively small changes in the distribution function, then

$$f_i(\mathbf{r} + \mathbf{c}_i\,\delta t,\ \mathbf{c}_i + \mathbf{F}_i\,\delta t,\ t + \delta t) = f_i(\mathbf{r}, \mathbf{c}_i, t) + \left[\mathbf{c}_i \cdot \frac{\partial f_i}{\partial \mathbf{r}}\,\delta t + \mathbf{F}_i \cdot \frac{\partial f_i}{\partial \mathbf{c}_i}\,\delta t + \frac{\partial f_i}{\partial t}\,\delta t\right] + \cdots \qquad (12.3)$$

Substitution of Eq. (12.3) into (12.2) and division by $\delta V\,\delta\Omega\,\delta t$ gives the *Boltzmann equation*,

$$\frac{\partial f_i}{\partial t} + \mathbf{c}_i \cdot \frac{\partial f_i}{\partial \mathbf{r}} + \mathbf{F}_i \cdot \frac{\partial f_i}{\partial \mathbf{c}_i} = \sum_{j=1}^{r} (\Gamma_{ij}^{(+)} - \Gamma_{ij}^{(-)}) \qquad (12.4)$$

We can combine the first two terms if we first observe that $\partial/\partial\mathbf{r}$ is, in reality, a notational convenience designating the gradient $\boldsymbol{\nabla}$. Then we can combine the local change of f_i with time, $\partial f_i/\partial t$, with the convective change of f_i, $\mathbf{c}_i \cdot \boldsymbol{\nabla} f_i$, to get

$$\left(\frac{\partial}{\partial t} + \mathbf{c}_i \cdot \boldsymbol{\nabla}\right) f_i = \text{the substantial derivative } \frac{Df_i}{Dt}$$

and

$$\frac{Df_i}{Dt} + \underbrace{\mathbf{F}_i \cdot \frac{\partial f_i}{\partial \mathbf{c}_i}}_{\substack{\text{change}\\ \text{in } f_i\\ \text{result-}\\ \text{ing from}\\ \text{external}\\ \text{forces}}} = \underbrace{\sum_j (\Gamma_{ij}^{(+)} - \Gamma_{ij}^{(-)})}_{\substack{\text{change in } f_i\\ \text{resulting from}\\ \text{collisions}}} \qquad (12.4a)$$

The term $\mathbf{F}_i \cdot \partial f_i / \partial \mathbf{c}_i$ really represents a kind of logical extension of the meaning of the substantial derivative. The left-hand side of Eq. (12.4a) now designates a net rate of change in the distribution of a swarm of particles. This change occurs directly in time; it also occurs by virtue of any change in location; and it occurs by virtue of external body forces. In the absence of collisions these changes would compensate and sum to zero. Accordingly, we define a new derivative in much the same spirit as the substantial derivative is defined:

$$\left(\frac{\partial f_i}{\partial t}\right)_{\text{coll}} \equiv \sum_j (\Gamma_{ij}{}^{(+)} - \Gamma_{ij}{}^{(-)}) \tag{12.4b}$$

Equation (12.4a) gives the equilibrium, or Maxwellian, distribution function when f_i does not depend upon t or $\mathbf{r}$, and when $\mathbf{F}_i$ is zero. This in turn results in the left-hand side of the equation being equal to zero. Thus we should be able to show that f_i is Maxwell's distribution, using[3]

$$\sum_j (\Gamma_{ij}{}^{(+)} - \Gamma_{ij}{}^{(-)}) = 0 \tag{12.5}$$

Our next task is that of writing the collision terms in terms of the distribution functions. This requires a more detailed look at the mechanics of molecular encounters than we have previously undertaken.

EXAMPLE 12.1 In Example 3.2 we obtained the steady-state distribution of particles in an isothermal atmosphere using statistical-mechanical methods. This result can be written as

$$f(\mathbf{C}, \mathbf{r}) = n(z)\left(\frac{m}{2\pi kT}\right)^{3/2} \exp\left(-\frac{mC^2}{2kT}\right)$$

where $n(z) = n_0 \exp(-mgz/kT)$. Is this result consistent with the Boltzmann integrodifferential equation?

The atmosphere is in local equilibrium; therefore, the collision terms vanish. The derivative, $\partial f/\partial t$, also vanishes, because the system is steady, and Eq. (12.4) reduces to

$$\mathbf{C} \cdot \frac{\partial f}{\partial z} + (-g)\frac{\partial f}{\partial \mathbf{C}} = 0$$

where the body force, $-g$, is negative because it acts opposite the direction of increasing z. Substituting the distribution function, above, in this expression gives

$$-\frac{mg\mathbf{C}}{kT} f + \frac{mg\mathbf{C}}{kT} f = 0$$

[3]This result is a special case of the "principle of detailed balancing." See E. H. Kennard, *Kinetic Theory of Gases*, McGraw-Hill, Inc., New York, 1938, p. 34, for a discussion of the principle.

which is identically true. Thus an equilibrium result can be considered a legitimate solution of the Boltzmann equation. This notion is examined further in Problems 12.8 and 12.9 and in the context of Eq. (12.94).

12.2 FORMULATION OF THE COLLISION TERM

MECHANICS OF A BINARY ENCOUNTER

Figure 12.2 shows two particles that have moved into each other's force fields and are suffering a "collision." Section 9.3 showed that the intermolecular forces, $\mathfrak{F}_i$, usually attract at a distance and repel at short range. They generally far overbalance the body forces $m_i\mathbf{F}_i$, at close range; however, the latter can be just as important because they act at all times — not just during the very brief collisions.

It aids us in our consideration of a collision to identify $\mathbf{c}_i$ and $\mathbf{c}_j$ as the velocities before the collision begins, and $\mathbf{c}'_i$ and $\mathbf{c}'_j$ as the velocities after it is complete. The following definitions are also made:

$$m_0 \equiv m_i + m_j \qquad M_i \equiv \frac{m_i}{m_0} \qquad M_j \equiv \frac{m_j}{m_0}$$

so

$$M_i + M_j = 1 \tag{12.6}$$

Then from conservation of momentum,

$$m_0\mathbf{G} = m_i\mathbf{c}_i + m_j\mathbf{c}_j = m_i\mathbf{c}'_i + m_j\mathbf{c}'_j \tag{12.7}$$

Fig 12.2 A binary encounter between particles i and j.

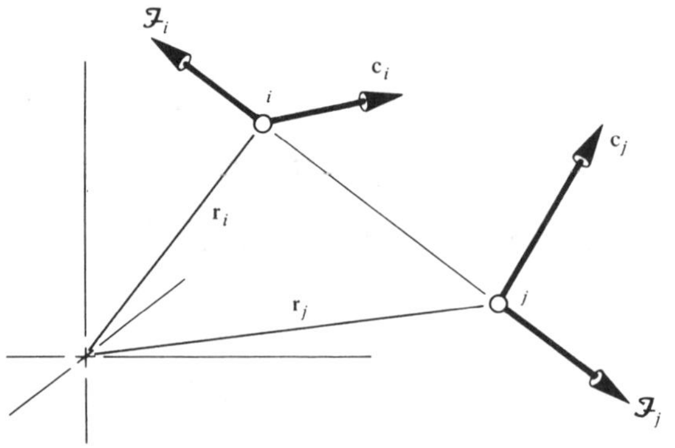

Furthermore, $\mathbf{G}$, the velocity of the center of mass of i and j, is a constant during a collision, and

$$\begin{aligned} \mathbf{g}_{ij} &\equiv \mathbf{c}_i - \mathbf{c}_j = -\mathbf{g}_{ji} \\ \mathbf{g}'_{ij} &\equiv \mathbf{c}'_i - \mathbf{c}'_j = -\mathbf{g}'_{ji} \\ g &\equiv |\mathbf{g}_{ij}| = |\mathbf{g}_{ji}| \end{aligned}$$

From these definitions the following results can easily be obtained (Problem 12.1):

$$\begin{aligned} \mathbf{c}_i &= \mathbf{G} + M_j\mathbf{g}_{ij} \\ \mathbf{c}'_i &= \mathbf{G} + M_j\mathbf{g}'_{ij} \\ \mathbf{c}_j &= \mathbf{G} + M_i\mathbf{g}_{ji} \\ \mathbf{c}'_j &= \mathbf{G} + M_i\mathbf{g}'_{ji} \end{aligned} \tag{12.8}$$

At this point we wish to introduce the assumption that translational kinetic energy is conserved during a collision. This is a more restrictive form of the ideal-gas requirement that we embraced in Sec. 2.1. It means that collisions are elastic and that no energy is transferred between the translational and internal modes of storage during collisions. Equation (12.4) remains a perfectly general equation, but under this assumption the form of the term $\sum_j (\Gamma_{ij}^{(+)} - \Gamma_{ij}^{(-)})$ is limited for use with ideal monatomic gases. As a result of the assumption we can easily show that the relative speeds of the particles are the same before and after the collision. Thus

$$g = g' \tag{12.9}$$

The conservation-of-momentum requirement, Eq. (12.7), can also be expressed according to Newton's second law, as

$$m_i\ddot{\mathbf{r}}_i = \mathfrak{F}_i = -\mathfrak{F}_j = -m_j\ddot{\mathbf{r}}_j \tag{12.10}$$

It follows easily that

$$m_i m_j(\ddot{\mathbf{r}}_i - \ddot{\mathbf{r}}_j) = \mathfrak{F}_i(m_j + m_i) \tag{12.11}$$

or

$$\frac{d^2}{dt^2}(\mathbf{r}_i - \mathbf{r}_j) = \mathfrak{F}_i\left(\frac{m_i + m_j}{m_i m_j}\right) \tag{12.11a}$$

From this it can be shown (Problem 12.2) that

$$(\mathbf{r}_i - \mathbf{r}_j) \times (\mathbf{c}_i - \mathbf{c}_j) = \text{a constant vector } \mathbf{K} \tag{12.12}$$

The vector $\mathbf{K}$ must be perpendicular to both vectors $(\mathbf{r}_i - \mathbf{r}_j)$ and $(\mathbf{c}_i - \mathbf{c}_j)$, whose cross product it represents. Since $\mathbf{K}$ is a constant, the plane of collision defined by these vectors is also constant. Conservation of momentum therefore requires that binary collisions be *coplanar*.

CHARACTER OF THE COLLISION

Figures 12.3 and 12.4 show the character of an actual collision. Figure 12.3 is a three-dimensional representation of a single binary encounter. Figure 12.4 shows two such encounters in plan view. Both figures should be viewed as though they were translating with velocity $\mathbf{c}_i$ so that i appears to be stationary while the other particle approaches it with speed g. The terms, b, ϵ, apse, and χ, which appear in the figures have the following meanings:

Fig. 12.3 Three-dimensional representation of a binary encounter.

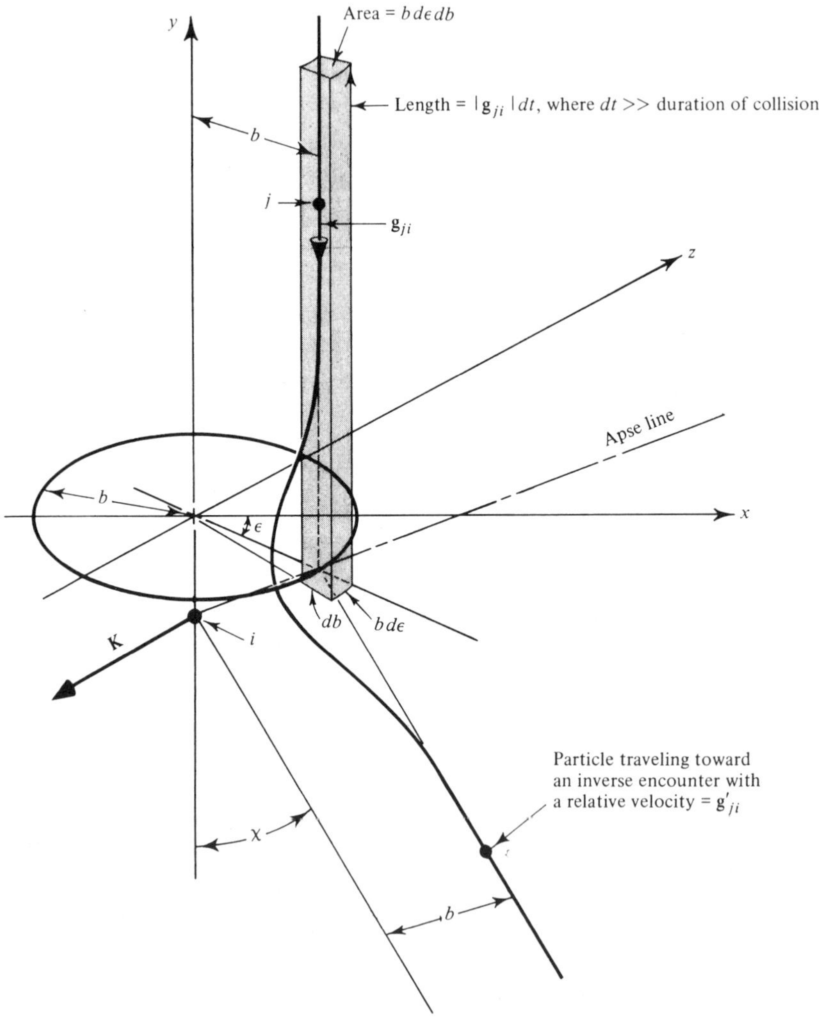

Fig. 12.4 Plan view of two binary collisions.

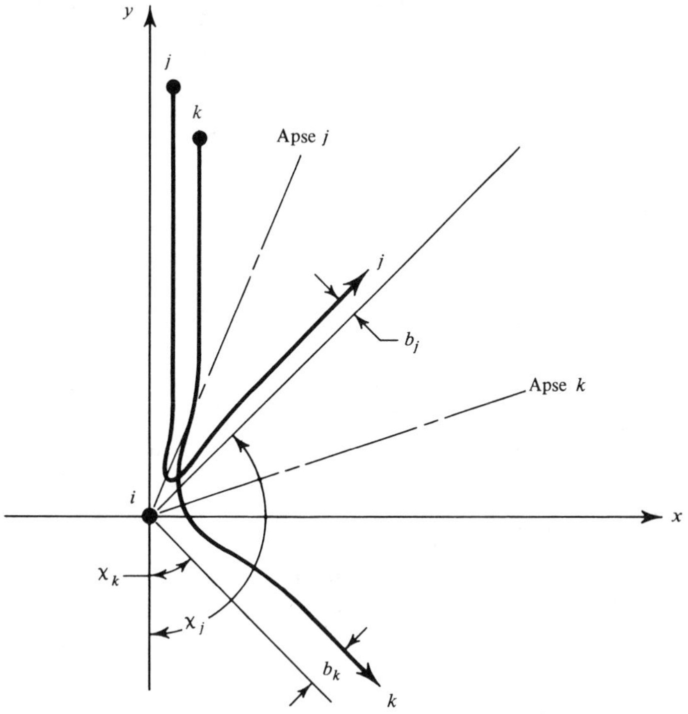

b = the *impact parameter*, or distance between the asymptotic paths of the two particles

ϵ = coordinate angle identifying the plane of the collision; since collisions are coplanar, ϵ is constant

Apse = center line of symmetry of the collision in the moving coordinates

χ = angle through which the approaching particle is deflected relative to particle i; depends upon b, g, and the law of molecular interaction

STATISTICS OF A BINARY ENCOUNTER

With this conceptualization of an encounter in mind, we seek to account for the effect of collisions in adding and removing particles from the group under consideration. The number of particles of the jth type approaching particles of the ith type during the time interval dt, within the volume increment $dV = gb\,d\epsilon\,db\,dt$, and in the velocity range $[\mathbf{c}_j, \mathbf{c}_j + d\mathbf{c}_j)$ can be written as

$$f_j(\mathbf{r}, \mathbf{c}_j, t)\,dV\,d\Omega = f_j gb\,d\epsilon\,db\,dt\,d\Omega_j$$

because the principle of equal a priori probabilities specifies that the velocities are randomly distributed in $dV\,d\Omega$.

Similarly, the number of i particles that are liable to be hit is $f_i gb\,d\epsilon\,db\,dt\,d\Omega_i$. It follows that in a neighborhood $dV\,d\Omega_i\,dt$ of position, velocity, and time, the number of collisions that remove particles from consideration is

$$\Gamma_{ij}^{(-)}dV\,d\Omega_i\,dt = \int_{b,\epsilon,\Omega_j} (f_i\,dV\,d\Omega_i)(f_j gb\,d\epsilon\,db\,dt\,d\Omega_j) \qquad (12.13)$$

The differentials $dV\,d\Omega_i\,dt$ are independent of b, ϵ, or $\mathbf{c}_j$ and can be factored out of the integral on the right-hand side. Equation (12.13) then becomes

$$\Gamma_{ij}^{(-)} = \int_{b=0}^{\infty}\int_{\epsilon=0}^{2\pi}\int_{\Omega_j} f_i f_j gb\,db\,d\epsilon\,d\Omega_j \qquad (12.13a)$$

An upper limit of infinity on the integral with respect to b assumes that long-range attraction forces decay rapidly enough to assure convergence. There are certain models for intermolecular attraction that do not converge and for which an arbitrary finite upper bound has to be used in the integration. Such models are, of course, implausible in this respect, although they might otherwise be quite useful.

The computation of $\Gamma_{ij}^{(+)}$ requires the introduction of an elegant conceptual device—that of the *inverse encounter*. We need some way of counting the exterior collisions that add particles to the range of interest. This counting must be accomplished from *within* the element of interest, however. To this end, let us consider the reversal of a collision as shown in Fig. 12.5. If the positions of two particles were to be shifted after a collision, they could be made to collide in such a way as to restore the original configuration.

Fig. 12.5 Restoration of original velocities by an inverse encounter.

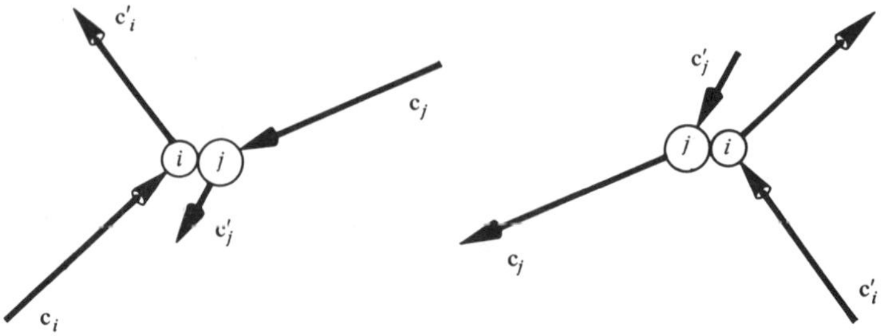

(a) a direct encounter between two particles

(b) an inverse encounter between two particles

With the help of this idea we can devise a scheme for counting the particles added to the region of interest by external collisions, even though we do not step outside the region to execute the count. For every collision that occurs within the region, there exists the restorative velocity pair ($\mathbf{c}'_i$, $\mathbf{c}'_j$) in the surroundings. The primed velocities possess distribution functions f'_i and f'_j, and these distribution functions can be used to count collisions that restore particles to the region. Thus we can write the following expression, just as we wrote Eq. (12.13):

$$\Gamma_{ij}^{(+)}\, dV\, d\Omega_i\, dt = \int_{b',\epsilon,\Omega'_j} f'_i f'_j g' b'\, db'\, d\epsilon\, dt\, d\Omega_j'\, d\Omega'_i\, dV' \qquad (12.14)$$

The terms g', b', and dV' are the same as g, b, and dV, so the primes on them can be dropped. Since collisions generally change the distribution functions f_i and f_j to f'_i and f'_j, these primes must be retained. Finally, the primes on $d\Omega'_j$ and $d\Omega'_i$ can be eliminated with the help of the Jacobian of transformation:

$$d\Omega'_j\, d\Omega'_i = \left| J\left(\frac{\Omega'_j, \Omega'_i}{\Omega_j, \Omega_i}\right)\right| d\Omega_j\, d\Omega_i \qquad (12.15)$$

It turns out that although J is negative because of the changes in direction with collisions, it is numerically equal to unity by virtue of Liouville's theorem.[4] Thus Eq. (12.14) becomes

$$\Gamma_{ij}^{(+)} = \int_{b=0}^{\infty} \int_{\epsilon=0}^{2\pi} \int_{\Omega_j} f'_i f'_j g b\, db\, d\epsilon\, d\Omega_j \qquad (12.14a)$$

Returning to Boltzmann's equation, Eq. (12.4), with Eqs. (12.13a) and (12.14a) we obtain Boltzmann's nonlinear integrodifferential equation for elastic particles:

$$\frac{\partial f_i}{\partial t} + \mathbf{c}_i \cdot \frac{\partial f_i}{\partial \mathbf{r}} + \mathbf{F}_i \cdot \frac{\partial f_i}{\partial \mathbf{c}_i} = \sum_j \iiint (f'_i f'_j - f_i f_j) g b\, db\, d\epsilon\, d\Omega_j \qquad (12.16)$$

For a single-component gas the Boltzmann equation becomes

$$\frac{\partial f}{\partial t} + \mathbf{c} \cdot \frac{\partial f}{\partial \mathbf{r}} + \mathbf{F} \cdot \frac{\partial f}{\partial \mathbf{c}} = \iiint (f' f'_1 - f f_1) g b\, db\, d\epsilon\, d\Omega_1 \qquad (12.17)$$

In Eq. (12.17) we have retained the distinction between the distribution functions f, for particles considered to be stationary, and f_1, for particles considered to be approaching the stationary ones. This distinction is necessary, because the integrations on the right-hand

[4]See, for example, S. Chapman and T. G. Cowling, *The Mathematical Theory of Non-Uniform Gases*, Cambridge University Press, New York, 1960, sec. 3.52.

side are only to be done over f'_1 and f_1, not over f' and f. The distribution f_1 — or f_j in Eq. (12.16) — is, in effect, a dummy variable because it depends upon the variable of integration, Ω_j.

MEAN FREE PATH

We found in Sec. 11.2 that a simple argument shows that the mean free path of a spherical particle moving among stationary spherical particles would be $(n\pi\sigma^2)^{-1}$, where σ is the diameter of the particles. We alleged at that point that this result would be reduced to $(\sqrt{2}n\pi\sigma^2)^{-1}$ when all particles move about with a Maxwellian velocity distribution. Equation (12.13) now provides the means for developing this expression.

Figure 12.6 shows the collision between two spherical but dissimilar molecules. The equivalent diameter of the particles is $\sigma_{12} = (\sigma_1 + \sigma_2)/2$. We have shown in connection with Eq. (12.13) that the number of collisions per unit volume and time in the range $d\Omega_1$, $d\Omega_2$, db, and $d\epsilon$ is

$$f_1 f_2 g b \, db \, d\epsilon \, d\Omega_1 \, d\Omega_2$$

By introducing $b = \sigma_{12} \sin \psi$, or $b \, db = \sigma_{12}^2 \cos \psi \sin \psi \, d\psi$ (where $0 \le \psi \le \pi/2$) and integrating, we can obtain the volume rate of collisions, N_{12}:

$$N_{12} = \int_{\Omega_1} \int_{\Omega_2} f_1 f_2 g \sigma_{12}^2 \, d\Omega_1 \, d\Omega_2 \int_0^{\pi/2} \cos \psi \sin \psi \, d\psi \int_0^{2\pi} d\epsilon \quad (12.18)$$

The introduction of Eq. (2.43), the Maxwell distribution, for f_1 and f_2 and integration of the result gives (see Problem 12.3)

$$N_{12} = 2n_1 n_2 \sigma_{12}^2 \left[\frac{2\pi kT(m_1 + m_2)}{m_1 m_2} \right]^{1/2} \quad (12.19)$$

Fig. 12.6 Collision between hard spherical molecules.

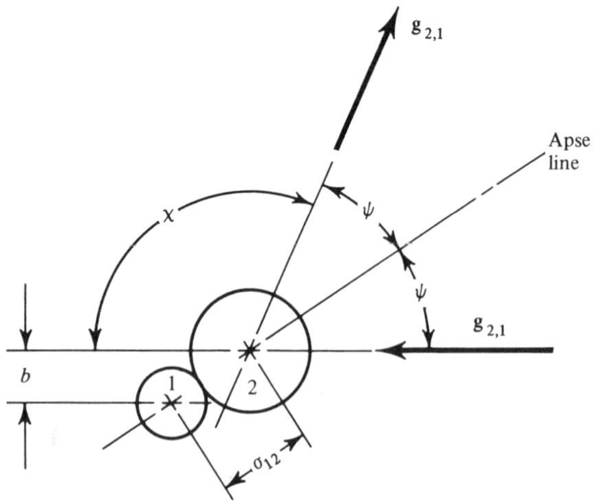

If there is only one molecular species, N_{11} can be obtained by replacing the subscript 2 with 1:

$$N_{11} = 4n_1{}^2\sigma_1{}^2\sqrt{\frac{\pi kT}{m_1}} \tag{12.20}$$

Actually this represents the result of counting each collision twice. It is therefore a correct representation of the volume rate at which *particles* undergo collision, but it is exactly double the number of *collisions*.

We must next determine τ, the average time elapsed between collisions of a given molecule [recall Eq. (11.54)]. The frequency with which the particle undergoes collisions is N_{11}/n_1, or, for several species,

$$\frac{1}{\tau_1} = \frac{N_{11}}{n_1} + \frac{N_{12}}{n_1} + \cdots = \frac{\sum_i N_{1i}}{n_1} \tag{12.21}$$

or

$$\tau_j = \frac{n_j}{\sum_i N_{ji}} \tag{12.21a}$$

where we pass to the more general subscript j. The mean free path of the jth component is then

$$l_j = \tau_j\bar{C}_j = \frac{n_j\bar{C}_j}{\sum_j N_{ji}} \tag{12.22}$$

Substituting the average speed for a Maxwellian gas, $\sqrt{8kT/\pi m}$ (recall Problem 2.11) and Eq. (12.19) for N_{ji}, we obtain

$$l_j = \frac{1}{\pi \sum_i (n_i\sigma_{ji}{}^2\sqrt{(m_j/m_i) + 1})} \tag{12.23}$$

For a single-component gas this becomes very nearly the result that we anticipated in Sec. 11.2,

$$l = \frac{1}{\sqrt{2}\pi\sigma^2 n} \tag{12.23a}$$

It is useful to refine this result a bit further. The use of the result of Eq. (12.18) to obtain Eq. (12.22) involves an implicit assumption that $l = \overline{\tau C} = \tau\bar{C}$. The factoring of the average in this case is a reasonable approximation but not entirely accurate. A more precise computation gives $l = \overline{\tau C} = 1.051\tau\bar{C}$, so

$$l = \frac{1.051}{\sqrt{2}\pi\sigma^2 n} \tag{11.29}$$

as we anticipated.

EXAMPLE 12.2 Find the mean free path, l_m, for molecules (m) and l_e, for electrons (e) in a "plasma". (Recall Example 11.1.)

Using Eq. (12.23) we obtain for l_e,

$$l_e = \left[\pi\left(n_e\sigma_{ee}^2\sqrt{2} + n_m\sigma_{em}^2\sqrt{\frac{m_e}{m_m} + 1}\right)\right]^{-1}$$

but σ_{ee}^2 is negligibly small and $\sigma_{em}^2 = \sigma_m^2/4$. Furthermore, $m_e/m_m \ll 1$, so

$$l_e = \frac{4}{\pi n_m \sigma_m^2}$$

which is what we found in Example 11.1. For l_m we get

$$l_m = \left[\pi\left(n_m\sigma_{mm}^2\sqrt{2} + n_e\sigma_{em}^2\sqrt{\frac{m_m}{m_e} + 1}\right)\right]^{-1}$$

This result is surprising at first glance. Since $m_m/m_e \gg 1$, the path is very short, unless n_e is quite small. The reason is that a slow-moving molecule does not get far before it is struck by one of the fast-moving electrons. If we neglect electron collisions as characterized by the second term, then this gives exactly Eq. (12.23a).

Before returning to the problem of actually solving Boltzmann's equation for the distribution function, we look at some other results that can be obtained from it. Solving the equation is a difficult business, as it happens, but much can be learned from it short of actual solution.

12.3 BOLTZMANN *H* THEOREM

COLLISION INVARIANTS OF A BINARY ENCOUNTER

A *collision invariant* Ψ (or *summation invariant* as it is frequently called), is any molecular property[5] Φ whose combined value for both i and j particles is conserved during a collision:

$$\Psi_i + \Psi_j = \Psi'_i + \Psi'_j \tag{12.24}$$

Mass, momentum, and energy are the three basic collision invariants that we are concerned with. Since $m_i = m'_i$ and $m_j = m'_j$,

$$m_i + m_j = m'_i + m'_j \tag{12.25}$$

We have already noted that

$$m_i\mathbf{c}_i + m_j\mathbf{c}_j = m_i\mathbf{c}'_i + m_j\mathbf{c}'_j \tag{12.7}$$

[5]The symbol Φ denotes any molecular property; Φ includes Ψ as a special case.

Finally, since we have restricted our discussion to particles that undergo elastic collisions,

$$\tfrac{1}{2}m_i(\mathbf{c}_i)^2 + \tfrac{1}{2}m_j(\mathbf{c}_j)^2 = \tfrac{1}{2}m_i(\mathbf{c}'_i)^2 + \tfrac{1}{2}m_j(\mathbf{c}'_j)^2 \tag{12.26}$$

All three of these equations are examples of Eq. (12.24).

It is possible to show that particle mass, momentum, and energy (or linear combinations of them) are the only collision invariants in a translational gas. Let us now prove the following useful result related to collision invariants:

$$\sum_{ij} \iiiint \Phi_i(f'_i f'_j - f_i f_j) gb\, db\, d\epsilon\, d\Omega_j\, d\Omega_i = 0 \qquad \text{if } \Phi_i = \Psi_i \tag{12.27}$$

This result is fundamental to our subsequent analysis of binary encounters. To prove it, we first write from symmetry,

$$\iiiint \Phi_i(f'_i f'_j - f_i f_j) gb\, db\, d\epsilon\, d\Omega_i\, d\Omega_j = \iiiint \Phi'_i(f_i f_j - f'_i f'_j) gb\, db\, d\epsilon\, d\Omega'_i\, d\Omega'_j$$

But $d\Omega_i\, d\Omega_j = d\Omega'_i\, d\Omega'_j$; therefore,

$$\iiiint \Phi_i(f'_i f'_j - f_i f_j) gb\, db\, d\epsilon\, d\Omega_i\, d\Omega_j = -\iiiint \Phi'_i(f'_i f'_j - f_i f_j) gb\, db\, d\epsilon\, d\Omega_i\, d\Omega_j$$

If $a = d$, then $a = a/2 + d/2$, so we can write

$$\iiiint \Phi_i(f'_i f'_j - f_i f_j) gb\, db\, d\epsilon\, d\Omega_i\, d\Omega_j = \tfrac{1}{2}\iiiint (\Phi_i - \Phi'_i)(f'_i f'_j - f_i f_j) gb\, db\, d\epsilon\, d\Omega_i\, d\Omega_j \tag{12.28}$$

Let us now restrict Φ to functions of $\mathbf{r}$, t, and $\mathbf{c}$. We then observe that interchanging $\mathbf{c}_i$ and $\mathbf{c}_j$ should not alter the right-hand side of Eq. (12.28), because the meaning of such an interchange would be no more than a change in the sequence of integration. Thus

$$\iiiint (\Phi_i - \Phi'_i)(f'_i f'_j - f_i f_j) gb\, db\, d\epsilon\, d\Omega_i\, d\Omega_j = \iiiint (\Phi_j - \Phi'_j)(f'_i f'_j - f_i f_j) gb\, db\, d\epsilon\, d\Omega_i\, d\Omega_j$$

It follows that

$$\sum_{i,j} \iiiint \Phi_i(f'_i f'_j - f_i f_j) gb\, db\, d\epsilon\, d\Omega_i\, d\Omega_j$$
$$= \tfrac{1}{4} \sum_{i,j} \iiiint (\Phi_i + \Phi_j - \Phi'_i - \Phi'_j)(f'_i f'_j - f_i f_j) gb\, db\, d\epsilon\, d\Omega_i\, d\Omega_j \tag{12.29}$$

The right-hand side of Eq. (12.29) will vanish when Φ is a summation invariant Ψ so the proof is complete.

DERIVATION AND DISCUSSION OF THE *H* THEOREM

In 1872 Boltzmann[6] undertook to show that dilute nonequilibrium systems would naturally approach equilibrium. Using the Boltzmann equation he showed how this approach takes place and that the end result is the Maxwell distribution function. He began by defining a quantity,[7] $H = H(t)$, for a spatially uniform, but nonequilibrium, gas:

$$H(t) \equiv \sum_i \int_{\Omega_i} f_i(\mathbf{c}_i, t) \ln f_i(\mathbf{c}_i, t)\, d\Omega_i \tag{12.30}$$

or

$$H(t) = \sum_i n_i \overline{\ln f_i} \tag{12.30a}$$

The importance of H is that it is a monotonic function of time. To show this we write

$$\frac{\partial H}{\partial t} = \sum_i \frac{\partial}{\partial t} \int_{\Omega_i} f_i \ln f_i\, d\Omega_i$$

Since t and Ω_i are independent, it is possible to differentiate under the integral. The result is

$$\frac{\partial H}{\partial t} = \sum_i \int_{\Omega_i} (1 + \ln f_i) \frac{\partial f_i}{\partial t}\, d\Omega_i \tag{12.31}$$

But, for $f_i = f_i(t, \mathbf{c}_i, \text{only})$ and for $\mathbf{F}_i = 0$, Eq. (12.16) — Boltzmann's equation — is

$$\frac{\partial f_i}{\partial t} = \sum_j \iiint (f'_i f'_j - f_i f_j) gb\, db\, d\epsilon\, d\Omega_j \tag{12.32}$$

[6]Boltzmann derived the *H* theorem in *Wien. Ber.*, **66,** 275 (1872). He obtained the Maxwell distribution four years earlier in *Wien Ber.*, **58,** 517 (1868). Ter Haar includes a careful discussion of implications of the *H* theorem in his book *Elements of Statistical Mechanics*, Holt, Rinehart and Winston, Inc., New York, 1960, App. 1.

[7]*H* in the present context is unrelated to the Hamiltonian function that we mentioned earlier.

Using Eq. (12.32) to eliminate $\partial f_i/\partial t$ from Eq. (12.31), we obtain

$$\frac{\partial H}{\partial t} = \sum_{i,j} \iiiint (1 + \ln f_i)(f'_i f'_j - f_i f_j) gb\, db\, d\epsilon\, d\Omega_i\, d\Omega_j \quad (12.33)$$

after appropriate interchanges of integration and summation. Finally, noting that $(1 + \ln f_i)$ can be regarded as a molecular property Φ_i, we can substitute Eq. (12.33) in Eq. (12.29) and obtain

$$\frac{\partial H}{\partial t} = -\tfrac{1}{4} \sum_{i,j} \iiiint \left(\ln \frac{f'_i f'_j}{f_i f_j}\right)(f'_i f'_j - f_i f_j) gb\, db\, d\epsilon\, d\Omega_i\, d\Omega_j \quad (12.34)$$

A moment's reflection will reveal that the quantity $[\ln (a/d)](a - d)$ can only be positive, whatever values a and d might be given. The quantity gb is also positive. It therefore follows from Eq. (12.34) that

$$\frac{\partial H}{\partial t} \leq 0 \quad (12.35)$$

Equation (12.35) is the substance of Boltzmann's H theorem. It tells us that H displays a quality not unlike entropy in that both are monotonic in time, although entropy increases. Like S, H tends to a limiting value, but Eq. (12.30) or (12.30a) shows that this value is negative instead of positive. We look further at the relation between H and S later in this section. First, we must consider a difficulty in our statement of the theorem.

When we discussed the statistics of a binary encounter, in the course of deriving the Boltzmann equation, we incorporated the principle of equal a priori probabilities[8] into the equation. This approximation, as it turns out, does not seriously limit the usefulness of Boltzmann's equation, although it does tend to smooth out the behavior that it predicts. Thus we might follow Huang[9] in constructing a plot such as is shown in Fig. 12.7.

Figure 12.7 compares the actual approach of a system to equilibrium (as it is influenced by the small-scale fluctuations that result from collisions) with the approach to equilibrium predicted by Eq. (12.33). We recognize that Eq. (12.33) includes the Boltzmann equation and therefore includes the smoothing effect that entered when we assumed molecular chaos to exist (footnote 8). Actually, *some* of the local maxima represent points at which molecular chaos is achieved momentarily. This is the case because (in accordance with the H theorem) spontaneous microfluctuations, in a forward

[8]We take for granted that use of the principle of equal a priori probabilities implies that a state of molecular chaos exists.
[9]K. Huang presents a searching discussion of these matters in *Statistical Mechanics*, John Wiley & Sons, Inc., New York, 1963, chap. 4.

Fig. 12.7 Character of the approach to equilibrium.

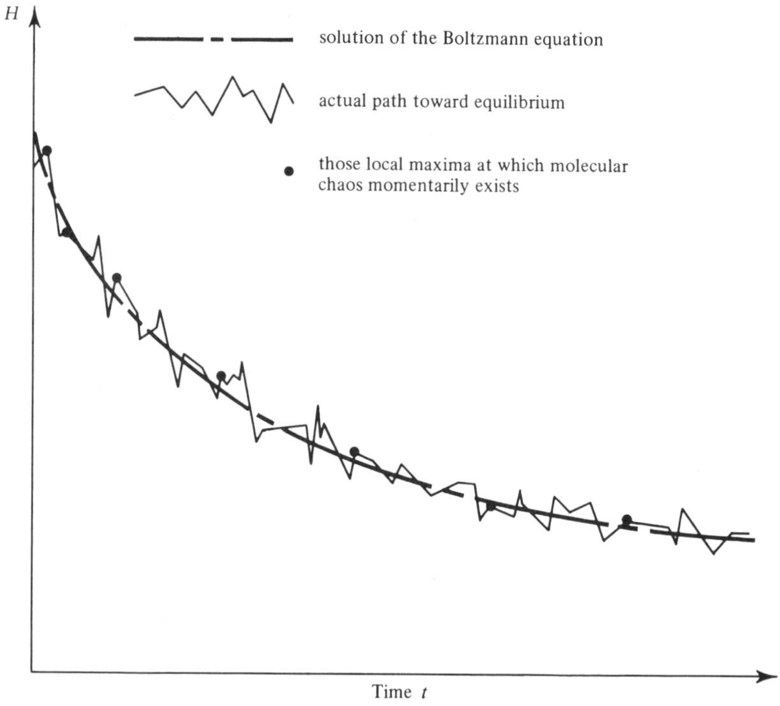

direction in time, can only carry a system to a lower value of H than that which exists in the state of molecular chaos. However, another gas in the same state, but with all molecular motions reversed, would go to a lower H in forward time, as well. Since this would be the way in which we might back up from the local maximum in question, we see that *if* any points of perfect molecular chaos exist, they must be local maxima.

The use of the idea of molecular chaos in Boltzmann's equation eliminates the possibility of backward microfluctuations and results in a smoothed curve (Fig. 12.7) that threads its way through these local maxima.

Of course, as the size of the system under consideration is increased, these fluctuations rapidly become too small to notice; but they never vanish. To allege that they did vanish would be to open the way for a variety of paradoxical difficulties. Consider, for example, Poincaré's theorem: *A system having finite energy and confined to a finite volume will, after a sufficiently long time, return to an arbitrarily small neighborhood in phase space of almost any initial state.*

This might at first seem to be blatant contradiction of the *H* theorem, which implies that a system can only move irrevocably to equilibrium. Actually the return to an initial state predicted by Poincaré would have to be the result of a possible, but very unlikely, fluctuation. If it is *possible*, it would also be *inevitable*, but only after an unreasonable wait. We must therefore allow the microfluctuation that would satisfy Poincaré's theorem even though we accept the *H* theorem in the main. But such a fluctuation is not worth considering any more than is the likelihood of getting back to pure salt and pepper in the example in Sec. 3.1.

Another paradox is our apparent creation of a function *H* that is irreversible in time out of a molecular model that is completely reversible. Here again, Fig. 12.7 provides the needed clarification. The function *H*, although time-irreversible in the main, is actually subject to minor reversals due to fluctuations. Only the use of Boltzmann's equation in writing Eq. (12.33) led to an absolutely irreversible approximation for *H*.

Nevertheless, it is a matter of great interest that the kinetic theory has now exposed the microscopic origins of the time-dependent quality of *irreversibility* — a quality that did not manifest itself in the equilibrium situations described by statistical mechanics.

MAXWELL DISTRIBUTION

Equilibrium will be established when $\partial H/\partial t$ vanishes. In accordance with Eq. (12.34), this occurs either when

$$\ln\left(\frac{f'_i f'_j}{f_i f_j}\right) = 0 \tag{12.36}$$

or when

$$f'_i f'_j - f_i f_j = 0 \tag{12.36a}$$

and these conditions are clearly equivalent. Equation (12.36a) results in the right-hand side of Eq. (12.16) vanishing. This condition [as we noted at Eq. (12.5)] should define the Maxwell distribution function. Let us now show that this is the case.

Equation (12.36a) tells us that in equilibrium, collisions result in no change of the distribution function. Expanding Eq. (12.36) we obtain

$$\ln f'_i + \ln f'_j = \ln f_i + \ln f_j$$

This tells us that $\ln f$ must be a collision invariant property.[10] But a collision invariant must in turn be a linear function of m, $m\mathbf{c}$, or

[10]We henceforth drop the subscripts i and j in this section. Actually we could just as well carry either subscript.

$m\mathbf{c}^2/2$, because they are the only quantities for which we have conservation principles. Thus

$$\ln f = am + \mathbf{b}\cdot(m\mathbf{c}) - \beta\left(\frac{m\mathbf{c}^2}{2}\right) \tag{12.37}$$

where a, $\mathbf{b}$, and β are constants that must be evaluated. By completing the square we obtain

$$\ln f = \ln \alpha - \frac{m\beta}{2}\left(\mathbf{c} - \frac{\mathbf{b}}{\beta}\right)^2 \tag{12.37a}$$

where

$$\ln \alpha \equiv am + \frac{m\beta}{2}\left(\frac{\mathbf{b}}{\beta}\right)^2 \tag{12.38}$$

It follows that

$$f = \alpha \exp\left[-\frac{m\beta}{2}\left(\mathbf{c} - \frac{\mathbf{b}}{\beta}\right)^2\right] \tag{12.39}$$

Equation (12.39) looks something like a Maxwell distribution function, but we still must eliminate α, β, and $\mathbf{b}$ in favor of the basic macroscopic parameters: n or ρ/m, T, and $\mathbf{c}_0$. Three physical constraints serve to do this:

$$n = \int_\Omega f\, d\Omega \tag{12.40}$$

$$\frac{3}{2}kT = \frac{m}{2n}\int_\Omega c^2 f\, d\Omega \tag{12.41}$$

$$\rho\mathbf{c}_0 = \int_\Omega m\mathbf{c}_i f\, d\Omega \tag{12.42}$$

The substitution of Eq. (12.37a) in Eqs. (12.40), (12.41), and (12.42) leads, after straightforward manipulation (Problem 12.4), to

$$\alpha = n\left(\frac{m\beta}{2\pi}\right)^{3/2} \tag{12.43}$$

$$\beta = \frac{1}{kT} \tag{2.34}$$

$$\mathbf{b} = \beta\mathbf{c}_0 \tag{12.44}$$

We note that the peculiar velocity $\mathbf{C}$ is equal to $(\mathbf{c} - \mathbf{c}_0)$ and we return to Eq. (12.39) with these constants. The result is the Maxwell distribution:

$$f = n\left(\frac{m}{2\pi kT}\right)^{3/2} \exp\left(-\frac{m\mathbf{C}^2}{2kT}\right) \tag{2.43}$$

H FUNCTION AND ENTROPY

In a single-component gas,

$$H = \int_{\Omega} f \ln f \, d\Omega = n \overline{\ln f}$$

but when the gas is in equilibrium,

$$\ln f = \ln n + \tfrac{3}{2} \ln \left(\frac{m}{2\pi kT}\right) - \frac{m\mathbf{C}^2}{2kT}$$

so if we compute the average of $\ln f$ over $\mathbf{C}$, we obtain

$$\overline{\ln f} = \ln n + \tfrac{3}{2}\left[\ln \left(\frac{m}{2\pi kT}\right) - 1\right]$$

The equilibrium value of H is then

$$H_e = n\left(\ln n + \tfrac{3}{2}\left[\ln \left(\frac{m}{2\pi kT}\right) - 1\right]\right) \tag{12.45}$$

Now let us compare this with the macroscopic expression for the entropy of N_0 moles of an ideal gas,

$$S = \int N_0\left(c_v \frac{dT}{T} + R^0 \frac{dV}{V}\right) + S_0 \tag{12.46}$$

But

$$\frac{dV}{V} = -\frac{dn}{n} \qquad c_v = \frac{3k}{2m} M \qquad \frac{R^0}{M} = \frac{k}{m}$$

so Eq. (12.46) can be rewritten as

$$S = N_0 \int \frac{k}{m}\left(\frac{3}{2} d \ln T - d \ln n\right) + S_0 \tag{12.47}$$

Multiplying Eq. (12.45) by kV, adding it to Eq. (12.47), and noting that $N_0 N_A/n$ is V, we obtain

$$S = -kVH_e + Vnk\left\{\ln (T^{3/2}/n) + \ln n + \tfrac{3}{2}\left[\ln \left(\frac{m}{2\pi kT}\right) - 1\right]\right\} + \text{constant}_1$$

or

$$S = -kVH_e + \left\{\frac{3N_0 k}{2}\left[\ln \left(\frac{m}{2\pi k}\right) - 1\right] + \text{constant}_1\right\}$$

The expression in braces *does not depend upon the state of the gas.* Thus we finally obtain

$$S = -kVH_e + \text{constant}_2 \tag{12.48}$$

The equilibrium value of H is therefore a direct function of the entropy. However, there are many limitations to making a very broad interpretation of S in terms of H or vice versa. Entropy is defined irrespective of any substance, but only for equilibrium. The H function, on the other hand, is defined only for a very restrictive molecular model (or "kinetic hypothesis"), but it is not limited to equilibrium. Although the two can be related only for an equilibrium ideal gas, there is still great value in the comparison. Both serve as "time's arrow" by pointing to a basic unidirectionality of processes in time. Entropy does so by increasing between the equilibrium end points that mark the beginning and end of an irreversible process. The H function does so by decreasing in the direction of increasing time *during* the process.

EXAMPLE 12.3 Compare $\partial H/\partial t$ with $\partial S/\partial t$ in a monatomic ideal gas that is being heated very slowly and uniformly.

Such a gas can be considered in quasistatic equilibrium. Therefore, in accordance with the Sackur–Tetrode equation [Eq. (7.15a)],

$$\frac{\partial S}{\partial t} = Nk\frac{3}{2T}\frac{\partial T}{\partial t} = \frac{3knV}{2T}\frac{\partial T}{\partial t}$$

Combining this with Eq. (12.45) gives

$$\frac{\partial H_e}{\partial t} = -\frac{3n}{2T}\frac{\partial T}{\partial t} = -\frac{1}{kV}\frac{\partial S}{\partial t}$$

which is consistent with Eq. (12.48).

12.4 FUNDAMENTAL EQUATIONS OF FLUID MECHANICS

In providing the means for obtaining the nonequilibrium distribution function from a detailed description of molecular behavior, the Boltzmann equation provides a basis for obtaining the macroscopic equations of fluid mechanics. Not only can we obtain the equations but we can also get the transport coefficients that appear in them by applying the rules of averaging with the distribution function.

GENERAL EQUATION OF CHANGE OF MOLECULAR PROPERTIES

Let us first obtain from the Boltzmann equation an equation for the transport of those molecular properties that are conserved during collisions—the so-called collision invariants $\Psi_i(\mathbf{r}, \mathbf{c}_i, t)$. To do this we multiply Eq. (12.16) by Ψ_i and integrate it over $d\Omega_i$. Then we sum the result over the several components present,

$$\sum_i \int \Psi_i\left(\frac{\partial f_i}{\partial t} + \mathbf{c}_i\cdot\nabla f_i + \mathbf{F}_i\cdot\frac{\partial f_i}{\partial \mathbf{c}_i}\right) d\Omega_i$$

$$= \sum_{i,j} \iiiint \Psi_i(f'_i f'_j - f_i f_j) g b \, db \, d\epsilon \, d\Omega_i \, d\Omega_j$$

But [recall Eq. (12.27)] the right-hand side has to vanish, so

$$\sum_i \int \Psi_i\left(\frac{\partial f_i}{\partial t} + \mathbf{c}_i\cdot\nabla f_i + \mathbf{F}_i\cdot\frac{\partial f_i}{\partial \mathbf{c}_i}\right) d\Omega_i = 0 \tag{12.49}$$

Equation (12.49) expresses the conservation of Ψ_i — mass, momentum, or energy — for a nonequilibrium gas. To put it in more useful form we must next complete the indicated integrations by interpreting them as appropriate averages. The first term, for example, becomes

$$\int \Psi_i \frac{\partial f_i}{\partial t} d\Omega_i = \frac{\partial}{\partial t}\int \Psi_i f_i \, d\Omega_i - \int f_i \frac{\partial \Psi_i}{\partial t} d\Omega_i = \frac{\partial}{\partial t}(n_i\overline{\Psi}_i) - n_i \overline{\frac{\partial \Psi_i}{\partial t}} \tag{12.50}$$

The second term becomes

$$\int \Psi_i \mathbf{c}_i\cdot\nabla f_i \, d\Omega_i = \nabla\cdot\int \Psi_i \mathbf{c}_i f_i \, d\Omega_i - \int f_i \nabla\cdot(\Psi_i \mathbf{c}_i) \, d\Omega_i$$

$$= \nabla\cdot(n_i\overline{\Psi_i \mathbf{c}_i}) - n_i(\overline{\mathbf{c}_i\cdot\nabla\Psi_i}) \tag{12.51}$$

To simplify the third term we must first recognize that $\mathbf{F}_i$ is not dependent upon velocity. Thus

$$\int \Psi_i \mathbf{F}_i\cdot\frac{\partial f_i}{\partial \mathbf{c}_i} d\Omega_i = \mathbf{F}_i\cdot\int \Psi_i \frac{\partial f_i}{\partial \mathbf{c}_i} d\Omega_i = \mathbf{F}_i\cdot\left[\left(\Psi_i f_i\right)_{-\mathbf{c}_i=-\infty}^{\mathbf{c}_i=\infty} - \int f_i \frac{\partial \Psi_i}{\partial \mathbf{c}_i} d\Omega_i\right]$$

But f_i vanishes as $|\mathbf{c}_i| \to \infty$. Thus

$$\int \Psi_i \mathbf{F}_i\cdot\frac{\partial f_i}{\partial \mathbf{c}_i} d\Omega_i = -n_i \mathbf{F}_i\cdot\overline{\frac{\partial \Psi_i}{\partial \mathbf{c}_i}} \tag{12.52}$$

Combining Eqs. (12.50), (12.51), and (12.52) in Eq. (12.49) we obtain the *general equation of change*, or the *Boltzmann conservation equation:*

$$\sum_i \left[\frac{\partial}{\partial t}(n_i\overline{\Psi}_i) + \nabla\cdot(n_i\overline{\Psi_i \mathbf{c}_i}) - n_i\left(\overline{\frac{\partial \Psi_i}{\partial t}} + \overline{\mathbf{c}_i\cdot\nabla\Psi_i} + \mathbf{F}_i\cdot\overline{\frac{\partial \Psi_i}{\partial \mathbf{c}_i}}\right)\right] = 0 \tag{12.53}$$

Substitution of the appropriate Ψ_i in the general equation of change should then lead to the continuity, momentum, or energy equation.

EXAMPLE 12.4 Suppose that a stationary, single-component gas is spatially uniform. What additional conditions are implied?

The general equation of change becomes

$$\frac{\partial}{\partial t}(n\overline{\Psi}) - n\overline{\frac{\partial \Psi}{\partial t}} - n\mathbf{F}\cdot\overline{\frac{\partial \Psi}{\partial \mathbf{C}}} = 0$$

or

$$\overline{\Psi}\frac{\partial n}{\partial t} - n\mathbf{F}\cdot\overline{\frac{\partial \Psi}{\partial \mathbf{C}}} = 0$$

Now let us enter this equation with the three basic collision invariant quantities. First, for $\Psi = m$, the mass of the molecules, we obtain

$$\frac{\partial \rho}{\partial t} = 0 \qquad \text{or} \qquad \rho = \rho(\mathbf{r}) = \text{constant}$$

and for $\Psi = m\mathbf{C}$, the momentum of the particles,

$$\overline{\mathbf{C}}\frac{\partial \rho}{\partial t} = \rho\mathbf{F}$$

and finally for $\Psi = mC^2/2$, the kinetic energy of the particles,

$$\frac{\overline{\mathbf{C}^2}}{\overline{\mathbf{C}}}\frac{\partial \rho}{\partial t} = 2\rho\mathbf{F}$$

The last two conditions, when combined with the first, both give $\mathbf{F} = 0$.

Thus a stationary single-component gas cannot be spatially uniform unless it is free of body force fields, and its density (or number density) must be constant in time as well as space. The earth's atmosphere, for example, sustains a gravity body force, but it must vary in density in the direction of the field.

CONTINUITY EQUATION

We now develop full statements of the three conservation equations, beginning with continuity or conservation of mass. It will help if we first restate and extend some definitions. First, the partial and total densities ρ_i and ρ are

$$\rho_i \equiv n_i m_i \qquad \text{and} \qquad \rho \equiv \sum_i \rho_i = \sum n_i m_i \tag{12.54}$$

The "diffusion velocity," or average peculiar velocity $\overline{\mathbf{C}_i}$, no longer vanishes in a multicomponent system as it did in a single-component system:

$$\overline{\mathbf{C}_i} \equiv \frac{1}{n_i}\int \mathbf{C}_i f_i \, d\Omega = \frac{1}{n_i}\int (\mathbf{c}_i - \mathbf{c}_0) f_i \, d\Omega \tag{12.55}$$

The gross velocity $\mathbf{c}_0$ is defined with respect to the entire system,

$$\mathbf{c}_0 \equiv \frac{1}{\rho}\sum_i \rho_i \overline{\mathbf{c}_i} = \frac{1}{\rho}\sum_i n_i m_i \overline{\mathbf{c}_i} \tag{12.56}$$

so Eq. (12.55) becomes

$$\overline{\mathbf{C}}_i = \overline{\mathbf{c}}_i - \mathbf{c}_0 \tag{2.23a}$$

The mass flux of the ith component across a surface moving at a velocity $\mathbf{c}_0$ in the gas is

$$\mathbf{J}_{m_i} = \int m_i \mathbf{C}_i f_i \, d\Omega = \rho_i(\overline{\mathbf{c}}_i - \mathbf{c}_0) \tag{12.57}$$

J_{m_i} thus does not vanish. The total mass flux with respect to $\mathbf{c}_0$,

$$\sum_i \mathbf{J}_{m_i} = \sum_i \rho_i(\overline{\mathbf{c}}_i - \mathbf{c}_0) = \rho\mathbf{c}_0 - \rho\mathbf{c}_0 = 0$$

of course *does* vanish.

The first statement of conservation of mass that we wish to extract from Eq. (12.53) is the so-called *global continuity equation*, or simultaneous conservation of all species. We begin by writing Eq. (12.53) for $\Psi = m_i$:

$$\sum_i \left[\frac{\partial(n_i m_i)}{\partial t} + \nabla \cdot (n_i m_i \overline{\mathbf{c}}_i) - n_i \left(\frac{\partial m_i}{\partial t} + \overline{\mathbf{c}_i \cdot \nabla m_i} + \overline{\mathbf{F}_i \cdot \frac{\partial m_i}{\partial \mathbf{c}_i}} \right) \right] = 0$$

The last term in this expression is zero, and the first two terms become the global continuity equation

$$\frac{\partial}{\partial t}\sum_i \rho_i + \nabla \cdot \sum_i \rho_i \mathbf{c}_0 = 0$$

or

$$\frac{\partial \rho}{\partial t} + \nabla \cdot \rho \mathbf{c}_0 = 0 \tag{12.58}$$

This is the conventional continuity equation. It is *global* in the sense that it applies to properties that have been averaged over all species.

An expression of continuity of the component molecular species can also be established. To do so we must first show that for $\Psi_i = m_i$, the ith terms in Eq. (12.49) vanish identically. This proof is left as an exercise (Problem 12.5). It follows that the derivation of Eq. (12.58) can be repeated without carrying the summation sign and

$$\frac{\partial \rho_i}{\partial t} + \nabla \cdot \rho_i \overline{\mathbf{c}}_i = 0 \tag{12.59}$$

Equation (12.58) is the conventional continuity equation, and it shows that the processes of averaging that we used to eliminate the distribution function have yielded the gross mass relationship that we would anticipate. Equation (12.59) is interesting in that it shows that the same conservation law must also be satisfied for each component. It can also be cast in the form (Problem 12.6)

$$\frac{D\rho_i}{Dt} + (\boldsymbol{\nabla}\cdot\mathbf{J}_{m_i} + \rho_i\boldsymbol{\nabla}\cdot\mathbf{c}_0) = 0 \tag{12.60}$$

where D/Dt is the substantial derivative. Thus we find that the increase of the density of a moving particle at a point is balanced by the divergence of the mass flux at that point and a dilational effect.

If a chemical reaction takes place in a gas mixture, then a rate-of-creation-of-species term r_i (equal to the mass of the ith component created per unit volume per unit time), must be added to Eq. (12.59) to balance the equation,

$$\frac{\partial\rho_i}{\partial t} + \boldsymbol{\nabla}\cdot\rho_i\overline{\mathbf{c}_i} = r_i \tag{12.61}$$

Summing Eq. (12.61) over all components results in

$$\sum_i r_i = 0 \tag{12.62}$$

because the left-hand side sums to Eq. (12.58). Equation (12.62) is just the stoichiometric equation for the chemical reaction.

CONSERVATION OF MOMENTUM

Now let us return to the general equation of change [Eq. (12.53)] and obtain the macroscopic equation for the conservation of momentum. This time $\Psi_i = m_i\mathbf{c}_i$. The five terms in the equation must be simplified into tractable form. The first term becomes

$$\sum_i \frac{\partial}{\partial t}(n_i m_i\overline{\mathbf{c}_i}) = \frac{\partial\rho\mathbf{c}_0}{\partial t} \tag{12.63}$$

The second term is more complicated,[11]

$$\begin{aligned}\sum_i \boldsymbol{\nabla}\cdot[n_i m_i\overline{(\mathbf{C}_i + \mathbf{c}_0)(\mathbf{C}_i + \mathbf{c}_0)}] &= \boldsymbol{\nabla}\cdot\Big(\sum_i n_i m_i\overline{\mathbf{C}_i\mathbf{C}_i} + 2\sum_i n_i m_i\overline{\mathbf{C}_i}\mathbf{c}_0 \\ &\qquad + \sum_i n_i m_i\mathbf{c}_0\mathbf{c}_0\Big) \\ &= \boldsymbol{\nabla}\cdot\Big(\sum_i \rho_i\overline{\mathbf{C}_i\mathbf{C}_i} + 2\mathbf{c}_0\sum_i \mathbf{J}_{m_i} + \rho\mathbf{c}_0\mathbf{c}_0\Big)\end{aligned}$$

The sum $\sum_i \mathbf{J}_{m_i}$ is zero as long as there are no chemical reactions, and $\mathbf{c}_0\mathbf{c}_0$ and $\overline{\mathbf{C}_i\mathbf{C}_i}$ must be recognized as a dyadic whose array

[11]Our vector and tensor notation in this section follows that of J. O. Hirschfelder, C. F. Curtiss, and R. B. Bird, *Molecular Theory of Gases and Liquids*, John Wiley and Sons Inc., New York, 1954; and of Chapman and Cowling, footnote 4.

is a second-order tensor. In fact, $\rho_i\overline{\mathbf{C}_i\mathbf{C}_i}$ is the stress tensor, p_{kl} [recall Eq. (2.29)], or the momentum flux, J_{mom}. Hence the second term reduces to

$$\sum_i \nabla\cdot[\rho_i\overline{(\mathbf{C}_i+\mathbf{c}_0)(\mathbf{C}_i+\mathbf{c}_0)}] = \nabla\cdot\left[\sum_i p_{kl_i} + \rho\mathbf{c}_0\mathbf{c}_0\right] \qquad (12.64)$$

The third and fourth terms vanish because Ψ_i (or $m_i\mathbf{c}_i$) is independent of $\mathbf{r}$ and t, and the fifth term becomes

$$\mathbf{F}_i\,\overline{\frac{\partial(m_i\mathbf{C}_i)}{\partial\mathbf{C}_i}} = m_i\mathbf{F}_i \qquad (12.65)$$

Equations (12.63), (12.64), and (12.65) can then be substituted back into Eq. (12.53). The result is

$$\sum_i\left[\frac{\partial(\rho_i\mathbf{c}_0)}{\partial t} + \nabla\cdot p_{kl_i} + \nabla\cdot(\rho\mathbf{c}_0\mathbf{c}_0) - n_im_i\mathbf{F}_i\right] = 0$$

or

$$\left(\rho\frac{\partial\mathbf{c}_0}{\partial t} + \mathbf{c}_0\frac{\partial\rho}{\partial t}\right) + \nabla\cdot p_{kl} + \rho(\mathbf{c}_0\cdot\nabla)\mathbf{c}_0 + \mathbf{c}_0(\nabla\cdot\rho\mathbf{c}_0) - \sum_i \rho_i\mathbf{F}_i = 0$$

Equation (12.58), the global continuity statement, can be used to eliminate two terms from this equation. The result is

$$\rho\frac{D\mathbf{c}_0}{Dt} = -\nabla\cdot p_{kl} + \sum_i \rho_i\mathbf{F}_i \qquad (12.66)$$

Equation (12.66) is the basis for the Navier–Stokes equation of motion of a viscous fluid, but the stress term has not yet been expressed in terms of rates of strain — of fluid velocities. We know from a strictly continuum derivation of the Navier–Stokes equation that

$$p_{kl} = \begin{cases} p - 2\mu\dfrac{\partial u_k}{\partial x_l} + \dfrac{2}{3}\mu\nabla\cdot\mathbf{c}_0 - \kappa\nabla\cdot\mathbf{c}_0 & k = l \\[2ex] -\mu\left(\dfrac{\partial u_k}{\partial x_l} + \dfrac{\partial u_l}{\partial x_k}\right) & k \neq l \end{cases} \qquad (12.67)$$

In the continuum derivation, the shear and bulk coefficients of viscosity, μ and κ, have to be obtained from physical measurements. The bulk coefficient characterizes the resistance of a fluid to a pure dilational motion — the motion of the fluid within a balloon when the exterior pressure is reduced, for example. It assumes importance only in cases of very rapid dilations, such as might occur in a shock wave, and it is often ignored.

In the microscopic derivation we must remember that the term ∇p_{kl} was obtained from Eq. (12.51), which included the integration of f_i. Once f_i has been obtained — not an easy task — this integration can be carried out. The result will include the transport coefficients explicitly and clarify such matters as the relative importance of κ.

CONSERVATION OF ENERGY

Finally, we write Eq. (12.53) in terms of the remaining collision invariant energy. In this case $\Psi_i = m_i\mathbf{C}_i^2/2$. The first term in Eq. (12.53) is then

$$\sum_i \frac{\partial}{\partial t}\left(n_i \overline{\frac{m_i}{2}\mathbf{C}_i^2}\right) = \frac{\partial}{\partial t}(\rho E) \tag{12.68}$$

where E is the kinetic energy per unit mass in the fluid. The second term is

$$\begin{aligned}\sum_i \nabla\cdot\left(n_i \overline{\frac{m_i}{2}\mathbf{C}_i^2\mathbf{c}_i}\right) &= \nabla\cdot\sum_i n_i \frac{m_i}{2}\overline{\mathbf{C}_i^2\mathbf{C}_i} + \nabla\cdot\sum_i n_i \frac{m_i}{2}\overline{\mathbf{C}_i^2\mathbf{c}_0} \\ &= \nabla\cdot\mathbf{J}_E + \nabla\cdot\rho E\mathbf{c}_0 \end{aligned} \tag{12.69}$$

where $\mathbf{J}_E$ is the flux of kinetic energy consistent with Eq. (2.26).

The third term vanishes because $\mathbf{C}_i$ is independent of t. The fourth term is a little more complicated:

$$\begin{aligned}\sum_i n_i\overline{\mathbf{c}_i\cdot\nabla\left(\frac{m_i}{2}\mathbf{C}_i^2\right)} &= \sum_i n_i m_i\overline{\mathbf{c}_i\cdot\mathbf{C}_i\cdot\nabla(\mathbf{c}_i-\mathbf{c}_0)} \\ &= \sum_i n_i m_i[\overline{\mathbf{c}_i\cdot(\mathbf{C}_i\cdot\nabla\mathbf{c}_i)} - \overline{\mathbf{c}_i\cdot(\mathbf{C}_i\cdot\nabla\mathbf{c}_0)}]\end{aligned}$$

The first term on the right-hand side of this equation vanishes because $\mathbf{c}_i$ is independent of $\mathbf{r}$, and we are left with

$$\sum_i n_i\overline{\mathbf{c}_i\cdot\nabla\left(\frac{m_i}{2}\mathbf{C}_i^2\right)} = -\sum_i n_i m_i\overline{\mathbf{c}_i\cdot(\mathbf{C}_i\cdot\nabla\mathbf{c}_0)} = -J_{\text{mom}} : \nabla\mathbf{c}_0 \tag{12.70}$$

The symbol : denotes a tensor product defined as follows in Einstein and summational notation, respectively:

$$J_{\text{mom}} : \nabla\mathbf{c}_0 = p_{kl}\frac{\partial u_k}{\partial x_l} = \sum_k\sum_l p_{kl}\frac{\partial u_k}{\partial x_l}$$

The fifth term is

$$\sum_i n_i\overline{\mathbf{F}_i\cdot\frac{\partial}{\partial\mathbf{c}_i}\left(\frac{m_i}{2}\mathbf{C}_i^2\right)} = \sum_i n_i m_i\mathbf{F}_i\cdot\overline{\mathbf{C}_i} = \sum_i \mathbf{F}_i\cdot\mathbf{J}_{mi} \tag{12.71}$$

Combining Eqs. (12.68), (12.69), (12.70), and (12.71) in Eq. (12.53), we obtain the energy equation,

$$\frac{\partial}{\partial t}(\rho E) + \nabla\cdot(\rho E\mathbf{c}_0) + \nabla\cdot\mathbf{J}_E + p_{kl} : \nabla\mathbf{c}_0 - \sum_i \mathbf{F}_i\cdot\mathbf{J}_{mi} = 0$$

The first two terms can be combined into a substantial derivative by addition of the global continuity equation. Then

$$\underbrace{\rho\frac{DE}{Dt}}_{\text{unsteady and convection term}} = \underbrace{-\boldsymbol{\nabla}\cdot\mathbf{J}_E}_{\text{conduction term}} - \underbrace{p_{kl}:\boldsymbol{\nabla}\mathbf{c}_0}_{\text{dissipation term}} + \underbrace{\sum_i \mathbf{F}_i\cdot\mathbf{J}_{m_i}}_{\text{work done by external forces}} \quad (12.72)$$

Under each of the terms in Eq. (12.72) is a description of its physical meaning. The left-hand term, written as a substantial derivative, shows the change of energy of an element of the gas as it moves. The right-hand side shows the three basic effects that give rise to this change. The first is heat conduction into or out of the element. The second is the result of viscous dissipation of mechanical flow energy and should always contribute positively to the right-hand side. The third is external work done upon the fluid.

If we accept the empirical relation $\mathbf{J}_E = -\lambda\boldsymbol{\nabla}T$, Fourier's law, for the moment, and if we take $E = c_v(T - T_{\text{ref}})$, then Eq. (12.72) for a single- component incompressible fluid becomes

$$\rho c_v\frac{DT}{Dt} = \lambda\nabla^2 T - \text{dissipation}$$

But for an incompressible fluid $c_v = c_p \equiv c$, so that if we ignore dissipation and introduce the thermal diffusivity[12] $a \equiv \lambda/\rho c$ then

$$\frac{1}{a}\frac{DT}{Dt} = \nabla^2 T \quad (12.73)$$

If the fluid is stationary, Eq. (12.73) further reduces to the equation for heat conduction in a solid:

$$\frac{1}{a}\frac{\partial T}{\partial t} = \nabla^2 T \quad (12.74)$$

Like the Navier–Stokes equation, this result can only be considered valid if solution of the Boltzmann integrodifferential equation gives a distribution function that yields the correct transport coefficients. It turns out that Fourier's law with constant thermal conductivity, or Newton's law of shear with a constant viscosity coefficient, are only valid results when deviations from equilibrium are comparatively small. In a few highly nonequilibrium situations, such as in flow through a shock wave, Boltzmann's equation shows these relations to be quite inaccurate. Equations (12.66) and (12.72) do not include these transport laws and are general, however.

[12]The thermal diffusivity is actually defined as $\lambda/\rho c_p$. The development of an explanation as to why this is true is left as an exercise (Problem 12.7).

12.5 ON SOLVING THE BOLTZMANN EQUATION

We have extracted a good deal of material from Boltzmann's equation thus far without really solving it — without evaluating its dependent variable f_i. By and large, the kind of solutions that must be undertaken to solve today's important problems in nonequilibrium gas dynamics are both sophisticated and laborious in the extreme. They are also approximate in all cases.

Our task in this section is to show the direction of thinking required to arrive at a solution. We set up some of the better known approximations and obtain results from them, but we only present a small beginning of a very large subject.

COLLISIONS AND THE COLLISION INTEGRAL

The collision term $(\partial f_i/\partial t)_{\text{coll}}$ represents the influence of collisions in altering the time rate of change of the distribution function of a fluid element. If collisions are frequent it is large, and as the frequency approaches zero it vanishes.

Enskog perceived, in 1917, that this fact could be the starting point for a successive approximation method for solving Boltzmann's equation.[13] We begin our considerations by looking at the influence of collisions on the collision integral in the way Enskog did.

The Boltzmann equation (12.16) can first be written in the form

$$\left(\frac{\partial f_i}{\partial t}\right)_{\text{coll}} = \sum_j J(f_i f_j) \qquad (12.75)$$

where we have summarized the lengthy integrals on the right-hand side with the notation $J(f_i f_j)$. We then consider a quantity ϵ which we wish to introduce as a perturbation parameter. When the number of collisions is large ϵ (as well as the period τ between collisions) is small. When the number of collisions is small, ϵ (and τ) become large.

As the number of collisions decreases to zero, the right-hand side of Eq. (12.75) disappears and an equilibrium situation is ultimately established. When the number is large, on the other hand, any nonequilibrium f_i will very rapidly "relax" to an equilibrium value $f_i^{(0)}$.

[13]Enskog's solution is discussed by J. O. Hirschfelder, C. F. Curtiss, and R. B. Bird, *Molecular Theory of Gases and Liquids*, John Wiley & Sons, Inc., New York, 1954, sec. 7-3-b; and by S. Chapman and T. G. Cowling, *The Mathematical Theory of Non-Uniform Gases*, Cambridge University Press, New York, 1952, chap. 7.

The preceding considerations suggest that we might scale time in the Boltzmann equation, such that it always will correspond with the same rate of collisions, by introducing $t' = \epsilon t$. Thus

$$\left(\frac{\partial f_i}{\partial t'}\right)_{\text{coll}} = \frac{1}{\epsilon}\sum_j J(f_i f_j) \tag{12.76}$$

Next we note that the distribution function f_i should depend upon ϵ which, as we can now see, denotes the extent of the deviation from equilibrium. Thus

$$f_i(c_i, r, t', \epsilon) = f_i(c_i, r, t', 0) + \left.\frac{\partial f_i}{\partial \epsilon}\right|_{\epsilon=0} \epsilon + \left.\frac{\partial^2 f_i}{\partial \epsilon^2}\right|_{\epsilon=0} \frac{\epsilon^2}{2!} + \cdots \tag{12.77}$$

or

$$f_i = f_i^{(0)} + \epsilon f_i^{(1)} + \epsilon^2 f_i^{(2)} + \cdots \tag{12.77a}$$

where $f_i^{(0)}$ should turn out to be the equilibrium, or Maxwell, distribution function, $f_i^{(1)}$, is $(\partial f_i/\partial \epsilon)_{\epsilon=0}$, and so on.

We defer a consideration of Enskog's successive approximation method until we have first looked at a simpler approximate solution suggested by Eqs. (12.76) and (12.77a).

SIMPLE LINEAR APPROXIMATION TO THE BOLTZMANN EQUATION

Equation (12.77) can be recast in the form

$$\left[\left(\frac{\partial f_i}{\partial \epsilon}\right)_{\text{coll}}\right]_{\epsilon=0} = \frac{f_i - f_i^{(0)}}{\epsilon} \tag{12.77b}$$

for situations in which deviations from equilibrium are small. But ϵ is a quantity that increases with the period τ between collisions. We therefore make an assumption that takes a roughly similar form,

$$\left(\frac{\partial f_i}{\partial t}\right)_{\text{coll}} = -\frac{f_i - f_i^{(0)}}{\tau} \tag{12.78}$$

The right-hand side is now negative because, while $f_i - f_i^{(0)}$ increases with ϵ, it must decrease with time to satisfy the H theorem. This says that the total rate of change of the distribution function will be proportional to the deviation from equilibrium at any instant and that τ is the appropriate time constant for the change.

Equation (12.78) is then the approximate form of Boltzmann's equation that we wish to solve. The right-hand side replaces the complicated collision term, and indeed it does properly characterize the influence of collisions.

By way of testing Eq. (12.78), let us consider a steady shear flow in a single-component, almost-Maxwellian gas without body forces. We suppose that there is a gross velocity distribution $u_0 = u_0(y)$.

The deviation from equilibrium is small because the right-hand side of the equation contributes little and the degree of approximation is slight. Let us seek to compute the viscosity for this case.

The full form of Eq. (12.78) for this case reduces to

$$\tau V \frac{\partial f}{\partial y} = f - f^{(0)} \tag{12.79}$$

where we are using the velocity notation that was set up in the context of Eq. (2.23). We can approximate $(\partial f/\partial y)$ by $(\partial f^{(0)}/\partial u)(\partial u/\partial y)$, where $u = U + u_0(y)$. But $f^{(0)}$ depends only on U, V, W—not on y or u_0—and u depends on y only through u_0. Thus $\partial f/\partial y \simeq (\partial f^{(0)}/\partial U) \times (du_0/dy)$. Furthermore, v is equal to V in this case, so Eq. (12.79) takes the form

$$f = f^{(0)} + \tau V \frac{du_0}{dy} \frac{\partial f^{(0)}}{\partial U} \tag{12.80}$$

where the equilibrium distribution $f^{(0)}$ will be the Maxwell velocity distribution,

$$f^{(0)} = n\left(\frac{m}{2\pi kT}\right)^{3/2} \exp\left[-\frac{m(U^2 + V^2 + W^2)}{2kT}\right] \tag{2.43}$$

We can now write the shear stress p_{yx} in accordance with Eqs. (2.28) and (1.51) as

$$\mu \frac{du_0}{dy} = -m \int VUf \, d\Omega \tag{12.81}$$

The first term on the right-hand side of equation (12.80), $f^{(0)}$, does not contribute to the shear stress, and the velocity gradient can be divided out. Equation (12.81) then simplifies to

$$\mu = -m \int \tau U V^2 \frac{\partial f^{(0)}}{\partial U} \, d\Omega \tag{12.82}$$

Now suppose we approximate $\tau(\mathbf{c})$ with a suitable mean period obtained from the Maxwell distribution [recall Eq. (12.22)],

$$\tau \simeq \frac{l}{\bar{C}} \tag{12.83}$$

When we factor this out of Eq. (12.82) and complete the indicated integrations we obtain [Problem (12.10)]

$$\mu \simeq mn \frac{\overline{V^2}}{\bar{C}} l$$

but $\bar{C}^2 \simeq (8/\pi)\overline{V^2}$, the Maxwell distribution value, so

$$\mu \simeq \frac{\pi}{8} \rho l \bar{C} \tag{12.84}$$

Equation (12.84) reveals exactly the same dependence of μ upon physical parameters that we obtained in Eq. (11.7) as a consequence of mean-free-path arguments. The constant $\pi/8$ would correspond with a value of $\alpha = 0.785$, which is a little low if we consider the persistence of velocities.

This simple result emphasizes the following point: The terrible complications of the Boltzmann equation can be fairly accurately simplified when deviations from equilibrium are not great. Fortunately, many physical processes in gases deviate only a little from equilibrium because the gross fluid motions are much slower than molecular motions. In the subsequent subsection we see how Enskog treated small deviations from equilibrium in a more formal way. The result is a solution that can be applied very broadly to problems of practical interest.

ENSKOG'S SUCCESSIVE-APPROXIMATION METHOD

We begin by substituting the series expansion for f_i, Eq. (12.77a), into Eq. (12.76). The result will be an equation in terms containing the factors $\epsilon^0, \epsilon^1, \epsilon^2, \epsilon^3, \ldots$. Collecting coefficients of like powers of ϵ, we obtain for ϵ^0,

$$0 = \sum_j J(f_i^{(0)} f_j^{(0)}) \tag{12.85}$$

for ϵ^1,

$$\left(\frac{\partial f_i^{(0)}}{\partial t'}\right)_{\text{coll}} = \sum_j [J(f_i^{(0)} f_j^{(1)}) + J(f_i^{(1)} f_j^{(0)})] \tag{12.86}$$

for ϵ^2,

$$\left(\frac{\partial f_i^{(1)}}{\partial t'}\right)_{\text{coll}} = \sum_j [J(f_i^{(0)} f_j^{(2)}) + J(f_i^{(1)} f_j^{(1)}) + J(f_i^{(2)} f_j^{(0)})] \tag{12.87}$$

and so forth (see Problem 12.11).

Equation (12.85) must be satisfied for all ϵ, no matter how small. That is because it is the equilibrium condition and its solution $f_i^{(0)}$ as we saw in Sec. 12.3, is the Maxwell distribution. For larger departures from equilibrium — for larger ϵ — the terms in ϵ must be retained and Eq. (12.86) must be solved for $f_i^{(1)}$. And $f_i^{(1)}$ will correct f_i for small deviations from equilibrium. It is then possible — at least in principle — to return to Eq. (12.87) with $f_i^{(0)}$ and $f_i^{(1)}$ and calculate the next correction, $f_i^{(2)}$. Additional equations would correct f_i further, adding one term to the series (12.77a) each time.

The solution to Eq. (12.85) has already been obtained in Sec. 12.3. The result for $f_i^{(0)}$ is the Maxwell distribution in form; but, because of the nonequilibrium constraints that it must satisfy, it actually differs slightly from the Maxwell distribution. The *full* distribution function f_i for nonequilibrium must satisfy three constraints: the conservation of particles,

$$\int_{\Omega_i} f_i \, d\Omega_i = n_i(\mathbf{r}, t) \tag{12.88}$$

the conservation of momentum,

$$\sum_i m_i \int_{\Omega_i} \mathbf{c}_i f_i \, d\Omega_i = \rho(\mathbf{r}, t)\mathbf{c}_0(\mathbf{r}, t) \tag{12.89}$$

and the conservation of thermal energy,

$$\tfrac{1}{2} \sum_i m_i \int_{\Omega_i} \mathbf{C}_i^2 f_i \, d\Omega_i = \tfrac{3}{2} nkT(\mathbf{r}, t) \tag{12.90}$$

[Recall Eqs. (12.40), (12.41), and (12.42).]

But $f_i^{(0)}$ *alone* satisfies each of these. Thus we require that

$$\int_{\Omega_i} f_i^{(r)} \, d\Omega_i = 0 \qquad r \neq 0 \tag{12.91}$$

$$\sum_i m_i \int_{\Omega_i} \mathbf{c}_i f_i^{(r)} \, d\Omega_i = 0 \qquad r \neq 0 \tag{12.92}$$

and

$$\tfrac{1}{2} \sum_i m_i \int_{\Omega_i} \mathbf{C}_i^2 f_i^{(r)} \, d\Omega_i = 0 \qquad r \neq 0 \tag{12.93}$$

It follows that the $\mathbf{r}$ and t dependence of n, ρ, $\mathbf{c}_0$, and T must be carried into $f_i^{(0)}$. Thus the equilibrium solution should actually be

$$f_i^{(0)} = n_i(\mathbf{r}, t)\left[\frac{m_i}{2\pi kT(\mathbf{r}, t)}\right]^{3/2} \exp\left[-\frac{m_i \mathbf{C}_i^2}{2kT(\mathbf{r}, t)}\right] \tag{12.94}$$

where $\mathbf{C}_i = \mathbf{c}_i - \mathbf{c}_0(\mathbf{r}, t)$.

Equation (12.94) is thus a space- and time-dependent variation on the Maxwell distribution. It shows us that when deviations from equilibrium are slight (or when gradients and time derivatives of n_i, $\mathbf{c}_0$, and T are small), the velocity distribution is *locally* Maxwellian.

It is this feature of ideal-gas behavior that allowed Maxwell to develop a barometric formula by considering the atmosphere to be isothermal and locally Maxwellian. We have already suggested that this is possible in Examples 3.2 and 12.1 and Problem 5.13. Problems 12.8 and 12.9 outline a more satisfactory derivation in the light of Eq. (12.94).

The second approximation to f_i is obtained by writing

$$f_i^{(1)} = f_i^{(0)}(\mathbf{r}, \mathbf{c}_i, t)\phi_i(\mathbf{r}, \mathbf{c}_i, t) \tag{12.95}$$

so

$$f_i \simeq f_i^{(0)}(1 + \epsilon\phi_i) \tag{12.96}$$

Substitution of this result in Eq. (12.86) gives[14] (Problem 12.15)

$$\frac{\partial f_i^{(0)}}{\partial t'} + \mathbf{c}_i \cdot \nabla f_i^{(0)} + \mathbf{F}_i \cdot \frac{\partial f_i^{(0)}}{\partial \mathbf{c}_i}$$
$$= \sum_j \iiint f_i^{(0)} f_j^{(0)} (\phi'_i - \phi'_j - \phi_i - \phi_j) g b \, db \, d\epsilon \, d\Omega_j \quad (12.97)$$

where not only is t' scaled so that $t' = \epsilon t$, but so, too, are the time dimensions in $\mathbf{c}_i$ and $\mathbf{F}_i$. Equation (12.97), of course, can be rewritten as

$$\left(\frac{\partial f_i^{(0)}}{\partial t}\right)_{\text{coll}} = \sum_j \iiint f_i^{(0)} f_j^{(0)} (\epsilon\phi'_i + \epsilon\phi'_j - \epsilon\phi_i - \epsilon\phi_j) g b \, db \, d\epsilon \, d\Omega_j$$
(12.97a)

Our objective is now to solve Eq. (12.97a) for $\epsilon\phi_i$ subject to the constraints (12.91), (12.92), and (12.93).

Equation (12.97) is an integral equation in ϕ_i, and solving it is a complicated business. The method outlined by Hirschfelder, Curtiss, and Bird[15] involves mathematical tools that are not at the immediate disposal of many technical people. We must therefore be content merely to write down the results of the computation, and to point out some qualitative ramifications of it. The solution for $\epsilon\phi_i$ takes the form

$$\epsilon\phi_i = -\mathcal{A}_i \cdot \nabla \ln T - \mathcal{B}_i : \nabla \mathbf{c}_0 + n \sum_j \vec{\mathcal{C}}_i^{\,j} \cdot \mathbf{d}_i \quad (12.98)$$

where

$$\mathbf{d}_i \equiv \nabla \left(\frac{n_i}{n}\right) + \left(\frac{n_i}{n} - \frac{\rho_i}{\rho}\right) \nabla \ln p - \frac{\rho_i}{\rho p} \left(\frac{\rho}{m_i} \mathbf{F}_i - \sum_j n_j \mathbf{F}_j\right) \quad (12.99)$$

and $\mathcal{A}_i$, $\mathcal{B}_i$, and $\vec{\mathcal{C}}_i^{\,j}$ are complicated functions of n_i, T, and $\mathbf{c}_0$.

TRANSPORT OF MASS, MOMENTUM, AND ENERGY

The mass flux, $\mathbf{J}_{m_i}$ or $\rho_i \overline{\mathbf{C}}_i$, is calculated as follows:

$$\rho_i \overline{\mathbf{C}}_i = \int m_i \mathbf{C}_i f_i \, d\Omega_i \simeq m_i \int \mathbf{C}_i f_i^{(0)} (1 + \phi_i) \, d\Omega_i$$

The resultant diffusion velocity then takes the form

$$\overline{\mathbf{C}}_i = \frac{n^2}{\rho n_i} \sum_j m_j D_{ij} \mathbf{d}_j - \frac{1}{\rho_i} D_i^T \nabla \ln T \quad (12.100)$$

[14]The symbol ϵ is serving here to designate both the angular coordinate in a molecular impact and the perturbation parameter. The meaning is clear in context, so we carry the double notation.
[15]See footnote 13.

where D_{ij} and D_i^T are diffusion coefficients that depend upon n_i, T, and $\mathbf{c}_0$. If we disallow any gradients of pressure or temperature, then this result reduces to Fick's law and D_{ij} is the same diffusion coefficient that we discussed in chapter 11.

By the same token, the energy flux is

$$\mathbf{q} = -\lambda' \nabla T + \tfrac{5}{2} kT \sum_j n_j \overline{\mathbf{C}}_j - nkT \sum_j \frac{1}{\rho_j} D_j^T \, \mathbf{d}_j \qquad (12.101)$$

where

$$\lambda' = \lambda + \text{function of } D_i^T\text{'s and } D_{ij}\text{'s} \qquad (12.102)$$

The added function in Eq. (12.102) is small, but it points to a curious interlinking between mass and heat diffusion. Let us consider this further.

If we impose a temperature gradient on an isobaric, homogeneous mixture of gases, there will be a flow of *heat* [in accordance with Eq. (12.101) or in accordance with common sense]. But Eq. (12.100) reveals that there will also be a flow of mass given by $\rho_i \overline{\mathbf{C}}_i = -D_i^T \nabla \ln T$, where D_i^T is the so-called coefficient of thermal diffusion. Of course, once mass diffusion begins, the flow of heat is altered by the second term on the right-hand side of Eq. (12.102) and by a concentration-gradient effect that enters through $\mathbf{d}_i$.

The same kind of complications would result if we imposed a concentration gradient on an initially isothermal mixture. We would once again find that $\mathbf{q}$ and $\rho_i \overline{\mathbf{C}}_i$ were necessarily interrelated in a mixture. This coupling of fluxes was predicted analytically before it was identified experimentally. More recently it has become the major subject of study in the comparatively new subject of irreversible thermodynamics.

Finally, it is possible to write the flux of momentum — the stress tensor p_{kl} at a point — as

$$\begin{aligned} p_{kl} &= -\sum_i m_j \int \mathbf{C}_j \mathbf{C}_j f_j \, d\Omega_j \\ &= p\delta_{kl} - \mu\left(\frac{\partial c_{0l}}{\partial x_k} + \frac{\partial c_{0k}}{\partial x_l}\right) + \tfrac{2}{3}\mu \nabla \mathbf{c}_0 \delta_{kl} \end{aligned} \qquad (12.103)$$

where δ_{kl} is the Kronecker delta and μ is another coefficient, which we recognize as the viscosity.

Equation (12.103) is very nearly Equation (12.67). It differs only in that κ, the bulk coefficient of viscosity, is absent. Thus the second approximation to f_i has yielded a form of the Navier–Stokes stress tensor that is known to give good results in all but the most extreme deviations from equilibrium. If the bulk coefficient of viscosity, κ, is to exert any influence it can only do so in a gas that suffers large deviations from equilibrium.

Subsequent analytical work has shed more light on this and other matters related to very strong deviations from equilibrium. Much of this work has been directed toward calculating $f_i^{(2)}$. More fruitful than this have been a variety of entirely different techniques for approximating and solving the Boltzmann equation. These matters are well beyond the scope of this introduction, however.

Problems 12.1 Derive Eqs. (12.7), (12.8), and (12.9).

12.2 Verify Eq. (12.12).

12.3 Verify Eq. (12.19).

12.4 Verify Eqs. (12.43), (2.34), and (12.44).

12.5 Prove that the terms in the summation (12.49) vanish identically when $\Psi_i = m_i$, and that Eq. (12.59) follows.

12.6 Verify Eq. (12.60).

12.7 The thermal diffusivity a is defined as $\lambda/\rho c_p$ for compressible flow—not as $\lambda/\rho c_v$, as the text above Eq. (12.73) might imply. To explain why this is true, first add Dp/Dt to both sides of Eq. (12.72) and simplify the equation. Then reduce the result to an equivalent form of Eq. (12.73). Discuss.

12.8 Consider an isothermal single component Maxwellian gas at rest in a force field [i.e., $\mathbf{F} = \mathbf{F}(\mathbf{r})$]. In this case, the distribution function f will be unchanged from that obtained for $\mathbf{F} = 0$, except that $n = n(\mathbf{r})$. Write the appropriate form of the Boltzmann integro-differential equation for this case and insert the space-dependent $f^{(0)}$ in it. Solve this result for $n(\mathbf{r})$ and obtain from it

$$f = n_0\left(\frac{m}{2\pi kT}\right)^{3/2} \exp\left[-\left(\frac{mC^2}{2kT} + \frac{mU}{kT}\right)\right]$$

in which $U(\mathbf{r})$ is the potential of the force field defined by $\mathbf{F} \equiv -\nabla U$ and $n(\mathbf{r}) = n_0$, for $U \to 0$.

12.9 Obtain distribution functions and the density $\rho(\mathbf{r})$ using the result of Problem 12.8 (cf. Examples 3.2 and 12.1 and Problem 5.13) (a) for the earth's isothermal atmosphere, and (b) for a Maxwellian gas in an isothermal centrifuge with angular velocity, ω. (Note that f will depend upon both C and $C_t \equiv \omega r$.)

12.10 Complete all the missing steps in the derivation leading to Eq. (12.84).

12.11 Obtain Eqs. (12.85), (12.86), and (12.87) and one more equation for the coefficients of ϵ^3.

12.12 A flat disk of unit area is placed in a dilute gas at rest with initial temperature T. Face A of the disk is at temperature T and face B is at temperature $T_1 > T$. Molecules striking face A reflect elastically. Molecules striking face B are absorbed by the disk only to reemerge from the same face with a Maxwellian distribution of temperature T_1.

(a) Assume that the mean free path in the gas is much smaller than the dimension of the disk. Present an argument to show that after a few collision times the gas can be described by the fluid mechanical equations, with face *B* replaced by a boundary condition for the temperature.

(b) Write down the first-order fluid mechanical equations for (a), neglecting the flow of the gas. Show that there is no net force acting on the disk.

(c) Assume that the mean free path is much larger than the dimensions of the disk, and find the net force acting on the disk.

12.13 Show that the Boltzmann equation for a gas of relatively low molecular mass interacting with a gas of very heavy molecules may be obtained as

$$\left(\frac{\partial f}{\partial t}\right)_{\text{coll}} = n_H \int (f' - f) g_{ij} b \, db \, d\epsilon$$

where n_H is the number density of heavy molecules. Indicate all assumptions made.

12.14 A dilute gas, infinite in extent and composed of charged molecules, each of charge e and mass m, comes to equilibrium in an infinite lattice of fixed ions. In the absence of an external electric field the equilibrium distribution function is

$$f^{(0)}(\mathbf{c}) = n\left(\frac{2\pi kT}{m}\right)^{3/2} \exp\left(-\frac{m\mathbf{c}^2}{kT}\right)$$

where n and T are constants. A weak uniform electric field $\mathbf{E}$ is then turned on, leading to a new equilibrium distribution function. Assume that Eq. (12.79) adequately takes into account the effect of collisions among molecules and between molecules and lattice. Calculate (a) the new equilibrium distribution function f to the first order, and (b) the electric conductivity σ defined by the relation $ne\overline{\mathbf{c}} = \sigma\mathbf{E}$.

12.15 Verify Eq. (12.97a).

12.16 One of the first things that students of viscous flow learn is that symmetry of the stress tensor requires any shear stress in the horizontal plane to be balanced by a shear in the vertical plane. Thus the shear τ_{yx} in a simple Couette or Poiseuille flow must be balanced by a τ_{xy}. Explain this balance using an argument based upon the microscopic behavior of an ideal gas. Derive a quantitative expression for τ_{xy} and compare it with the microscopic expression for τ_{yx}.

appendix A Stirling's approximation

The numerical evaluation of $n!$ is a seriously complicated task when n becomes large. The development of a convenient approximation for $n!$ is an absolute necessity when n approaches numbers on the order of Avogadro's number.

The gamma function $\Gamma(n+1)$ provides a closed-form expression that helps us to make such an approximation:

$$\Gamma(n+1) \equiv \int_0^\infty t^n e^{-t}\, dt \tag{A.1}$$

Equation (A.1) can be integrated by parts using

$$d(t^n e^{-t}) = e^{-t}(nt^{n-1}\, dt) - t^n(e^{-t}\, dt)$$

so that

$$\Gamma(n+1) = -(t^n e^{-t})_0^\infty + n\int_0^\infty t^{n-1} e^{-t}\, dt$$

The first term on the right-hand side vanishes and the integral is $\Gamma(n)$. Thus

$$\Gamma(n+1) = n\Gamma(n)$$

It follows by induction that

$$n! = \Gamma(n+1) \tag{A.2}$$

The gamma function therefore provides the means by which we can write $n!$ as an integral. Under the change of variable $x = t/n$, Eqs. (A.1) and (A.2) give

$$n! = n^{n+1} \int_0^\infty x^n e^{-nx}\, dx \tag{A.3}$$

The integral in Eq. (A.3) can be simplified in the following approximate way. First divide the integral

$$n^{n+1} \int_0^\infty e^{-nx} x^n\, dx = n^{n+1} e^{-n} \left[\int_0^2 e^{n(1-x+\ln x)}\, dx + \int_2^\infty e^{-n(x-1-\ln x)}\, dx \right] \tag{A.4}$$

and note that, since $(x - 1 - \ln x) \geq 0.306$, the second integral on the right-hand side will be negligible for large n. Furthermore, employ the change of variable, $x = 1 + \epsilon$, where $-1 \leq \epsilon \leq 1$, in the first integral on the right-hand side. Under this transformation, and the series expansion

$$\ln(1+\epsilon) = \epsilon - \frac{\epsilon^2}{2} + \frac{\epsilon^3}{3} - \cdots$$

Eq. (A.4) permits us to write Eq. (A.3) as

$$n! \sim n^{n+1} e^{-n} \int_{-1}^{1} e^{(n-\epsilon^2/2+\epsilon^3/3-\cdots)}\, d\epsilon$$

Once more we observe that, for large n, $\epsilon^2/2$ contributes far more than any subsequent term. Accordingly,

$$n! \sim n^{n+1} e^{-n} \int_{-1}^{1} e^{-\epsilon^2 n/2}\, d\epsilon \simeq n^{n+1} e^{-n} \sqrt{\frac{2}{n}} \int_{-\infty}^{\infty} e^{-\epsilon^2 n/2}\, d\left(\epsilon \sqrt{\frac{n}{2}}\right)$$

or

$$n! \simeq n^{n+1} e^{-n} \sqrt{\frac{2\pi}{n}} = \sqrt{2\pi n} \left(\frac{n}{e}\right)^n \tag{A.5}$$

or

$$\ln n! \simeq (n + \tfrac{1}{2}) \ln n - n + \tfrac{1}{2} \ln(2\pi) \tag{A.5a}$$

Equation (A.5) is the lead term in Stirling's series.[1] The next

[1]See, for example, E. T. Whittaker and G. N. Watson, *A Course of Modern Analysis*, 4th ed., Cambridge University Press, New York, 1963, sec. 12.33.

term is on the order of $(1/12n)$ times the first term and is thus completely negligible for large n. Indeed, Eq. (A.5) can be even further simplified for large n,

$$\ln n! \simeq n \ln n - n \tag{A.6}$$

Equation (A.6) is the expression we usually designate as *Stirling's approximation* and which we have most occasion to use subsequently. Figure A.1 shows the percentage error of the approximation for insufficiently large n. It is interesting to note that, although the percentage error drops, the absolute error $\ln n! - (n \ln n - n)$, increases monotonically with n. The error in Eq. (A.5a) is, of course, far less than this.

Problem A.1 Plot, against n, the error of Eq. (A.5a) and the error of $\ln n! \simeq n \ln n$, as well as the error of Eq. (A.6). Discuss the results.

Fig. A.1 Percent error in Stirling's approximation, for low n.

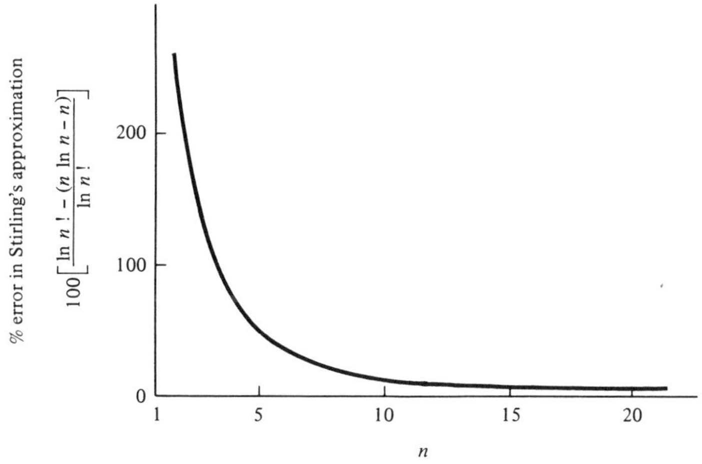

appendix B Lagrange's method of undetermined multipliers

Suppose we wish to maximize or minimize a function of several variables, $P(X_1, X_2, \ldots, X_n)$. P might be the power output of an engine, the cost of manufacturing a device, or perhaps the amount of material required in a product. Let us also suppose that P is a continuous function and that it has derivatives. It is necessary that the derivative of P with respect to any of its independent variables be zero when it is maximum. Thus

$$dP = 0 = \sum_{i=1}^{n} \frac{\partial P}{\partial X_i} dX_i \tag{B.1}$$

In general, the function $P(X_1, X_2, \ldots, X_n)$ relates to some physical device or system. The variables, $X_1, X_2, \ldots, X_n$, characterize the device (that is, its length, width, area, thickness, etc.) and are not all independent. There will usually exist certain equations, called *constraints*, which relate some or all of these variables to one another. The constraints can be written in the form

$$\begin{aligned} \psi_1(X_1, X_2, \ldots, X_n) &= 0 \\ \psi_2(X_1, X_2, \ldots, X_n) &= 0 \\ &\vdots \\ \psi_m(X_1, X_2, \ldots, X_n) &= 0 \end{aligned} \tag{B.2}$$

If m represents the number of equations of constraint and n represents the number of independent variables (the X_i's), then the number of "degrees of freedom" is $n - m$. If $n = m$, there is no freedom; all the variables are determined and P can have but one value. When $m < n$, the freedom can be used to select some of the variables in such a way as to maximize or minimize P.

Only $n - m$ of the dX_i's in Eq. (B.1) can be varied arbitrarily; the remaining m of them must be varied in such a way as to satisfy the differential form of the equations of constraint (B.2),

$$d\psi_j = 0 = \sum_{i=1}^{n} \frac{\partial \psi_j}{\partial X_i} dX_i \qquad j = 1, 2, \ldots, m \tag{B.3}$$

Rather than try to decide which m of these dX_i's should be dependent and which $n - m$ of them should be independent, we shall use a method invented by Lagrange for this kind of problem. Lagrange's method consists of introducing arbitrary functions, $\lambda_1, \lambda_2, \ldots, \lambda_m$, as follows:

We multiply $d\psi_1$ by λ_1, $d\psi_2$ by λ_2, and so on, so that

$$\sum_{i=1}^{n} \frac{\partial P}{\partial X_i} dX_i = 0 \tag{B.1}$$

$$\begin{aligned} \lambda_1 \sum_{i=1}^{n} \frac{\partial \psi_1}{\partial X_i} dX_i &= 0 \\ &\;\vdots \\ \lambda_m \sum_{i=1}^{n} \frac{\partial \psi_m}{\partial X_i} dX_i &= 0 \end{aligned} \tag{B.4}$$

The λ's, which we call *Lagrangian multipliers*, are arbitrary functions of the X_i's, which can be assigned as we please. Adding Eqs. (B.1) and (B.4) gives

$$\sum_{i=1}^{n} \left(\frac{\partial P}{\partial X_i} + \lambda_1 \frac{\partial \psi_1}{\partial X_i} + \cdots + \lambda_m \frac{\partial \psi_m}{\partial X_i} \right) dX_i = 0 \tag{B.5}$$

The multipliers, $\lambda_1, \lambda_2, \ldots, \lambda_m$, have to be chosen so that the parentheses will be zero for every dX_i. We therefore obtain n equations of the form

$$\frac{\partial P}{\partial X_i} + \lambda_1 \frac{\partial \psi_1}{\partial X_i} + \cdots + \lambda_m \frac{\partial \psi_m}{\partial X_i} = 0 \qquad i = 1, 2, \ldots, n \tag{B.6}$$

Since we have n equations involving the m unknown λ's, we can solve for these λ's using m equations and substitute the λ's in the $n - m$ remaining equations. These $n - m$ equations plus the original m equations give us n equations, which will be sufficient to determine the n values of X_i which define P_{max}.

Lagrange's method of undetermined multipliers may be summarized as follows:

1. Write down the differential of the function, and equate it to zero.
2. Take the differential of each equation of constraint, and multiply by as many different Lagrangian multipliers as there are equations of constraint.
3. Add all the equations, factoring the sum so that each differential appears only once.
4. Equate the coefficient of each differential to zero.
5. Solve for the m values of λ_i.
6. Substitute these λ_i's in the n differential coefficients and solve for the n maximizing values of X_i.

EXAMPLE B.1 Maximize the function, $P(x, y, z) = 8xyz$, subject to the constraint $\psi_1(x, y, z) = x^2 + y^2 + z^2 - 1 = 0$. ($n = 3$ and $m = 1$.)

Step 1: $dP = 8yz\,dx + 8xz\,dy + 8xy\,dz = 0$

Step 2: $\lambda_1\,d\psi_1 = 2\lambda_1 x\,dx + 2\lambda_1 y\,dy + 2\lambda_1 z\,dz = 0$

Step 3: $(8yz + 2\lambda_1 x)\,dx + (8xz + 2\lambda_1 y)\,dy + (8xy + 2\lambda_1 z)\,dz = 0$

Step 4: $8yz + 2\lambda_1 x = 0$; $8xz + 2\lambda_1 y = 0$; $8xy + 2\lambda_1 z = 0$

Step 5: $\lambda_1 = -4yz/x = -4xz/y = -4xy/z$

Step 6: $8yz - 8x^2z/y = 0$; therefore, $y = x$

$8xz - 8xy^2/z = 0$; therefore, $z = y$

$8xy - 8yz^2/x = 0$; therefore, $x = z$

Thus

$$-\frac{\lambda_1}{4} = x = y = z = \frac{1}{\sqrt{3}}$$

Thus we have shown that $8xyz$ is maximal, subject to the requirement $x^2 + y^2 + z^2 = 1$, when $x = y = z = 1/\sqrt{3}$.
Finally,

$$P_{\max} = \frac{8}{3\sqrt{3}}$$

Problems B.1 Metal cans of volume $V = V_0$ are made by welding a top and bottom on a cylinder of length L and diameter D. The cylinder has one vertical seam. Metal costs C_m cents/ft^2 and weld costs C_w cents/ft. (a) Find the L/D that gives minimum area, and (b) find the relation between L and D for minimum cost.

B.2 Sketch a geometrical interpretation of the example problem. Discuss your interpretation fully.

appendix C Sturm-Liouville system[1]

The Sturm–Liouville system includes all linear, second-order, ordinary differential equations with two homogeneous boundary conditions. A great deal can be said about the solutions to this system without actually solving it. The system is

$$\frac{d}{dx}\left[r(x)\frac{dX}{dx}\right] + [q(x) + \lambda p(x)]X = 0 \tag{C.1}$$

with boundary conditions

$$c_1X(a) + c_2\left.\frac{dX}{dx}\right|_{x=a} = 0 \tag{C.2}$$

$$c_3X(b) + c_4\left.\frac{dX}{dx}\right|_{x=b} = 0$$

Before stating the theorems governing the solutions $X = X(x)$ of this system we must develop the concept of orthogonality of functions.

[1]See, for example, R. V. Churchill, *Fourier Series and Boundary Value Problems*, McGraw-Hill, Inc., New York, 1941, chap. III. See also G. Birkhoff and G. C. Rota, *Ordinary Differential Equations*, Ginn & Company, Boston, 1962, chap. X.

ORTHOGONAL FUNCTIONS

Two vectors, $\mathbf{A} = (a_1, a_2, \ldots, a_i, \ldots, a_n)$ and $\mathbf{B} = (b_1, b_2, \ldots, b_i, \ldots, b_n)$, are orthogonal if their dot product vanishes:

$$\mathbf{A} \cdot \mathbf{B} = \sum_{i=1}^{n} a_i b_i = 0$$

Now let us consider a set of mutually orthogonal vectors: $\mathbf{A}_1, \mathbf{A}_2, \ldots, \mathbf{A}_n, \ldots$. We *normalize* the set, where normalization means reduction of the set to a new set of unit vectors, $\mathbf{\Phi}_n$:

$$\mathbf{\Phi}_n = \frac{\mathbf{A}_n}{|\mathbf{A}_n|}$$

The set, $\mathbf{\Phi}_1, \mathbf{\Phi}_2, \ldots, \mathbf{\Phi}_n, \ldots$, is called an *orthonormal* set. It has the property that

$$\mathbf{\Phi}_n \cdot \mathbf{\Phi}_m = \sum_i \phi_{i_n} \phi_{i_m} = \delta_{nm} \tag{C.3}$$

where the ϕ's are the components and δ_{nm} is the Kronecker delta. A typical example of an orthonormal set is given by the three Cartesian unit vectors, **i**, **j**, **k**.

A vector **f** in the multidimensional vector space described by $\mathbf{\Phi}_1, \mathbf{\Phi}_2, \ldots$, can always be expressed as a linear combination of these unit vectors:

$$\mathbf{f} = C_1 \mathbf{\Phi}_1 + C_2 \mathbf{\Phi}_2 + \cdots + C_n \mathbf{\Phi}_n + \cdots \tag{C.4}$$

The Cartesian position vector, $\mathbf{r} = \mathbf{i}x + \mathbf{j}y + \mathbf{k}z$, is an example of a vector expressed as linear combination of a set of orthonormal vectors. It follows from Eq. (C.4) that we can write

$$\mathbf{f} \cdot \mathbf{\Phi}_n = C_n \tag{C.5}$$

or

$$\mathbf{f} = \sum_n (\mathbf{f} \cdot \mathbf{\Phi}_n) \mathbf{\Phi}_n \tag{C.6}$$

Equation (C.6) is called an expansion of **f** in the $\mathbf{\Phi}_n$.

Now let us suppose that each of a set of *functions* $\Phi_n(x)$ is considered to be a *vector* in the infinite space that is formed by regarding the value of $\Phi_n(x)$ at *each point* in the interval $[a, b]$ as being a component. Then if for any two such functions

$$\int_a^b \Phi_n(x) \Phi_m(x)\, dx = \delta_{nm} \tag{C.7}$$

the functions $\Phi_n(x)$ are called orthonormal by analogy with Eq. (C.3). The integral in Eq. (C.7) is, of course, the appropriate extension of the summation, $\sum_i \phi_{i_n} \phi_{i_m}$, in Eq. (C.3).

It follows that we should be able to express any well-behaved function $f(x)$ — that is to say, some other "vector" in the same infinite space — as a linear function of the "vectors" in the orthonormal set. Thus

$$f(x) = C_1\Phi_1(x) + C_2\Phi_2(x) + \cdots + C_n\Phi_n(x) + \cdots$$

Then the appropriate extensions of Eqs. (C.5) and (C.6) are

$$C_n = \int_a^b f(x)\Phi_n(x)\,dx$$

and

$$f(x) = \sum_n \Phi_n(x) \int_a^b f(\xi)\Phi_n(\xi)\,d\xi \tag{C.8}$$

The restrictions upon the expansion (C.8) are that the function $f(x)$ be "well behaved" and that the functions $\Phi_n(x)$ be "complete." A treatment of the attributes "well behaved" and "complete" is far beyond the present discussion. "Well behaved" includes such features as piecewise continuity and finite maxima and minima. The requirement of "completeness" can be illustrated by suggesting that the Cartesian position vector be expressed in terms of **i** and **j**, without using **k**. This would be impossible because the set of orthonormal vectors is incomplete. Likewise, if no function is orthonormal to every $\Phi_n(x)$, in the space considered, then the system of $\Phi_n(x)$'s is complete.

The functions $X_n(x)$ and $X_m(x)$ are said to be *orthogonal* (as opposed to orthonormal) in $[a, b]$ if

$$\int_a^b X_n(x)X_m(x)\,dx = \delta_{nm}\int_a^b X_n^2(x)\,dx \tag{C.9}$$

They are said to be *orthogonal with respect to a weight function* $p(x)$ if

$$\int_a^b p(x)X_n(x)X_m(x)\,dx = \delta_{nm}\int_a^b p(x)X_n^2(x)\,dx \tag{C.10}$$

and they are said to be *orthogonal in the Hermitian sense* if

$$\int_a^b X_n(x)X^*_m(x)\,dx = \delta_{nm}\int_a^b |X_n(x)|^2\,dx \tag{C.11}$$

where X^*_m is the complex conjugate of X_m. The wave function Ψ is orthogonal in the Hermitian sense.

THEOREMS ON STURM–LIOUVILLE SYSTEMS

Proofs of the following theorems on Sturm–Liouville systems are fairly direct and are not repeated here. The terms *eigenvalues* and *eigenfunctions*[2] denote values of the constant λ in Eq. (C.1) and the corresponding solutions $X(x)$, respectively.

The first theorem (which we shall not state formally) says that for any *finite* interval $[a, b]$ and under rather general conditions on $p(x)$, $q(x)$, and $r(x)$, there exists a discrete set of eigenvalues, $\lambda = \lambda_1, \lambda_2, \ldots, \lambda_n, \ldots$, for which eigenfunctions, $X = X_1, X_2, \ldots, X_n, \ldots$, exist. This peculiarity of the Sturm–Liouville system is of great importance in quantum mechanics.

The second theorem says that the eigenfunctions (or in our case the standing-wave functions) should be orthogonal to one another

THEOREM Let $p(x)$,$q(x)$, and $r(x)$ in the Sturm–Liouville system be continuous in the interval $[a, b]$, and let λ_m, λ_n be any two distinct eigenvalues and $X_m(x)$, $X_n(x)$ be the corresponding eigenfunctions for which dX_m/dx and dX_n/dx are continuous. Then $X_m(x)$ and $X_n(x)$ are orthogonal in $[a, b]$ with respect to the weight function, $p(x)$. Furthermore, if $r(a) = 0$, the first boundary condition (C.2) can be dropped from the problem, and if $r(b) = 0$ the second one can be dropped. If $r(a) = r(b)$, both conditions can be replaced with the periodic condition $X(a) = X(b)$; $(dX/dx)_a = (dX/dx)_b$.

The last theorem is important because it results in the energy of quantum systems being *real* in every case that we encounter.

THEOREM If, in addition to the conditions stated in the preceding theorem, $p(x)$ does not change sign in $[a, b]$, then the eigenvalues are all real.

EXAMPLE C.1 In chapter 5 the Schrödinger equation for the stationary-wave function ψ of a rigid rotor is solved by separation of variables. One of the resulting systems is

$$\frac{1}{\sin\theta}\frac{d}{d\theta}\left(\sin\theta\frac{dP}{d\theta}\right) + \left(\frac{2I\epsilon}{\hbar^2} - \frac{m^2}{\sin^2\theta}\right)P = 0 \qquad (5.13)$$

$$P(\theta = 0) = P(\theta = \pi) \qquad \text{and} \qquad \left(\frac{dP}{d\theta}\right)_{\theta=0} = \left(\frac{dP}{d\theta}\right)_{\theta=\pi}$$

Analyze this system from the viewpoint of the Sturm–Liouville theory.

In this case the given equation is of the Sturm–Liouville form with $r = \sin\theta$; $\lambda =$ the energy, ϵ; $p = 2I\sin\theta/\hbar^2$; $q = -m^2/\sin\theta$; $a = 0$; and $b = \pi$. Since $r(\theta = a = 0) = r(\theta = b = \pi)$, the given boundary conditions are also legitimate members of the Sturm–Liouville system, in accordance with the last portion of the second theorem.

[2]These words are hybrids of the German *eigenwerte* and *eigenfunktion* and their literal translations, *characteristic value* and *characteristic function.*

The quantity $p = 2I \sin \theta/\hbar^2$ is always positive; thus the third theorem requires that the eigenvalues — the energies — must all be real. The first theorem says that these energies exist in a discrete set of values. And the second theorem requires that the solutions (the Legendre polynomials) must obey the following normalization condition:

$$\frac{2I}{\hbar^2}\int_0^\pi P^m(\theta)P^n(\theta) \sin\theta \, d\theta = \frac{2I}{\hbar^2}\,\delta_{mn}\int_0^\pi [P^m(\theta)]^2 \sin\theta \, d\theta$$

The same comments generally apply to Eq. (5.12) as well, but the eigenfunctions in this case are sines and cosines and the eigenvalues are $m = 0, \pm 1, \pm 2, \ldots$.

appendix D values of some physical constants and conversions factors

Å $= 10^{-8}$ cm	angstrom unit
atm $= 1.01325 \times 10^{6}$ dynes/cm^2	standard atmosphere
$c_l = 2.998 \times 10^{10}$ cm/sec	speed of light in a vacuum
$e = -1.60210 \times 10^{-19}$ Coulomb $= 4.8030 \times 10^{-10}$ esu	electronic charge (e is also used to designate the natural number, 2.718 . . .)
erg $= 10^{-7}$ Joule $= 0.2390 \times 10^{-7}$ cal $= 1$ dyne-cm $= 1$ g-cm^2/sec^2	
eV $= 1.6021 \times 10^{-12}$ erg	electron volt
$g_c = 32.174$ ft-lb$_m$/lb$_f$-sec^2 $=$ unity in cgs system	conversion factor in Newton's law
$h = 6.6256 \times 10^{-34}$ Joule-sec $= 6.6256 \times 10^{-27}$ erg-sec	Planck's constant
$\hbar = 1.05443 \times 10^{-27}$ erg-sec	$h/2\pi$
$hc_l/k = 1.4388$ cm-°K	second radiation constant
$\hbar^2/m_e e^2 = 5.29 \times 10^{-9}$ cm	first Bohr radius
$k = 1.3805 \times 10^{-23}$ Joule/°K $= 1.3805 \times 10^{-16}$ erg/°K	Boltzmann's constant

$L = (\pi k)^2/3e^2$
$= 2.45 \times 10^{-8}$ Watt-ohm/°C^2 ...... Lorenz number
$m_e = 9.1091 \times 10^{-28}$ g .... electron rest mass
$m_n = 1.67482 \times 10^{-24}$ g .... neutron rest mass
$N_A = 6.0225 \times 10^{23}$ molecules/g mole Avogadro's constant
ohm $= 8.9876$ cm/sec
$R = 109{,}737.3$ cm^{-1} .... Rydberg's constant
$R^0 = 8.3143 \times 10^7$ ergs/g mole-°K
$= 0.082054$ atm-liter/g mole-°K
$= 1544$ ft-lb$_f$/lb$_m$-mole-°R
$= 1.986$ Btu/lb$_m$-mole-°R
} ideal gas constant
$(\lambda T)_m = 0.28978$ cm-°K
$= 5215.6$ μ-°R
} .... Wien's displacement constant
$\sigma = 5.6697 \times 10^{-5}$ erg/cm^2-sec-°K^4
$= 1.714 \times 10^{-9}$ Btu/ft^2-hr-°R^4
} Stefan–Boltzmann constant

appendix E some useful formulas

1. Integrals

   (a)

$$\int_0^\infty x^n e^{-(ax)^m}\,dx = \frac{\Gamma\left(\dfrac{n+1}{m}\right)}{ma^{n+1}} \qquad n+1, a, m > 0$$

The following special cases, for $m = 2$, are of particular relevance:

$$\int_0^\infty x^n e^{-a^2x^2}\,dx = \frac{\Gamma\left(\dfrac{n+1}{2}\right)}{2a^{n+1}} = \begin{cases} \dfrac{\sqrt{\pi}}{2a} & \text{for } n = 0 \\ \dfrac{1}{2a^2} & \text{for } n = 1 \\ \dfrac{\sqrt{\pi}}{4a^3} & \text{for } n = 2 \\ \dfrac{1}{2a^4} & \text{for } n = 3 \end{cases}$$

$$\int_{-\infty}^\infty x^n e^{-a^2x^2}\,dx = \begin{cases} 2\displaystyle\int_0^\infty x^n e^{-a^2x^2}\,dx & \text{for } n = 0, 2, 4, \ldots \\ 0 & \text{for } n = 1, 3, 5, \ldots \end{cases}$$

(b) The gamma function, $\Gamma(n) \equiv \int_0^\infty x^{n-1}e^{-x}\,dx$. The following identities can be proved for the gamma function:

$$\Gamma(n+1) = n\Gamma(n)$$
$$\Gamma(n+1) = n! \qquad \text{when } n \text{ is a positive integer}$$
$$\Gamma(1) = \Gamma(2) = 1$$
$$\Gamma(\tfrac{1}{2}) = \sqrt{\pi} \qquad \Gamma(\tfrac{3}{2}) = \sqrt{\pi}/2$$

(c) The error function, $\text{erf}\,(x) \equiv \frac{2}{\sqrt{\pi}}\int_0^x e^{-\xi^2}\,d\xi$. The error function does not have an explicit solution. Numerical values of erf (x) are given as a function of x in Table E.1.

TABLE E.1 Values of the Error Function

x	erf(x)	x	erf(x)	x	erf(x)
0	0	0.70	0.677801	1.8	0.989091
0.05	0.056372	0.75	0.711156	1.9	0.992790
0.10	0.112463	0.80	0.742101	2.0	0.995322
0.15	0.167996	0.85	0.770668	2.1	0.997021
0.20	0.222703	0.90	0.796908	2.2	0.998137
0.25	0.276326	0.95	0.820891	2.3	0.998857
0.30	0.328627	1.0	0.842701	2.4	0.999311
0.35	0.379382	1.1	0.880205	2.5	0.999593
0.40	0.428392	1.2	0.910314	2.6	0.999764
0.45	0.475482	1.3	0.934008	2.7	0.999866
0.50	0.520500	1.4	0.952285	2.8	0.999925
0.55	0.563323	1.5	0.966105	2.9	0.999959
0.60	0.603856	1.6	0.976348	3.0	0.999978
0.65	0.642029	1.7	0.983790		

(d) The Debye function, $D(X) \equiv \frac{3}{X^3}\int_0^X [x^3/(e^x - 1)]\,dx$, is tabulated in Table 10.3. A related integral of some interest is the following:

(e)

$$\int_0^\infty \frac{x^{p-1}}{e^{ax} - 1}\,dx = \frac{\Gamma(p)}{a^p}\,\zeta(p) \qquad a, p > 0$$

where $\zeta(p)$ is the Riemann zeta function of p[see 2(b)]. Some special cases of the result for $a = 1$ are

$$\int_0^\infty \frac{x}{e^x - 1}\,dx = \frac{\pi^2}{6}$$

$$\int_0^\infty \frac{x^2}{e^x - 1}\,dx = 2\zeta(3) = 2.404114$$

and

$$\int_0^\infty \frac{x^3}{e^x - 1}\,dx = \frac{\pi^4}{15}$$

2. Series

(a) The binomial expansion is

$$(1 + x)^n = 1 + nx + \frac{n(n-1)}{2!}x^2 + \cdots + \frac{n!}{(n-r)!r!}x^r + \cdots$$

This series contains a finite number of terms for n = a positive integer. If n is a positive noninteger, it converges for $x^2 \leq 1$. If n is a negative noninteger, it converges for $x^2 < 1$.

(b) The zeta function is $\zeta(a) \equiv \sum_{i=1}^{\infty} i^{-a}$. This function is related to the integral given in Sec. 1(e). Some values of $\zeta(a)$ are:

a	$\zeta(a)$
1	∞
2	$\pi^2/6$
3	1.202057
4	$\pi^4/90$
5	1.036928
6	$\pi^6/945$
7	1.008349
8	$\pi^8/9450$

(c) Some transcendental functions (valid for all finite x unless noted otherwise) are

$$\sin x = x - \frac{x^3}{3!} + \frac{x^5}{5!} - \cdots \qquad \cos x = 1 - \frac{x^2}{2!} + \frac{x^4}{4!} - \cdots$$

$$e^x = 1 + x + \frac{x^2}{2!} + \cdots = \sum_{n=0}^{\infty} \frac{x^n}{n!}$$

$$\ln(1 + x) = x - \frac{x^2}{2} + \frac{x^3}{3} - \frac{x^4}{4} \cdots \qquad x^2 < 1 \text{ and } x = 1$$

$$\sinh x = x + \frac{x^3}{3!} + \frac{x^5}{5!} + \cdots \qquad \cosh x = 1 + \frac{x^2}{2!} + \frac{x^4}{4!} + \cdots$$

index